research perspectives in

hydraulics and water resources engineering

research perspectives in
hydraulics and water resources engineering

edited by **Rama Prasad & S Vedula**

Indian Institute of Science, Bangalore

World Scientific

New Jersey • London • Singapore • Hong Kong

Published by

World Scientific Publishing Co. Pte. Ltd.

P O Box 128, Farrer Road, Singapore 912805

USA office: Suite 1B, 1060 Main Street, River Edge, NJ 07661

UK office: 57 Shelton Street, Covent Garden, London WC2H 9HE

British Library Cataloguing-in-Publication Data
A catalogue record for this book is available from the British Library.

RESEARCH PERSPECTIVES IN HYDRAULICS AND WATER RESOURCES ENGINEERING

ISBN 981-02-4929-2

Printed in Singapore by FuIsland Offset Printing

DEDICATED TO
THE MEMORY OF

Prof. N. S. GOVINDA RAO

ARCHITECT OF THE GROWTH OF
CIVIL ENGINEERING DEPARTMENT
INDIAN INSTITUTE OF SCIENCE
BANGALORE

PREFACE

In the year 1950, the Civil and Hydraulic Engineering Section was established as part of the Department of Power Engineering at the Indian Institute of Science, Bangalore under the leadership of Prof. N. S. Govinda Rao. A number of research problems of national importance in hydraulic engineering as well as in structural and geotechnical engineering were studied by the Section. Gradually, in response to the needs of the society at large, problems in water resources engineering were taken up. The Section grew into a full fledged department, which was rechristened as the Department of Civil Engineering in 1972. In recognition of the excellence in its performance, the Department has been designated as a Centre for Advanced Studies by the University Grants Commission.

The Department celebrated the Golden Jubilee during the year July 2000–July 2001 with a variety of academic activities including special lectures, workshops and an international conference. As part of the celebrations, it was decided to bring out a publication consisting of invited state-of-art review articles on specific research topics in Civil Engineering by alumni of the Department. This Volume is the result of the efforts to this end.

We thank the authors for the contributions and for graciously agreeing for a peer review. We place on record our grateful thanks to Professors Donald Dean Adrian, Jean Berlamont, Subhash Chander, M. Hanif Chaudhry, M. C. Deo, Rema Devi, K. Elango, T. Gangadharaiah, R. J. Garde, S. N. Ghosh, N. J. D. Graham, S. K. Jain, S. Kumaraswamy, P. P. Nageswara Rao, S. Narasimhan, B. S. Pani, N. Rajaratnam, K. G. Ranga Raju, S. M. Seth, Ramesh P. Singh, K. Srinivasan, R. Srivastava and N. V. C. Swamy, whose reviews enhanced the quality of the Publication.

We gratefully acknowledge the excellent cooperation extended to us by the Chairman, Faculty, Staff and Students of the Department in bringing out this Volume.

Rama Prasad and S. Vedula

CONTENTS

TURBULENT JETS: APPLICATION OF POINT SOURCE CONCEPT

B. S. PANI
Dept. of Civil Engg., I.I.T. Bombay, Mumbai - 400 076, INDIA
E-mail: bspani@civil.iitb.ernet.in

S. B. DUGAD
Dept. of Civil Engg., I.I.T. Bombay, Mumbai – 400 076, INDIA
E-mail: rsdugad@civil.iitb.ernet.in

A theoretical distribution for the streamwise mean velocity of a three-dimensional turbulent free jet can be derived by following Reichardt's inductive theory applied to an elemental source. Using an integral type approach, the momentum and heat flux can be predicted at any point in the downstream region of a three-dimensional outlet by superposition. Further, it is possible to predict both the mean momentum flux and velocity profiles for three-dimensional multiple jets. The mean flow characteristics of three-dimensional wall jets can be predicted by modifying the free jet equations. Experimental observations are in agreement with predicted values.

1 Introduction

In a number of engineering applications, a high velocity stream meets an ambient fluid, which could be at rest or is in slow motion. A so called surface of separation results which is highly unstable. Hence, eddies move in a random fashion both along and across the stream bringing about an exchange of momentum, heat and constituents between adjoining layers. As a result, a region of finite thickness with a continuous distribution of velocity, temperature and species concentration is formed which is termed as a shear layer. Abramovich [1] calls this layer the "turbulent jet boundary layer". As the Reynolds number of flow is mostly higher than few thousand, it is justified to classify the shear layer as a turbulent shear layer. The type of turbulent jet that has been studied most is the one diffusing through an ambient fluid at rest. The thickening of the shear layer results on the one hand in the increase in the cross-section of the jet and on the other, to a gradual 'eating up' of its nonviscous core. The velocity in the potential core remains constant and the static pressure is constant throughout the flow cross-section.

A short distance beyond the potential-core region, the jet becomes similar in appearance to a flow of fluid from a source of infinitely small thickness. Based on the geometrical shape of the jet nozzle and the way the flow field develops, jets can be axisymmetric, radial, two-dimensional or three-dimensional in nature. Further, depending on the presence or absence of solid boundaries, a wall shear layer or a free shear layer is formed respectively. A frequently encountered example of the wall shear layer, which is also termed as a wall jet, is the one with a jet of fluid moving tangentially on a wall.

This article demonstrates the application of jet flows in Civil Engineering. Nontheless, it does highlight the basic principles, which are common to other branches of engineering.

2 Submerged Outlets

In tidal power schemes the basin outlets are submerged. Sometimes even control gate structures, encountered in irrigation schemes, operate under deeply submerged tailwater conditions. Submerged jets are also commonly used for disposal of sewage and waste-water into streams, lakes and the sea. The cooling water system of nuclear and thermal power plants often involves the discharge of warm water jets from submerged multiport diffusers. As a consequence of the adverse environmental impact of sewage as well as heat, dumped into the natural water body, the subject has gained immense importance. An efficient disposal system aims at achieving a thorough mixing of the jet with the ambient fluid as quickly as possible, thereby bringing down the concentration or the temperature excess of the jet fluid to within acceptable limits.

As pointed out earlier, flow from outlets frequently occurs very close to a boundary and the shear layer moves tangential to the channel bed. For such flows, a wall jet type of analysis is warranted. In case the outlet occupies the full width of the conveyance channel, it could be treated as a plane wall jet. It has been found that the classical plane wall jet serves as an useful model to describe flow below sluice gates [25]. However, in many instances, the jet emanates from outlets of finite aspect ratios which occupy only a part of the width of the channel. Narain [16] divided the flow field of the three-dimensional wall jets into 6 zones and used the classical momentum integral approach to solve the basic equations with the known boundary conditions. There could be a single outlet or a number of them placed in a line. Their flow analysis will be somewhat more involved because of the three-dimensional nature of flow.

If the boundaries are sufficiently away, having negligible influence on the flow, the shear layer can be treated as a free jet. A fair amount of information is available on the mean flow and turbulence characteristics of plane, axially symmetric and three-dimensional incompressible jets [28]. The flow field downstream of an array of jets has been predicted using the principle of superposition [11].

3 Transport Process

This is one of the central issues in studies of turbulent jets. Theories based upon the concept of 'mixing length' introduced by Taylor and independently by Prandtl have been quite successful. Though the theories are not able to describe the mechanism of turbulent-transport process in detail, the resulting semiempirical relations have been of great practical value. The velocity profiles are assumed to exhibit similarity i.e. prove universal when plotted in dimensionless co-ordinates.

$$\frac{u}{u_m} = f\left(\frac{y}{b}\right) \tag{1}$$

where, u is the longitudinal velocity, u_m is the maximum value of u on the axis, y is the transverse co-ordinate and b is a characteristic width. This has been found to be in agreement with experimental observations.

The application of Prandtl's theory of momentum transport to derive the mean velocity distribution in free jets and wakes has resulted in velocity profiles which are in fair agreement with measurements. However, one of the fundamental objections to the theory is that no account is taken of the effect of pressure fluctuations on momentum transport [9].

Based on Prandtl's mixing length hypothesis, Tollmien investigated the velocity distribution in free jets. Using the principle of conservation of 'momentum flux', the variation of the mean velocity component along the axis of a plane jet was found as,

$$\frac{u_m}{u_0} = \frac{1.2}{\sqrt{\dfrac{ax}{b_0}}} \tag{2}$$

where, u_m is the velocity on the axis, u_0 is the efflux velocity of the jet, b_0 is one half the thickness of the initial cross section of the jet, x is the axial distance from the nozzle and 'a' is a constant. The constant 'a' based on velocity decrease has an average value of approximately 0.10 [1].

Similarly, in case of axially symmetric turbulent jets, the maximum velocity decay relationship was obtained as,

$$\frac{u_m}{u_0} = \frac{0.96}{a\left(\dfrac{x}{r_0}\right)} \tag{3}$$

where, r_0 is the radius of the nozzle and value of 'a' varies between 0.066 to 0.076 depending on the shape of the initial velocity field.

Prandtl's mixing length hypothesis i.e. old theory can also be applied to find the distribution of temperature difference and concentration difference in transverse cross sections of plane and axially symmetric jets. The solution leads to the same distribution as was derived by Tollmien for the dimensionless velocity profile. However, experiments show that the temperature difference and velocity profiles do not coincide even though the temperature difference and concentration difference profiles are identical. The variations of the maximum temperature difference ΔT_m and concentration difference ΔC_m along the jet axis are also subject to the same law.

Observed values agree quite well with the theoretical values. The discrepancy between the temperature and velocity distributions found in experiments can be avoided by adopting Taylor's free turbulence theory.

Prandtl's new theory of free turbulence, introduced in 1942, assumed that the coefficient of turbulent viscosity is constant over the cross section of the jet and not the mixing length [1]. On the basis of similarity of flow at various cross sections, an expression for ε, the eddy viscosity can be obtained as

$$\varepsilon = \chi b(u_m - u_a) = \chi b \Delta u_m \tag{4}$$

where, χ is an empirical constant, b is the width of the mixing zone, u_m is the maximum velocity in the transverse section and u_a is the flow velocity of the ambient fluid. In case of jets discharging into a stationary ambient, $u_a = 0$ and

$$\varepsilon = \chi b u_m \tag{5}$$

Goertler made use of the new theory and obtained the dimensionless velocity distribution in the fully developed region of plane and axially symmetric jets. It may be noted that in case of axially symmetric jets, the kinematic eddy viscosity is constant over the entire field of flow. In other words, the governing equation would reduce to the same form as for an axially symmetric laminar jet, the two flows differing only by the magnitude of the viscosity coefficient. Otherwise, the computational formulas are the same. The dimensionless velocity components calculated from Goertler's formula for plane as well as the axially symmetric jets agree very well with Reichardt's experimental data [1]. In fact, solutions of Goertler and Tollmien give the velocity distribution with approximately the same accuracy almost over the entire transverse section. Tollmien's theory results in a jet of finite thickness while Goertler's solution is asymptotic in nature.

Though the mechanism of heat and constituent diffusion is the same in both the theories, Taylor's theory does not lead to a similarity of the velocity field. One can prove that

$$\frac{\Delta c}{\Delta c_m} = \frac{\Delta T}{\Delta T_m} = \sqrt{\frac{u}{u_m}} \tag{6}$$

4 Reichardt's Hypothesis

In all the methods discussed above, a deductive method was followed, i.e. the velocity distributions were deduced based on a hypothesis concerning turbulent shear stress or the turbulence diffusion coefficient. In contrast, Reichardt [27] proposed a inductive method. It was assumed that a turbulent transport process is a statistical process exactly analogous to molecular processes. The governing equations for molecular diffusion on heat conduction yield a Gaussian distribution for the concentration and the temperature, respectively. In a free jet the distribution of the total momentum flux, i.e. the total momentum resulting from the mean and fluctuating components of the velocity, has been observed to follow the Gaussian curve as well. Hence, the differential equation for $\rho \overline{U^2}$ must be identical with the equation for molecular diffusion or heat conduction. To transform the equation of motion into such a fashion so that the above requirement is met with, a separate hypothesis is needed.

Reichardt too followed a gradient type of transport process and assumed that the lateral transport of momentum is proportional to the transverse gradient of the horizontal component of momentum. If U and V are the instantaneous velocities in the x and y directions respectively, then

$$\overline{UV} = -\Lambda(x)\frac{\partial \overline{U^2}}{\partial y^2} \tag{7}$$

where, Λ is a length characterising the mixing, analogous to Prandtl's mixing length. Reichardt assumed Λ to be constant over each cross section and to depend only on the variable x. In applying Reichardt's equation to solve the cases of plane and axially symmetric jets one must further assume that the fields of the time averaged values of the dymamic pressure are universal, i.e.

$$\frac{\overline{U^2}}{\overline{U_m^2}} = f\left(\frac{y}{b}\right) = f(\xi) \tag{8}$$

where, b is a certain characteristic width of the layer and U_m is the instantaneous velocity on the axis of the jet.

Though Reichardt's hypothesis sounds reasonable, there are some serious limitations [9]. However, the fact that the theoretical results based on his hypothesis are in good agreement with experimental results, strongly support his theory for acceptance. The observed agreement between theory and experiments may not be surprising if one recalls that the equations of motion have been transformed artificially so as to agree with experimental results.

As a result of the governing equation being linear in $\overline{U}^2$ a great advantage accrues, not only mathematically but also from a engineering application point of view. Hence, the principle of superposition of elementary solutions can be adopted. The main purpose of this article is to demonstrate how a variety of practical problems can be worked out using the elementary solution as a 'building block'.

5 Axially Symmetric Jet Solution

It is assumed that the flow is quasi-steady, longitudinal pressure gradients are negligible, the amount of sensible heat formed by conversion from kinetic energy is negligible and that the molecular transport is insignificant. Further, considering an incompressible flow, the momentum equation for an axially symmetric flow can be written as

$$\frac{\partial}{\partial x}\left(\overline{U^2}\right)+\frac{1}{r}\cdot\frac{\partial}{\partial r}\left(r\cdot\overline{UV}\right)=0 \tag{9}$$

where U and V are the instantaneous velocities in x and r directions respectively and bars denote averaging with respect to time. Using Reichardt's hypothesis, (Eq. 7) in the above expression one can obtain

$$\frac{\partial}{\partial x}\left(\overline{U^2}\right)-\Lambda\left[\frac{\partial^2}{\partial r^2}\left(\overline{U^2}\right)+\frac{1}{r}\cdot\frac{\partial}{\partial r}\overline{U^2}\right]=0 \tag{10}$$

Introducing the principle of similarity of the dynamic pressure, Eq. 8, one can further modify the governing equation and write the following

$$2f+\xi f=-\frac{\Lambda}{b}\cdot\frac{dx}{db}\left[f''+\frac{f'}{\xi}\right] \tag{11}$$

As the above expression should be a function of ξ alone, set

$$\Lambda=b\frac{db}{dx} \tag{12}$$

Since the width of the jet depends on x, it follows from Eq. 12 that Λ must also be a function of x.

Thus, Eq. 11 can be re-written as

$$2f+f'\left(\xi+\frac{1}{\xi}\right)+f''=0 \tag{13}$$

The boundary conditions to be satisfied are; at $\xi = 0$; $f = 1$ and for $\xi \to \infty$; $f \to 0$. The solution of the above equation is

$$f=e^{-\frac{\xi^2}{2}} \tag{14}$$

or,

$$\frac{\overline{U}^2}{\overline{U}_m^2}=e^{-\frac{\xi^2}{2}} \tag{15}$$

As the forward momentum gets conserved, the following can be written,

$$\int\rho\overline{U}^2(2\pi r dr)=\rho\overline{U}_0^2A \tag{16}$$

where, A = area of the source and U_0 is the efflux velocity in plane of the source. From Eq. 15 and Eq. 16 one can get,

$$\frac{\overline{U}_m^2}{\overline{U}_0^2}=\frac{A}{2\pi b^2} \tag{17}$$

The time-averaged value of the instantaneous dynamic head can be expressed in terms of the mean velocity component u and the fluctuating component u′. Thus, $\overline{U}^2 = u^2 + \overline{u'^2}$. Since the quantity $\overline{u'^2}$ is an order of magnitude smaller than u^2, specifically in the central region of the jet, it seems reasonable to drop $\overline{u'^2}$ as an approximation.

Hence, from Eq. 15, the square of the mean velocity u^2 can be written in the following form,

$$\frac{u^2}{u_m^2} = e^{-\frac{\xi^2}{2}} \tag{18}$$

where, u_m is value of u on the axis of the jet. Also, Eq. 17 can be re-written as

$$\frac{u_m^2}{U_0^2} = \frac{A}{2\pi b^2} \tag{19}$$

If the jet is imagined to emanate from a elementary area dA, the variation of the dynamic pressure along the elemental jet axis can be found out from a modified form of the above equation written as,

$$\frac{u_m^2}{U_0^2} = \frac{dA}{2\pi b^2} \tag{20}$$

The principle of superposition can be adopted to find the resultant u_m^2 when there are more than one dA source of momentum flux.

6 Turbulent Jets from a Series of Holes

Knystautas [11] investigated theoretically and experimentally the flow field of a series of axisymmetric jets emanating from circular holes arranged in a line. Solutions of individual jets were superposed using Reichardt's theory to get the dynamic pressure profile and it was found that beyond certain downstream distance from the nozzle face, the flow field was akin to that of a plane jet. The theoretical predictions agreed reasonably well with experiments even to the extent of predicting the mean velocity profiles before attaining two-dimensionality. Momentum was observed to be very nearly conserved.

Dyban and Mazur [5,6] conducted experiments on the aerodynamic characteristics of a submerged jet formed by a system of axisymmetric jets issuing through a perforated surface. The individual jets degenerated at some distance into an axisymmetric jet. The main zone of the resulting jet was significantly different from the individual jets since turbulent mixing occurred not only on the outer surface of the equivalent jet but also within it because of entrainment by the constituent jets. Fluid entrained by the jet consumed 50 to 90 percent of the kinetic energy of the jet depending upon the hole spacing.

The flow field of multiple free jets resulting out of an array of rectangular nozzles was investigated by Marsters [13]. Turbulence characteristics were studied in detail for flow from an array of 20 nozzles. Both the mean flow features and the observed turbulence structure were not exhibiting similarity upto very large distances in the streamwise direction. Krothapalli *et al.* [12] conducted experiments on multiple three-dimensional jets. In confirmity with other's observations [29], three regions namely the potential core, the characteristic decay and the axially symmetric decay were found to exist. Because of the mutual interaction between the jets, the intensity of turbulence in the x-direction was lower than that of a single jet at corresponding locations.

To control the noise level of a jet, multiple nozzles arranged around a central axis were used and the jet velocity was 520 m/s [26]. At a distance of 14 times the nozzle diameter, the individual jets merged and beyond 56 times the diameter, the difference between multiple and single jet was not noticeable, either in respect of velocity or the extent of radial spread. Provision of an array of five nozzles offered the maximum advantage in reducing the noise level.

Fabris and Fejer [7] investigated the transfer of the kinetic energy from an array of primary jets to a surrounding secondary stream. An analytical technique, validated by experiments, was developed. The mixing efficiency of multiple jets exceeded the efficiency of an equivalent co-axial ducted jet by a significant amount.

7 Dual Mixing Lengths: Modified Reichardt's Hypothesis

7.1 Single Three Dimensional Outlet

Reichardt's hypothesis was extended by Dash [3] to find the velocity distribution and other bulk characteristics of three-dimensional jets. Based on boundary layer type of approximation involving instantaneous velocities, the governing equations of motion for a three-dimensional flow were simplified. For an incompressible quasi-steady flow, one can arrive at the following,

$$\frac{\partial}{\partial x}\left(\overline{U^2}\right)+\frac{\partial}{\partial y}\left(\overline{UV}\right)+\frac{\partial}{\partial z}\left(\overline{UW}\right)=0 \tag{21}$$

which shows the conservation of the momentum component in the x-direction.
Unlike an axisymmetric case, the growth rates of three-dimensional free jets in y and z directions are different. Hence, it was reasonable to select two different length parameters Λ_1 and Λ_2 to characterise the mixing in y and z directions respectively. This is similar to the approach adopted by Narain [16]. Adopting the momentum transfer law of Reichardt, one can write

$$\overline{UV} = -\Lambda_1 \frac{\partial}{\partial y} \overline{U}^2$$

and,
$$\overline{UW} = -\Lambda_2 \frac{\partial}{\partial z} \overline{U}^2 \tag{22}$$

where, Λ_1 and Λ_2 are functions of x alone. From Eqs. 21 and 22, we get the following governing equation

$$\frac{\partial}{\partial x} \overline{U}^2 = \Lambda_1 \frac{\partial^2}{\partial y^2} \overline{U}^2 + \Lambda_2 \frac{\partial^2}{\partial z^2} \overline{U}^2 \tag{23}$$

Further, the dynamic pressure profiles in the y and z directions are assumed to be similar. Thus, the universal dynamic pressure profiles can be expressed as

$$\frac{\overline{U}_m^2}{\overline{U}_{m0}^2} = f(\xi) \tag{24a}$$

and,
$$\frac{\overline{U}^2}{\overline{U}_m^2} = g(\zeta) \tag{24b}$$

where, $\xi = y/b_y$ and $\zeta = z/b_z$, b_y and b_z are the characteristic widths in the y and z directions respectively. U_m and U_{m0} are the maximum instantaneous velocities in the x-z plane and the x-y plane respectively, Fig. 1.

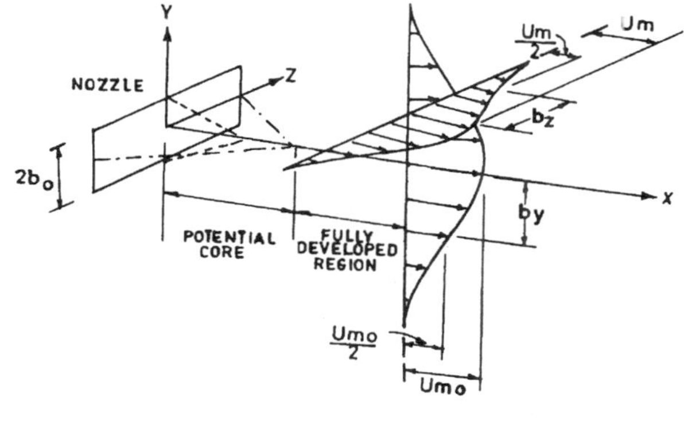

Figure 1. Definition sketch

A combination of the above two relationships in Eqs. 24a and 24b can be expressed as

$$\frac{\overline{U}^2}{\overline{U}_{m0}^2} = f \cdot g \tag{25}$$

Substituting Eq. 25 in Eq. 23 and following a procedure identical to the axisymmetric jet case, given in section 5, it can be proved that

$$\Lambda_1 = b_y \frac{db_y}{dy} \tag{26a}$$

and

$$\Lambda_2 = b_z \frac{db_z}{dz} \tag{26b}$$

For a negligible pressure gradient in the x-direction, the total forward momentum of the jet is conserved. Hence, one can show that

$$\overline{U^2_{m_0}} \alpha \frac{M_0}{b_y \cdot b_z} \tag{27}$$

where, M_0 is the momentum flux of the jet at the exit section of the outlet. Using Eqs. 25, 26a and 26b, the governing equation of motion in the x-direction can be written as follows,

$$g\frac{b_y'}{b_y}\left[f + \xi f' + f''\right] + f\frac{b_z'}{b_z}\left[g + \zeta g' + g''\right] = 0 \tag{28}$$

The differential equations in f and g can be solved independently using the following set of boundary conditions,
for $\xi = 0$; $f(\xi) = 1$ and $\xi = \infty$; $f(\xi) = 0$ and,
for $\zeta = 0$; $g(\zeta) = 1$ and $\zeta = \infty$, $g(\zeta) = 0$, the solution of Eq. 28 is given as

$$f = \frac{\overline{U^2_m}}{\overline{U^2_{m_0}}} = \exp\left(-\frac{\xi^2}{2}\right) \tag{29}$$

and

$$g = \frac{\overline{U^2}}{\overline{U^2_m}} = \exp\left(-\frac{\zeta^2}{2}\right) \tag{30}$$

Based on the distribution of the dynamic pressure distribution, given in Eqs. 29 and 30, one can arrive at the following relationship

$$\frac{\overline{U^2_{m_0}}}{\overline{U^2_0}} = \frac{A}{2\pi b_y \cdot b_z} \tag{31}$$

For axisymmetric flows b_y equals b_z and the above expression is identical to the one given in Eq. 17.

The contribution of the fluctuating velocity component being an order of magnitude smaller than that of the mean velocity, we may assume $\overline{U^2} = \overline{u^2}$, as has been done earlier. Thus, the mean velocity distribution in the y and z direction and the resulting decay of the maximum mean velocity u_{m0} can be expressed by the following:

$$\frac{u_m}{u_{m_0}} = \exp\left[-\frac{\xi^2}{4}\right] \tag{32}$$

$$\frac{u}{u_m} = \exp\left[-\frac{\zeta^2}{4}\right] \tag{33}$$

$$\frac{u}{u_{m_0}} = \exp\left[-\frac{\xi^2+\zeta^2}{4}\right] \tag{34}$$

and
$$\frac{u_{m_0}}{U_0} = \left[\frac{A}{2\pi\, b_y b_z}\right]^{\frac{1}{2}} \tag{35}$$

It is convenient to use the half-velocity scales, i.e. the half-widths $(b_y)_{0.5}$ and $(b_z)_{0.5}$ in analysing the experimental results. This change-over is simple, since a linear relationship exists between b_y, b_z and the corresponding half-widths. Using a scale factor of 1.665 we can re-write the above set of relationships in terms of the half-widths $(b_y)_{0.5}$ and $(b_z)_{0.5}$. For convenience of writing, the subscripts 0.5 could be dropped. In all future analysis and discussions, b_y and b_z will represent the half-widths. Thus, the final expressions in terms of the half-widths work out as,

$$\frac{u_m}{u_{m_0}} = \exp\left(-0.693\eta_y^2\right) \tag{36}$$

$$\frac{u}{u_m} = \exp\left(-0.693\eta_z^2\right) \tag{37}$$

$$\frac{u}{u_{m_0}} = \exp\left\{-0.693\left(\eta_y^2+\eta_z^2\right)\right\} \tag{38}$$

and
$$\frac{u_{m_0}}{U_0} = 0.664\left(\frac{A}{b_y \cdot b_z}\right)^{\frac{1}{2}} \tag{39}$$

In the axisymmetric velocity decay region of three-dimensional jets it has been established [24] that the half-widths b_y and b_z vary linearly with the distance x. Hence, let us write

$$b_y = c_1 x \tag{40a}$$

and
$$b_z = c_2 x \tag{40b}$$

From Eqs.35, 40a and 40b, we can express the variation of the maximum velocity as follows,

$$\frac{u_{m_0}}{U_0} = \frac{0.664}{c}\left[\frac{x}{\sqrt{A}}\right]^{-1} \tag{41}$$

where, the coefficient c = √(c_1.c_2).

Thus, from Eq. 41 it is clear that √A is a characteristic length parameter for evaluating the decay of the maximum velocity in the direction of flow. A similar observation was made based on dimensional analysis [18].

If more than one three-dimensional outlet exist in a row, the elementary solution for the dynamic profile for a single jet can be superposed to arrive at the resulting dynamic pressure field as in the case of axially symmetric jets.

Demissie and Maxwell [4] followed a similar approach by subdividing a three-dimensional outlet into square sized source elements to represent the initial distribution of momentum flux. Each element was assumed to behave like an axially symmetric source and the resulting momentum flux at downstream sections was obtained by linear superposition. Virtual origin corrections ranging between –3 to 0 times the slot width were used to bring closer agreement between theory and experiment pertaining to three-dimensional outlets of different aspect ratios like 4, 13 and 26.

7.2 Array of Three Dimensional Outlets

The results of the single three-dimensional jet can be used to predict the mean velocity field at various locations downstream of the nozzles. Thus, for N outlets of equal area A each, spaced uniformly along the z axis at a distances of s, centre to centre, placed symmetrically anbout the x axis, one can get

$$\frac{u_m^2}{U_0^2} = \frac{0.44A}{b_y \cdot b_z}\sum_{i=-j}^{j}\exp\left[-1.385\left\{\eta_y^2 + \left(\frac{s \cdot i}{b_z}\right)^2\right\}\right] \tag{42}$$

where, j = N/2.

Further, to get the maximum dynamic pressure along the x-axis, substitute η_y=0 in the above expression to yield,

$$\frac{u_{m_0}^2}{U_0^2} = \frac{0.44A}{b_y \cdot b_z}\sum\exp\left[-1.385\left(\frac{s \cdot i}{b_z}\right)^2\right] \tag{43}$$

In case of a large number of outlets in the series, the flow in the developed zone behaves as though it emanates from a plane slot having an equivalent width $2b_e$. The equivalent width can found from,

$$\rho U_0^2 A = \rho U_0^2 (2b_e \cdot s) \tag{44}$$

For example, for circular outlets spaced at intervals of 3d, the equivalent width would work out to be $\frac{\pi d}{12}$. By making substitutions, like A = $2b_e s$, s = dz and s.i = z, it is feasible to integrate Eq. 43 in the limits z = ∞ to z = -∞ and get

$$\frac{u_{m_0}^2}{U_0^2} = 1.325\frac{b_e}{b_y}$$

i. e.

$$\frac{u_{m_0}^2}{U_0^2} = 1.325\frac{b_e}{c_1 x} \tag{45}$$

An interesting feature of Eq. 45 is that the velocity decay for plane jets can be predicted based on the growth rate of three-dimensional jets. However, in this analysis the location of the virtual origins based on the velocity scale and the length scales have not been considered, i.e. they are assumed to lie in the plane of the nozzle.

8 Three Dimensional Wall Jets

It has been found that the shear force acting on a smooth wall is a small fraction of the momentum flux in the direction of flow [18]. Hence, as an approximation, the effect of friction could be ignored while predicting the gross characteristics of a three-dimensional wall jet. Therefore, the smooth wall can be treated as a reflector of momentum while developing the required relationship by substituting 2A in place of A of Eq. 39. The maximum velocity u_{m0} occurring close to the wall will be as follows,

$$\frac{u_{m0}}{U_0} = 0.94\left[\frac{A}{b_y b_z}\right]^{\frac{1}{2}} \tag{46}$$

As in the case of free jets, the principle of superposition can be adopted to get the maximum dynamic pressure downstream of N outlets of area A each, uniformly spaced at distances of s. The resulting expression is as follows,

$$\frac{u_{m0}^2}{U^2{}_0} = \frac{0.884A}{b_y b_z}\sum_{i=-j}^{j}\exp\left[-1.385\left(\frac{s.i}{b_z}\right)^2\right] \tag{47}$$

For an infinitely large number of wall jets, one can get from Eq. 47 the following relationship

$$\frac{u_{m_0}^2}{U_0^2} = \frac{1.325b_e}{c_1 x} \tag{48}$$

which looks identical to Eq. 45 of the free jet. However, one should note the difference that the equivalent width is considered to be b_e in Eq. 48, whereas it is $2b_e$ for the free jet case.

9 Mean Velocity Field : Experimental Results

9.1 Single Free Jet Velocity Decay

The variation of the length scale coefficients c_1, c_2 and c depend on the aspect ratio 'e' of the outlet. A typical plot for e = 10 is shown in Fig. 2. There seems to be a lot of variation in the values of c_1 and c_2 in the characteristic decay (CD) region and as expected, both approach a constant value in the axisymmetric decay region.

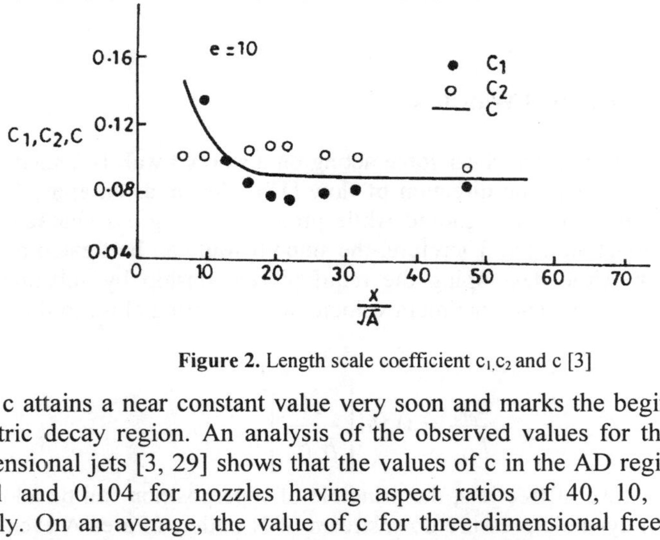

Figure 2. Length scale coefficient c_1, c_2 and c [3]

However, c attains a near constant value very soon and marks the beginning of the axisymmetric decay region. An analysis of the observed values for the growth of three-dimensional jets [3, 29] shows that the values of c in the AD region are 0.10, 0.09, 0.11 and 0.104 for nozzles having aspect ratios of 40, 10, 1 and 0.66 respectively. On an average, the value of c for three-dimensional free jets can be assumed as 0.10. Trentacoste and Sforza [31] and Rajaratnam [23] too have reported a value of c approximately equal to 0.10 for axially symmetric free jets emanating from circular nozzles. Thus, adopting a value of c = 0.10, the general expression for the decay of the maximum velocity, Eq. 41, can be written as,

$$\frac{u_{m0}}{U_0} = 6.64\left[\frac{x}{\sqrt{A}}\right]^{-1} \tag{49}$$

Originally, the above expression was suggested [18] as an empirical relationship. Figure 3 shows that for $x/\sqrt{A}$ greater than 10, Eq. 49 fits the data from different shapes of nozzle reasonably well.

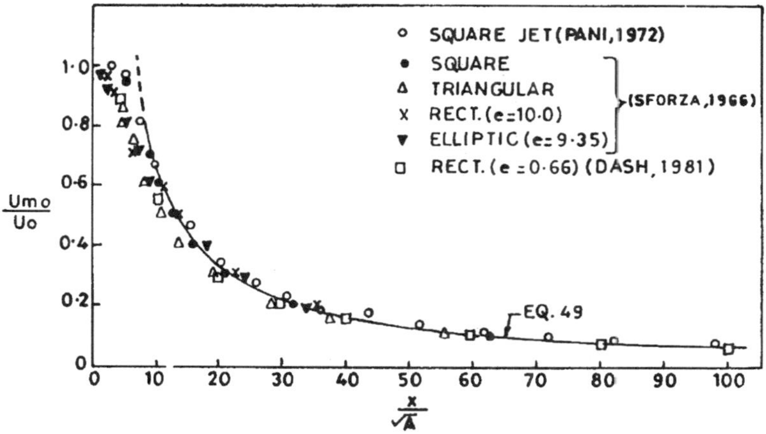

Figure 3. Decay of maximum velocity

Narain [17] presented solutions for the flux development of free jets issuing from orifices of different shapes by introducing two mixing lengths for the transfer of momentum in the Reichardt's theory. In order to bring agreement between the analytical expression for u_{m0} and observed values of other investigators, c was assumed as 0.08 in the developed zone. Suitable corrections for the location of the virtual origin were also introduced. For very slender jets (e = 0.025) c was assumed to be 0.118 in the potential core and the characteristic decay region, whereas, a value of c = 0.08 was selected for the axially symmetric region.

9.2 Velocity Decay of Multiple Three Dimensional Free Jets

The observed values of c_1, c_2 and c for a single rectangular outlet having an aspect ratio, e = 0.66, can be used to calculate the theoretical relationship between u_{m0}^2/U_0^2 and $x/\sqrt{A}$ for multiple jets as given in Eq. 47. This relationship has been plotted in Fig.4 as u_{m0}^2/U_0^2 versus $x/\sqrt{A}$ for N = 2, 3, 5 and 7. Observed values agree fairly well with the predicted curves. The plot of u_{m0}^2/U_0^2 versus $x/\sqrt{A}$ becomes linear as N becomes very large. It is also observed that the velocity decay curves, u_{m0}/U_0 versus $x/\sqrt{A}$ are distinctly different for varying values of N beyond the streamwise distance that marks the merger of the individual jets.

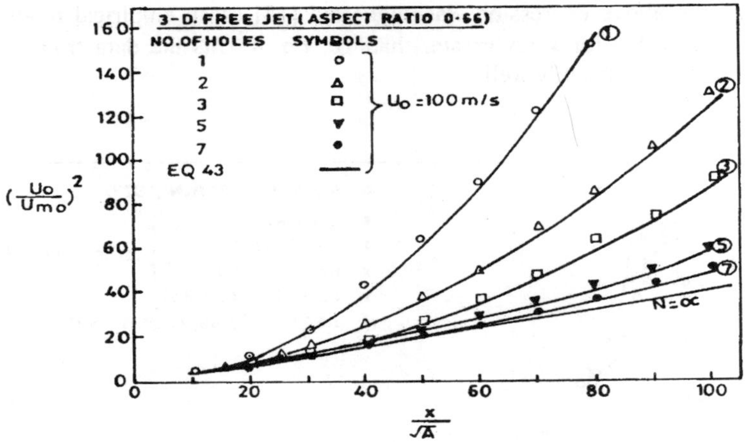

Figure 4. Variation of $\left(\dfrac{u_0}{u_{m0}}\right)^2$ with $\dfrac{x}{\sqrt{A}}$: Multiple free jets [3]

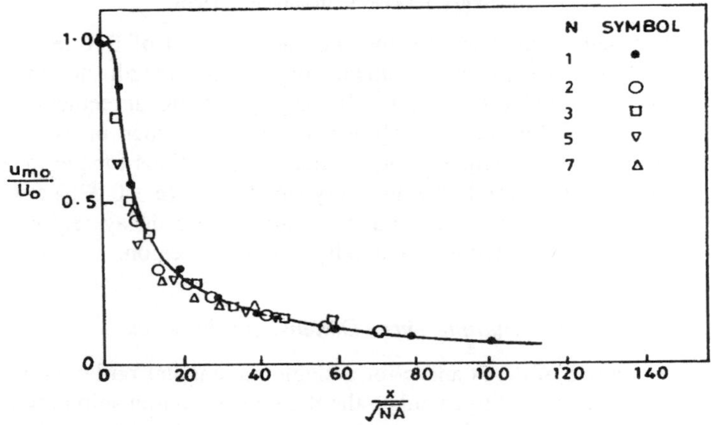

Figure 5. Decay of maximum velocity for multiple free jets: Unified plot [3]

Following the merger of the jets, the flow behaves as though it has come out of a single source having an equivalent area of NA. Hence, it is reasonable to expect that $\sqrt{NA}$ will be an appropriate length characteristic. The results are presented in Fig. 5 as u_{m0}/U_0 versus $x/\sqrt{NA}$ and can be described by a single curve for the range of N values used in the experiments.

Typical velocity distributions in the x-z plane of symmetry for N = 2, 3, 5 and 7 were computed and compared with the observed values [3]. A typical plot for N = 3 is shown in Fig. 6. The close agreement between the observed and predicted velocity distribution is evident beyond $x \cong 15\sqrt{A}$.

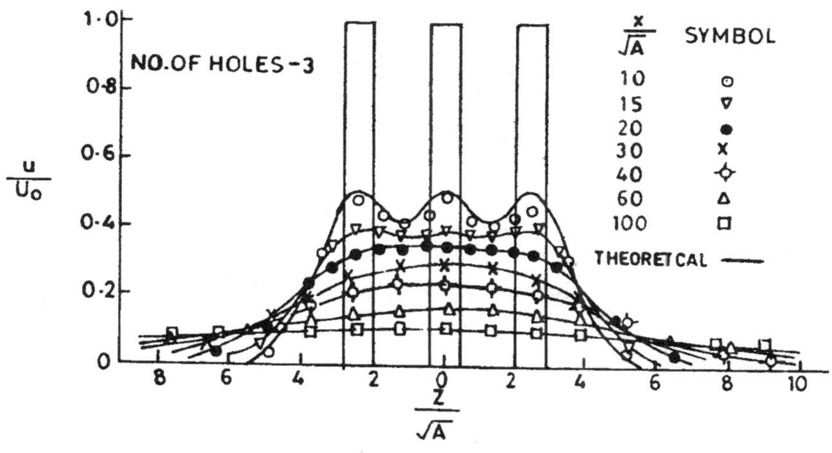

Figure 6. Velocity distribution in x-z plane of symmetry [3]

9.3 Three Dimensional Wall Jets

The non-dimensional velocity distribution in the x-y plane of symmetry normal to the wall is as shown in Fig. 7 for the case of a rectangular outlet of aspect ratio 1.5. The exponential curve given by Eq. 36 seems to be a reasonable fit, excepting in the proximity of the wall. The maximum value at $y \cong 0.16b_y$ is not very different than the one given by the Gaussian distribution. The velocity distribution parallel to the x-y plane at a distance of $\eta_y = 0.16$, is also shown in Fig.7.

The agreement of the observed values with that of the theoretical distribution can be seen for $x/\sqrt{A}$ exceeding 20. Hence, neglecting the wall shear effects, the decay of the maximum velocity u_{mo} was computed using c_1 and c_2 values found from experiments. The agreement between the observed and computed values u_{m0} in

Fig. 8 is good upto a value of $x/\sqrt{A} \cong 100$. This is not surprising since the corrections to account for the wall shear effect is very small in the initial reaches of the flow. Wall jets which issue from other shapes also behave in an identical manner [3].

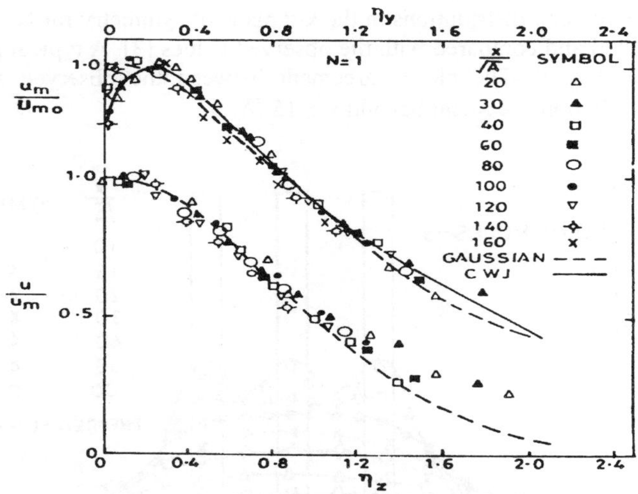

Figure 7. Non-dimensional velocity distribution: Single three-dimensional wall jet [3]

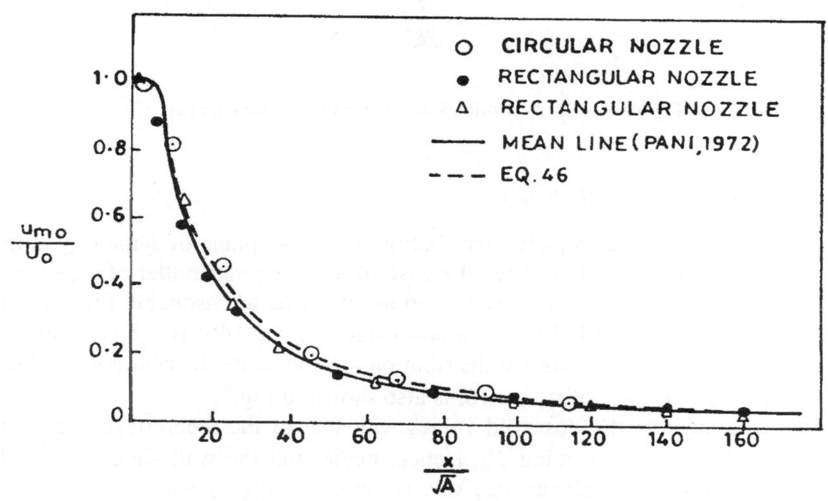

Figure 8. Decay of maximum velocity for three-dimensional wall jets [3]

9.4 Wall Jets Emanating from Multiple Outlets

The two length scales b_y and b_z for a single rectangular outlet can be used in Eq. 47 for various N values. The observed u_{m0} values follow closely the predicted relationship, Fig. 9.

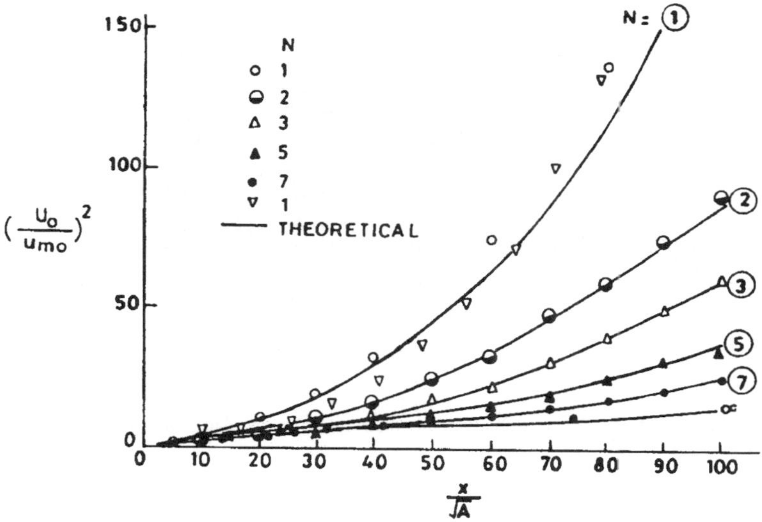

Figure 9. Variation of $\left(\dfrac{u_0}{u_{m0}}\right)^2$ with $\dfrac{x}{\sqrt{A}}$: Multiple wall jets [3]

As the value of N becomes large, the relationship between $(u_{m0}/U_0)^2$ and $x/\sqrt{A}$ tends to become linear and lies close to the plane wall jet type of diffusion. Further, the velocity distribution in the x-z plane (for $y = \delta$) shows that the profiles can be computed from theory with a fair degree of accuracy as shown in Fig. 10, which is a typical plot for N = 3. The velocity profiles normal to wall also exhibit similarity and are described by the classical wall jet profile. This is depicted in Fig. 11.

The behaviour of wall jet emanating from multiple circular outlets was also noticed to be identical to the rectangular outlet case. The spacing of the outlet has some influence on the decay of the maximum velocity as well as on the downstream distance at which merger of the jets takes place. As the spacing increases, the maximum velocity u_{m0} decays faster. A similar observation was made by Marsters [13] for free jets issuing from a linear array of rectangular outlets.

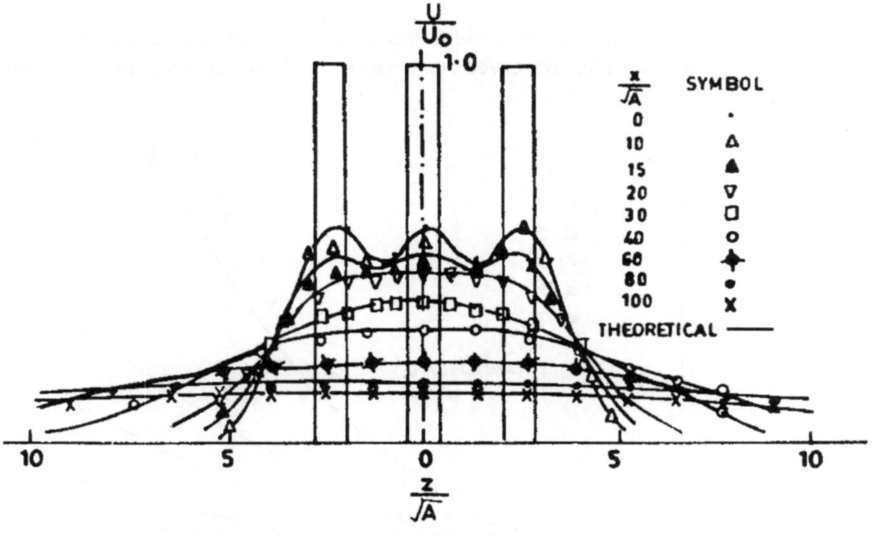

Figure 10. Non-dimensional velocity distribution in x-z plane: Multiple wall jets [3]

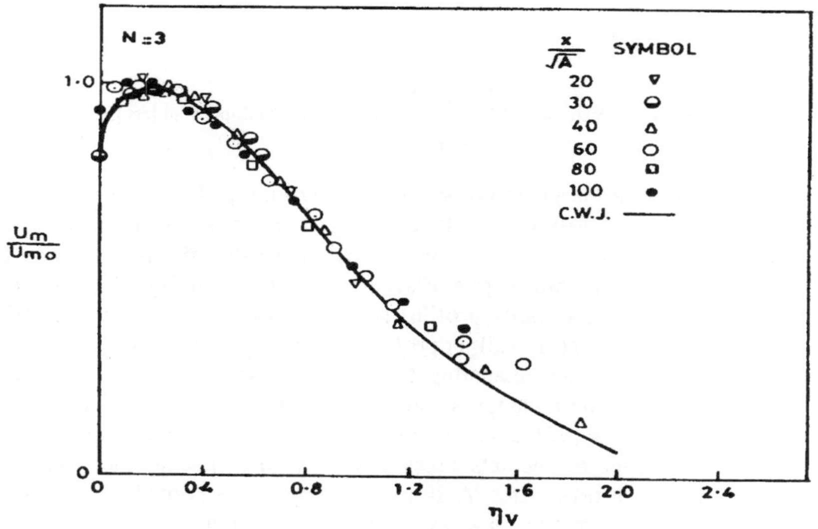

Figure 11. Non-dimensional velocity distribution in x-y plane: Multiple wall jets [3]

10 Use of Single Mixing Length for Three Dimensional Jets

10.1 Momentum Flux

So far, the different rates of spread in the x-y and the x-z planes of a three-dimensional jet were dealt with by introducing two mixing lengths in the y and z directions. The elementary solution for the distribution of the momentum flux in the developed zone of flow from an array of circular/rectangular outlets were superposed to derive the dynamic pressure field in the developed zone of multiple outlets. However, the flux distribution across the potential core and the characteristic decay region (jointly designated as the developing zone) of individual jets was not dealt with. The following sections describe how a single mixing length can be very effective in deriving the distribution of momentum flux and the mean velocity field without treating the flow field as different entities like the potential core, the characteristic decay and the axisymmetric decay region. The uncertainty in the location of different virtual origins, associated with the velocity decay and the growth of the shear layer in the y and z directions, can also be dispensed with. Recall that application of Reichardt's hypothesis for the case of an axially symmetric flow resulted in the following expression for the maximum dynamic pressure

$$\frac{u_m^2}{U_0^2} = \frac{dA}{2\pi b^2} \tag{20}$$

Further, the distribution of the momentum flux across the shear layer was also derived, which is re-written below as

$$\frac{u^2}{u_m^2} = \exp\left[-\frac{1}{2}\left(\frac{r}{b}\right)^2\right] \tag{50}$$

Consider a point P, Fig. 12, in any y-z plane at a downstream distance x from the plane of the rectangular outlet of size $2L_0$ x $2B_0$. Supposing the elemental area dA is located at distances of L and B from the x-axis in the y and z directions respectively. From Eqs. 20 and 50, the dynamic pressure at point P can be obtained by superposition. Thus, the total dynamic pressure resulting from flows through the outlet is given by

$$\frac{u^2}{U_0^{\,2}} = \int_{-L_0}^{L_0}\int_{-B_0}^{B_0} \frac{1}{2\pi b^2} \exp\left[-\frac{1}{2}\left\{\left(\frac{Z-L}{b}\right)^2 + \left(\frac{y-B}{b}\right)^2\right\}\right] dL.dB \tag{51}$$

The above expression can be integrated to yield the final result as,

$$\frac{u^2}{U_0^2} = \frac{1}{4}\left\{erf\left(\frac{y-B_0}{\sqrt{2}\cdot b}\right) - erf\left(\frac{y+B_0}{\sqrt{2}\cdot b}\right)\right\}\left\{erf\left(\frac{z-L_0}{\sqrt{2}\cdot b}\right) - erf\left(\frac{z+L_0}{\sqrt{2}\cdot b}\right)\right\} \quad ...(52)$$

To get the maximum dynamic pressure along the x-axis, select y = 0 and z = 0 in Eq. 52, resulting in the following relationship

$$\frac{u_{m_0}^2}{U_0^2} = erf\left(\frac{B_0}{\sqrt{2}\cdot b}\right)\cdot erf\left(\frac{L_0}{\sqrt{2}\cdot b}\right) \quad (53)$$

Hence, the velocity decay relationship works out as,

$$\frac{u_{m_0}}{U_0} = \left[erf\left(\frac{B_0}{\sqrt{2}\cdot b}\right)\cdot erf\left(\frac{L_0}{\sqrt{2}\cdot b}\right)\right]^{\frac{1}{2}} \quad (54)$$

At any distance x in the downstream direction, the velocity distribution u/u_{m0} can be obtatined by suitably combining Eqs. 52 and 53.

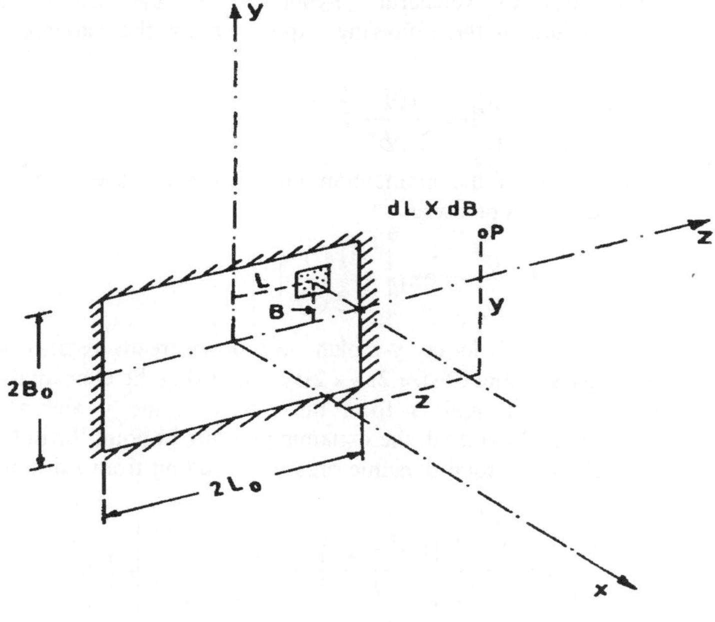

Figure 12. Point source definition sketch

10.2 Heat Flux

Let the instantaneous absolute temperature T be expressed as $T = T_0+\Delta T$, where T_0 is a reference temperature independent of x or r, and ΔT is the temperature excess which depends both on x and r. Using the turbulent transport equation for a scalar quantity and the continuity equation, one can get

$$\frac{\partial}{\partial x}\left(\overline{U\Delta T}\right)+\frac{1}{r}\cdot\frac{\partial}{\partial r}\left(r\overline{V\Delta T}\right)=0 \tag{55}$$

Taking recourse to Reichardt's hypothesis, the lateral transport of heat flux is given as

$$\overline{V\Delta T} = -\Lambda_t \frac{\partial}{\partial r}\left(\overline{U\Delta T}\right) \tag{56}$$

Combining Eqs. 55 and 56, and assuming that the distribution of the heat flux satisfies the similarity relationship

$$\frac{\overline{U\Delta T}}{\overline{U_m\Delta T_m}} = g\left(\frac{y}{b_t}\right) = g(\zeta) \tag{57}$$

it is possible to solve for the distribution function g. The solution works out to be

$$g = \frac{\overline{U\Delta T}}{\overline{U_m\Delta T_m}} = e^{-\frac{\zeta^2}{2}} \tag{58}$$

This is identical to that of the momentum flux distribution, excepting that the length scale b_t is different than b.

Based on the fact that the heat flux is conserved, it is possible to express the heat flux distribution for an elemental jet of area dA as

$$\frac{\overline{U_m\Delta T_m}}{\overline{U_0\Delta T_0}} = \frac{dA}{2\pi b_t^2} \tag{59}$$

or ,

$$\frac{\overline{U\cdot\Delta T}}{\overline{U_0\Delta T_0}} = \exp\frac{\left(\frac{-\zeta^2}{2}\right)\cdot dL\cdot dB}{2\pi b_t^2} \tag{60}$$

Hence, by summing up the contributions from various elemental areas, an expression for the distribution of heat flux for a three-dimensional jet can be shown to be as follows

$$\frac{\overline{U\cdot\Delta T}}{\overline{U_0\Delta T_0}} = \frac{1}{4}\left\{erf\frac{y-B_0}{\sqrt{2.}\cdot b_t} - erf\frac{y+B_0}{\sqrt{2.}\cdot b_t}\right\}\left\{erf\frac{z-L_0}{\sqrt{2.}\cdot b_t} - erf\frac{z+L_0}{\sqrt{2.}\cdot b_t}\right\} \tag{61}$$

As the temporal mean of the product of the fluctuating velocity and the fluctuating excess temperature is an order smaller than $u \cdot \overline{\Delta T}$, one can replace the left hand side of Eq. 61 by $\dfrac{u \cdot \overline{\Delta T}}{U_0 \overline{\Delta T_0}}$. With the help of the momentum distribution, given in Eq. 52, the following expression for the distribution of mean temperature excess can be written as

$$\frac{\overline{\Delta T}}{\overline{\Delta T_0}} = \frac{\frac{1}{4}\left\{erf\left(\frac{y-B_0}{\sqrt{2}\cdot b_t}\right) - erf\left(\frac{y+B_0}{\sqrt{2}\cdot b_t}\right)\right\}\left\{erf\left(\frac{z-L_0}{\sqrt{2}\cdot b_t}\right) - erf\left(\frac{z+L_0}{\sqrt{2}\cdot b_t}\right)\right\}}{\left[\frac{1}{4}\left\{erf\left(\frac{y-B_0}{\sqrt{2}\cdot b}\right) - erf\left(\frac{y+B_0}{\sqrt{2}\cdot b}\right)\right\}\left\{erf\left(\frac{z-L_0}{\sqrt{2}\cdot b}\right) - erf\left(\frac{z-L_0}{\sqrt{2}\cdot b}\right)\right\}\right]^{\frac{1}{2}}} \quad \ldots(62)$$

Error function tables can be used to evaluate the right hand side easily for various locations (y,z) over a cross section when the size of the nozzle (L_0, B_0) and the widths b and b_t are specified.

11 Co-flowing Jets

The point source concept and the principle of superposition can be extended to coflowing circular jets. Available experimental results of Landis and Shapiro [1] show the validity of the approach. On similar lines, the analysis can be extended to predict the mean velocity field of co-flowing three-dimensional jets, the definition sketch for which is given in Fig. 13.

Following the steps outlined earlier for a jet diffusing in an ambient fluid at rest, for an elemental point source it can be found that

$$\frac{u(u-U_1)}{U_0(U_0-U_1)} = \frac{dA}{2\pi b^2}\exp\left(\frac{\xi^2}{2}\right) \qquad (63)$$

By superposing the solution from elemental areas, r.dr.dϕ, spread over the cross-section of a circular nozzle of diameter D, the following expression for the excess momentum flux can be arrived at

$$\frac{u_m\,(u_m-U_1)}{U_0(U_0-U_1)} = 1-\exp\left\{-\frac{1}{2}\left(\frac{D}{cx}\right)^2\right\} \qquad (64)$$

Let us denote $u_m - U_1 = \Delta u_m$, and $U_0 - U_1 = \Delta u_0$. Hence, the above expression can be re-written in the following form

$$\frac{x}{D} = \frac{1}{2\sqrt{2}.c}\left\{-\ln\left(1-\frac{u_m}{U_0}\frac{\Delta u_m}{\Delta u_0}\right)\right\}^{-\frac{1}{2}} \tag{65}$$

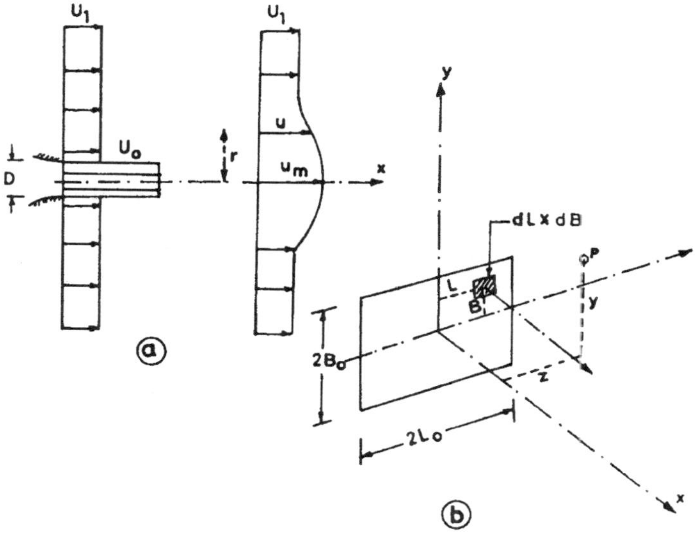

Figure 13. Definition sketch for co-flowing jets

Further, at any location x, let the ratio of the velocity excess Δu_m on the axis of jet and the velocity excess Δu_0 at the efflux section be denoted by q and the ratio U_1/U_0 be designated as β. Then, from Eq. 65 we can write

$$\frac{x}{D} = \frac{1}{2\sqrt{2}.c}\left[-\ln\left\{1-\beta\,(q-q^2)-q^2\right\}\right]^{-\frac{1}{2}} \tag{66}$$

The form of the relationship expressed in Eq. 66 is very convenient to evaluate the coefficient c from experimentally observed maximum excess velocity decay for specified values of β.

Riverting our attention to the co-flowing three-dimensional jets, it can be easily proved that on the axis of the jet

$$\frac{u_{m0}\Delta u_{m0}}{U_0 \Delta U_0} = erf\left(\frac{B_0}{\sqrt{2}\,cx}\right).erf\left(\frac{L_0}{\sqrt{2}\,cx}\right) \tag{67}$$

Thus, the expression shows that in general

$$\frac{u_{m0}\Delta u_{m0}}{U_0\Delta U_0} = \phi\left[\frac{cx}{2B_0}, e\right] \tag{68}$$

where, e is the aspect ratio of the rectangular outlet. The role of the velocity ratio U_1/U_0 appears implicitly and is reflected in the value of c.

Based on Eq. 67, a set of theoretical curves can be drawn as shown in Fig. 14, considering e as the third parameter. However, experimental results are not available to check the validity of the relationship.

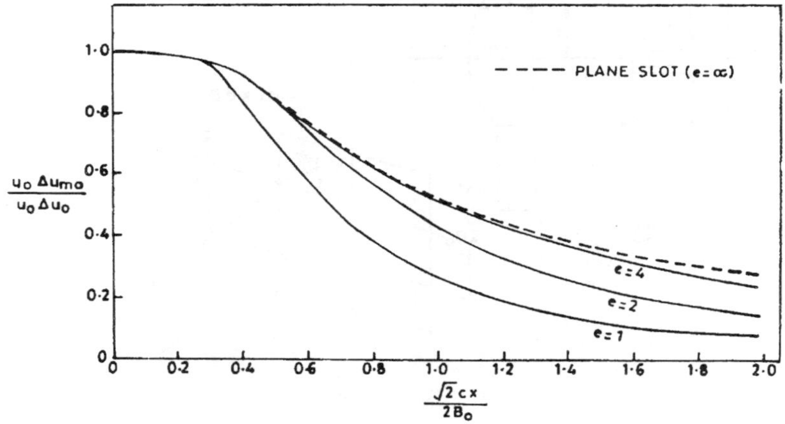

Figure 14. Variation of excess momentum flux for three-dimensional co-flowing jets [21]

12 Experimental Verification

12.1 Circular Jets

The variation of the centre line velocity u_m for a circular jet is shown in Fig. 15, wherein data from various sources are used. The theoretical curve of Maxwell [14] which is also based on the point source method can be seen to agree with the observed values for x/d greater than 12. Most investigators, including Maxwell, introduce a correction for the location of the virtual origin to help in bringing a closer agreement between the two distributions. However, such an approach is not warranted while applying the point source method of solution. The widely accepted relationship u_{m0}/U_0 = 6.2/ (x/d) fits very well with the data in the fully developed zone.

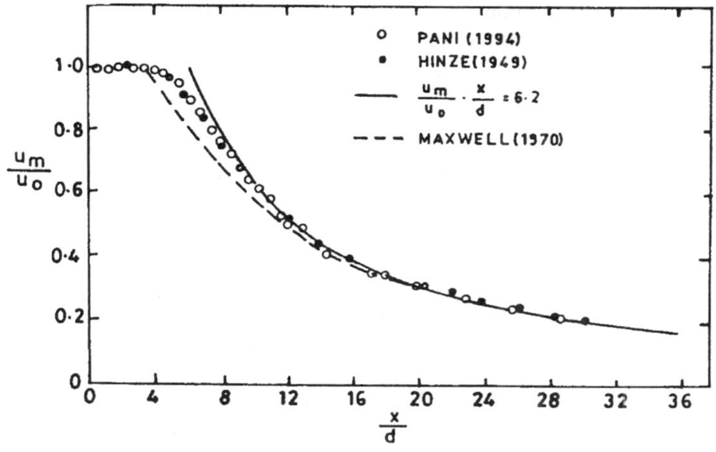

Figure 15. Axial velocity decay – Circular nozzle

The decay of the temperature excess along the axis of the jet is given in Fig. 16. Though there is reasonable agreement between the observed data and the theoretical curve of Abramovich [1] deviations are evident in the developing zone, very near the orifice. To check the present approach, a slightly modified route is followed as explained below.

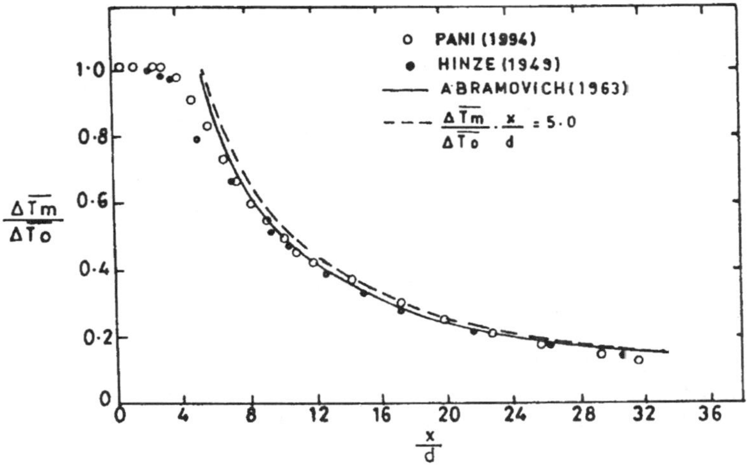

Figure 16. Axial temperature decay - Circular nozzle

For the case of axially symmetric jet, using linear superposition, Eqs. 20 and 59 will result in the following which involves the maximum velocity u_m and the maximum temperature excess ΔT_m. The expressions are,

$$\frac{x}{D} = \frac{1}{2\sqrt{2}.c}\left\{-\ln\left(1-\frac{u^2{}_m}{U^2{}_0}\right)\right\}^{-\frac{1}{2}} \tag{69}$$

and

$$\frac{x}{D} = \frac{1}{2(\sqrt{2}.c)\lambda}\left\{-\ln\left(1-\frac{\Delta T_m^2}{\Delta T_0^2}\right)\right\}^{-\frac{1}{2}} \tag{70}$$

A plot of the observed velocities and the temperature excess in the above forms is shown in Fig. 17 resulting in √2 c = 0.080 and the spread coefficient, λ = 1.18. These values compare well with the values suggested by many other investigators. The linear plot amply justified the validity of the basic equation derived from the point source concept. Making use of the above c and λ values, the lateral distribution of momentum flux and the heat flux can be found out for specified streamwise locations.

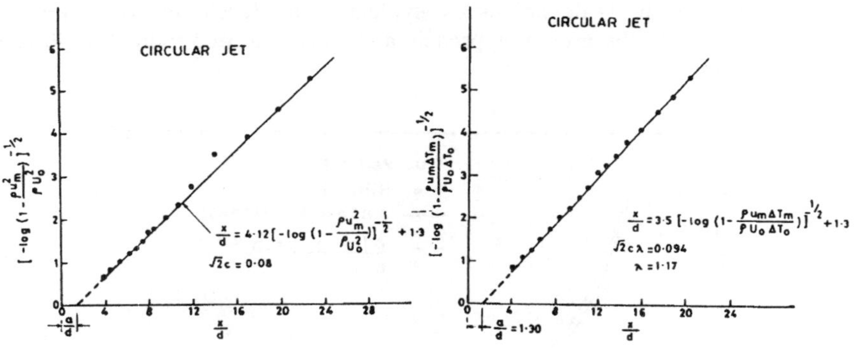

Figure 17. Plot for determination of c and λ - Circular jet [21]

A typical set of results is shown in Fig. 18 for x/√A = 6.0. The agreement between computed and observed values is quite remarkable. For large x/√A values the theoretical distributions approach the Gaussian curve and is in agreement with the observations of other investigators in the developing zone as well as in the developed zone of the jet. Thus, the present solution is able to predict the distributions.

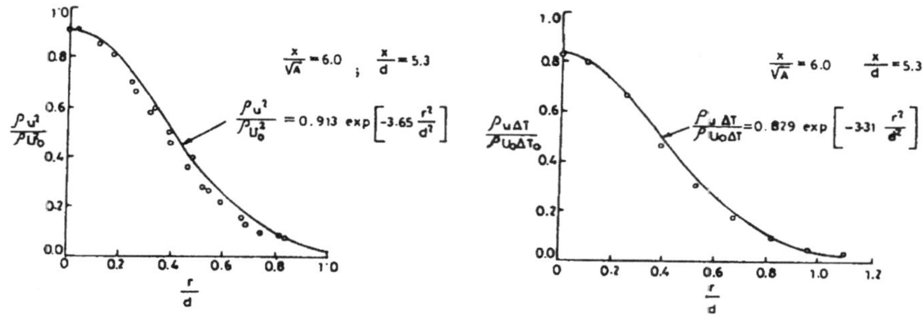

Figure 18. Lateral distribution of momentum and heat flux [21]

12.2 Three Dimensional Jets

Measured distributions of velocity and temperature excess [21] were analyzed for the case of a square nozzle resulted in $\sqrt{2}c = 0.085$ and $\lambda = 1.136$. These are somewhat different from the values found for circular nozzles indicating that the shape of the outlet could not be fully ignored. However, to avoid multiplicity of choice for values of $\sqrt{2}c$ and λ, the values pertaining to axially symmetric flows are used in the analysis of all three-dimensional jets that follow. Typical results for a rectangular jet having an aspect ratio of 3 are shown in Figs. 19a, 19b and 20a, 20b for the distribution of velocity and temperature in two mutually perpendicular planes. The agreement between the theoretical distribution and the observed values is superior in the x-y plane compared to the x-z plane. Similar observations can be made in Maxwell's [14] experiments concerning the distribution of momentum flux.

The decay of the maximum velocity and temperature excess along the axis of the jet is shown in Fig. 21. The observed temperature measurements show a better agreement with the predicted values, a fact which was also found to be true for jets having aspect ratios of 4 and 8 [21]. In the past Narain [17] attempted to bring better agreement between predicted and observed values by using different values for the spread coefficient c in the developing and the developed zone. However, such adjustments in c negate the generality of the point source concept applied to the case of three-dimensional jets and need to be avoided.

12.3 Co-flowing Circular Jets

The spread coefficient of coflowing jets depends primarily on the velocity ratio β. The results of Landis and Shapiro [1] on co-flowing circular jets were analysed by plotting the excess momentum flux along the axis of the jet as given in Eq. 69. The spread coefficient $\sqrt{2}$ c was found to be 0.08, 0.0476 and 0.0327 for $\beta = 0$, 0.25 and 0.46 respectively.

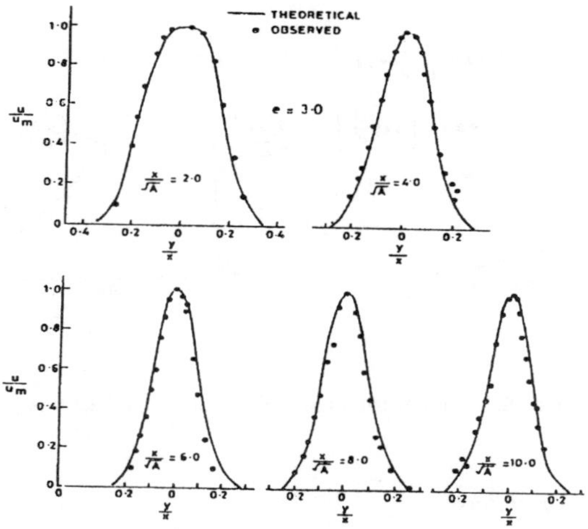

Figure 19(a). Velocity distribution in x-y plane: Point source method [21]

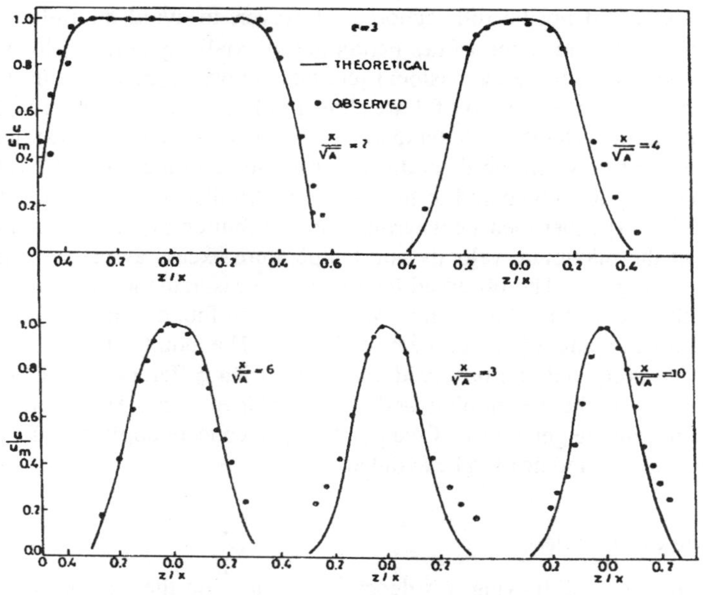

Figure 19(b). Velocity distribution in x-z plane: Point source method [21]

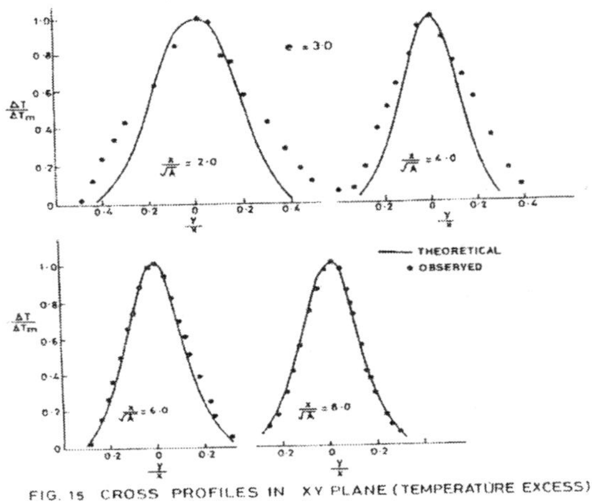

Figure 20(a). Temperature distribution in x-y plane: Point source method [21]

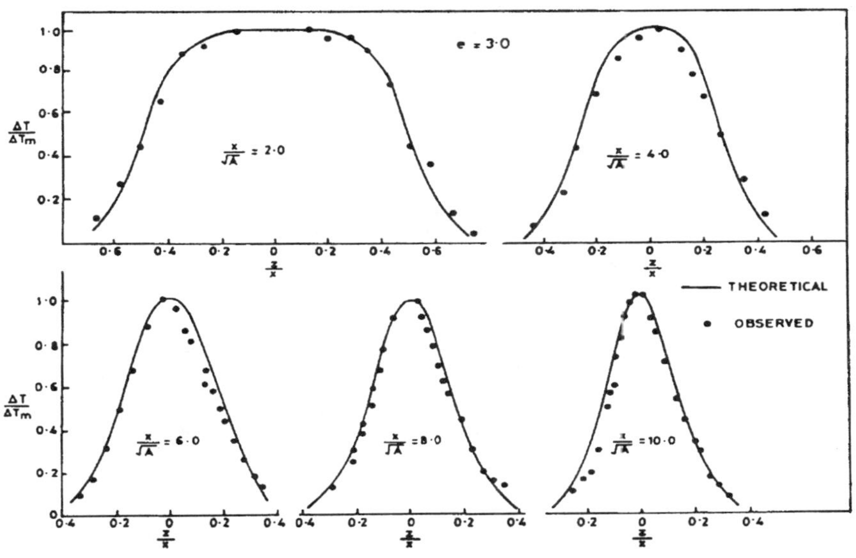

Figure 20(b). Temperature distribution in x-z plane: Point source method [21]

These values are quite close to those given by Abramovich's empirical relation

$$b = (\sqrt{2}\,c)_0\, x \frac{1-\beta}{1+\beta} \tag{71}$$

where, the bracketed quantity corresponds to β = 0 condition. The variation of the spread coefficient is shown in Fig. 22.

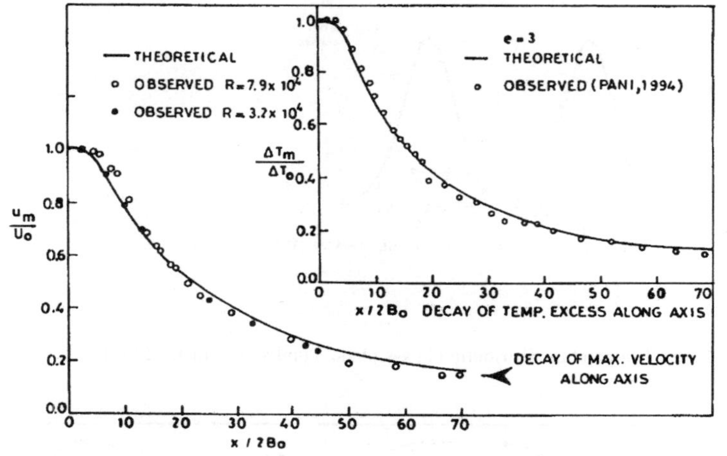

Figure 21. Decay of maximum velocity and temperature excess for a three-dimensional jet: Point source method [21]

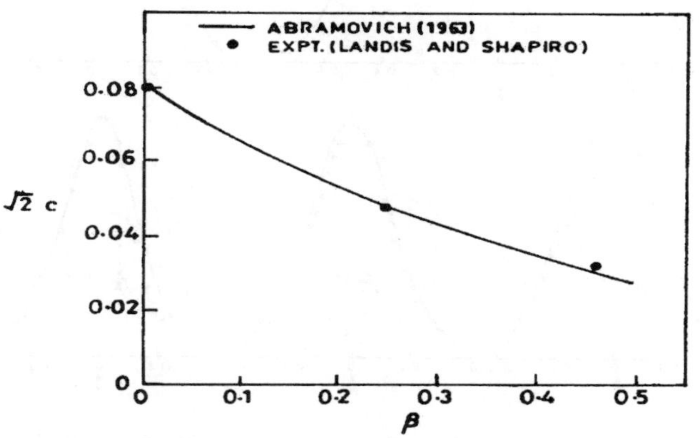

Figure 22. Variation of c with velocity ratio β

The decay of the excess momentum flux along the jet axis is found from Eq. 64 and the variation is shown in Fig. 23. The agreement between observed and predicted values seems to be reasonable for various β values. It was observed that though the lateral distribution of the excess momentum flux is Gaussian for large distances from the nozzle, in case of coflowing streams the actual distribution adjusts in a very gradual fashion to attain the normal distribution.

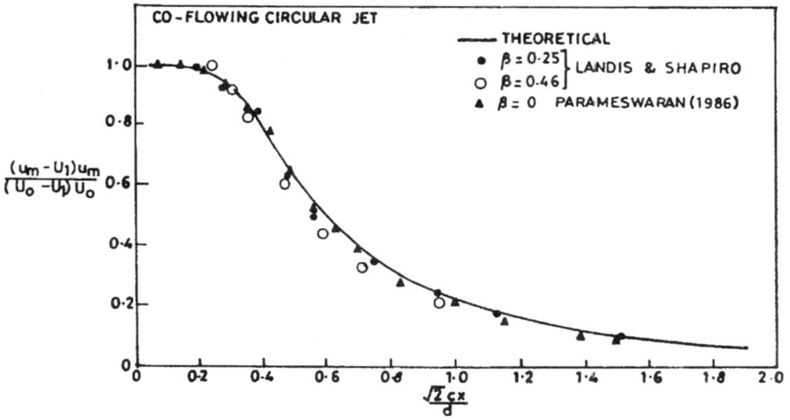

Figure 23. Momentum-excess flux along the jet axis

13 Diffusion of Jets in a Confined Space

In the analysis presented thus far, the jets were assumed to diffuse in ambient fluids, extents of which were not limited by any boundary. However, in some applications, turbulent jets diffuse in ambients which are bounded by solid boundaries. Two typical cases are presented below to demonstrate that the point source approach can be generalised and extended to estimate the momentum and energy fluxes, which are quite close to observed values.

13.1 Head Loss in Junction Boxes

Manholes or junction boxes are provided in storm sewer networks at regular intervals. The measured head loss coefficient K values are much smaller than the conventional value of 1.5 obtained from minor-loss formulas. Pedersen and Mark [22] applied the submerged jet theory to predict K for a through flow condition. The ratio of the longitudinal length of the junction box (D_m) to the inlet pipe diameter is normally less than six. Hence, any section of the jet located at a distance x from the efflux section comprises of two zones, namely, the potential core and the shear layer. The sum of the contributions from potential core and the shear layer yield the discharge and energy flux over the cross-section. Mudgal [15] adopted Reichardt's inductive theory for free turbulence and using the point source concept, evaluated the head loss in junction boxes. The variation of the momentum and energy flux in the core of the jet carrying the initial mass is shown in Fig. 24. The curves suggested by Abramovich [1] are also shown. Supported by experiments, it was found by Mudgal [1] that the submerged jet theory can be applied successfully to predict the head-loss coefficient K for junction boxes of different sizes.

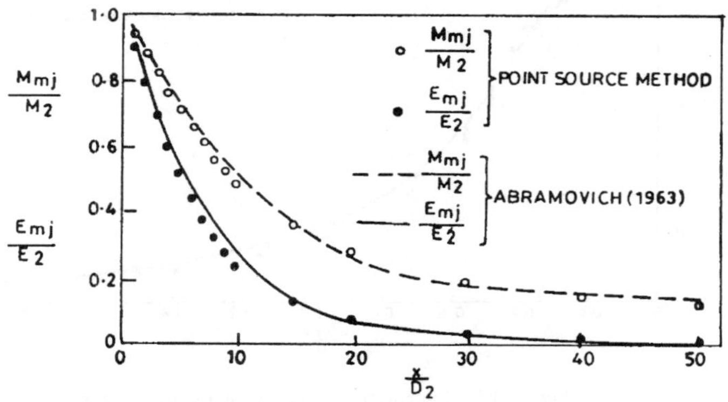

Figure 24. Momentum and energy flux of jets in a confined space [15]

13.2 Mixing in Flocculators

Removal of suspended solids from the raw water is an important operation in water treatment plants. Jets can be used for rapid mixing of coagulants with the water in the flocculation chamber. As jet flocculators do not have moving mechanical components, problems associated with malfunctioning and maintenance of the unit can be eliminated. The removal of suspended solids by jet flocculators is as effective as any other hydraulic flocculator. The intensity of mixing depends upon the temporal mean velocity gradient G. It can be measured or calculated in terms of the power input by the following expression [2].

$$G = \sqrt{\frac{P}{\mu \forall}} \tag{72}$$

where, P = total input of power in Watts, μ = absolute viscosity of water (N-S/m^2) and ∀ = volume of water to which power is supplied (m^3). It has been reported [8] that high G values result in break-up of flocs. The product G.T is a useful parameter for the design and operation of flocculators.

The power input i.e. the loss of energy P in the jet flocculation chamber was computed by Suresh Kumar [30] making use of the point source concept. The estimated values were in good agreement with the measured values.

Hence, alternate sizes of the flocculation chamber can be designed to have the G value in the recommended optimal range to effect good floc formation.

14 Conclusions

1. Reichardt's hypothesis can be used to linearise the equations of motion and the resulting solution is suitable for linear superposition.
2. The point source concept is a very general method of predicting the momentum and heat flux downstream of axially symmetric jets, three-dimensional and plane jets.
3. It suffices to use a single mixing length in the theoretical solution. The same solution helps to find the distribution of momentum and heat flux in the developed and developing flow regions of all types of jets. Thus, there is no need to divide the flow field into different zones and the results need no correction for the location of the virtual origin.
4. For jets issuing from outlets of finite size, having an area of cross section A, the lateral distribution of momentum and heat flux tends to become Gaussian in nature. This is generally true for downstream locations beyond $10\sqrt{A}$ from the plane of the outlet.
5. The point source method can be adopted to predict the flow field of multiple jets, circular or rectangular in cross-section, arranged in a row.
6. The spread coefficient c in case of coflowing jets depends on the ratio of the secondary stream velocity to that of the primary jet velocity prevailing at the nozzle exit. The distribution of momentum and heat flux for coflowing jets can be found out by superposing the point source solutions if an appropriate c value is used.
7. The decay of maximum velocity for three-dimensional wall jets can be predicted by considering the wall as a frictionless reflector. The results are good, specifically for $x < 100\sqrt{A}$.
8. The methodology of treating the cross-section of the jet as an agglomerate of a large number of point sources can also be used to predict the loss of energy due to diffusion of jets in confined spaces like junction boxes and flocculators.

References

1. Abramovich G. N., The Theory of Turbulent Jets (English edition translated by Scripta Technica, MIT Press, Cambridge, Mass., 1963).

2. Camp T. R. and Stein P. C., Velocity gradients and internal work in fluid motion *Jr. Boston Soc. Civil Eng.*, **30** (1943), pp. 219-237.
3. Dash R., Studies on single and multiple three-dimensional turbulent jets, (Ph. D. Thesis, Dept. of Civil Engg., Indian Institute of Technology, Bombay, India, 1981).
4. Demissie M. and Maxwell W. H. C., Three-dimensional slot jets, *Jr. of Hyd. Division*, **108** (1982), No. HY2, Tech. Note, pp. 247-251.
5. Dyban Ye. P. and Mazur A. I., A method for calculating the axial velocity in an array of axisymmetric jets issuing from a perforated plate, *Fluid Mechanics, Soviet Research*, **8** (1979), pp. 30-38.
6. Dyban Ye. P. and Mazur A. I., Certain features of an array of axisymmetric jets issuing from a perforated plate, *Fluid Mechanics Soviet Research,* **8** (1979), pp. 91-98.
7. Fabris G. and Fejer A. A., Confined mixing of multiple jets, *Jr. of Fluids Engg.,* **96** (1974), pp. 92-95.
8. Gemmel Robert S., Mixing and Sedimentation (in A Hand-book of Public Water Supplies by American Water Works Association Inc., McGraw-Hill Company, New York, 1971) pp. 123-157.
9. Hinze J. O., Turbulence (McGraw-Hill Series in Mechanical Engg., New York, 1975).
10. Hinze J. O., Zijnen B. G. and Van Der Hegge, Transfer of heat and matter in the turbulent mixing zone of an axially symmetric jet, *Jr. Appl. Sci. Res.*, **A1** (1949) pp. 435-461.
11. Knystautas R., The turbulent jet from a series of holes in line, *Aeronautical Quarterly*, **15** (1964) pp. 1-29.
12. Krothapalli A., Bagnoff D. and Karamchetty K., Development and structure of rectangular jets in a multiple jet configuration, *AIAA Jour.*, **18** (1980) pp. 945-950.
13. Marsters G. F., Measurements in the flow field of a linear array of rectangular nozzles, *Jr. of Aircraft*, **17** (1980) pp. 774-780.
14. Maxwell W. H. C., Flux development region in submerged jets, *Jr. of Engg. Mechanics Div.*, **96** (1986) No. EM6, pp. 1061-1079.
15. Mudgal B. V., Drag characteristics of sills in forced hydraulic jumps and energy loss in junction boxes (Ph. D. Thesis, Deptt. of Civil Engg., Indian Institute of Technology, Bombay, India, 1994).
16. Narain J. P., Three-dimensional turbulent wall jets, *Canadian Jr. of Chemical Engg.*, **53** (1975) pp. 245-251.
17. Narain J. P., Momentum flux development from three-dimensional turbulent free jets, *Jr. of Fluids Engg.*, **96** (1976).
18. Pani B. S., Three dimensional turbulent wall jets (Ph. D Thesis, Dept. of Civil Engg., University of Alberta, Edmonton, Alberta, Canada, 1972).
19. Pani B. S. and Dash R., Three-dimensional single and multiple free jets, *Jr. of Hyd. Engg.*, **109** (1983) pp. 254-269.

20. Pani B. S. and Parameswaran P. V., Momentum and heat flux characteristics of three dimensional jets based on point source concept, *Jr. Hyd. Res.*, IAHR, **32** (1994) pp. 53-66.
21. Parameswaran P. V., Characteristics of three-dimensional heated jets and swirling buoyant jets (Ph. D. Thesis, Deptt. of Civil Engg., Indian Institute of Technology, Bombay, India, 1986).
22. Pedersen F. B. and Mark O., Head loss in storm sewer manholes: submerged jet theory, *Jr. of Hyd. Engg.* **116**(1990) No. HY11, pp. 1317-1328.
23. Rajaratnam N., Turbulent Jets (Elsevier, Amsterdam, 1976).
24. Rajaratnam N. and Pani B. S., Three-dimensional turbulent wall jets, *Jr. of Hyd. Division*, **1** (1974) pp. 69-83.
25. Rajaratnam N. and Subramanya K., Diffusion of rectangular wall jets in channels, *Jr. Hyd. Res.*, IAHR, **5** (1967) pp. 281-294.
26. Ranganathan S. and Reid I. M., A study of multiple jets, Tech. Note, *AIAA Jour.*, **19** (1981) pp. 124-127.
27. Reichardt H., On a new theory of free turbulence, *Jr. Roy. Aero. Soc.*, **47** (1943) pp. 167-176.
28. Sfier A. A., Investigation of 3-D turbulent rectangular jets, *AIAA Jour.*, **17** (1979) pp. 1055-1060.
29. Sforza P. M., Steiger M. H. and Trentacoste N., Studies on three-dimensional viscous jets", AIAA Jour., **4** (1966) pp. 800-806.
30. Suresh Kumar N., Jet mixing in flocculators (Ph. D Thesis, Dept. of Civil Engg., Indian Institute of Technology, Bombay, India, 1997).
31. Trentacoste N. and Sforza P. M., Further experimental results for three-dimensional free jets, *AIAA Jour.*, **5** (1967) pp. 885-891

VELOCITY AND SHEAR DISTRIBUTIONS IN OPEN CHANNELS

KANDULA V. N. SARMA
Viswanadha Institute of Technology and Management
104, Kandula Vari Illu , 48-8-19, Dwaraka Nagar
Vishakhapatnam-16
INDIA

BOYINA V. RAMANA PRASAD
Department of Environmental and Energy Engineering
Rensselaer Polytechnic Institute
5049 JEC / 110 8th Street
Troy, NY, 12180-3590
USA

This article presents and discusses critically the attempts made so far by different investigators in studying the velocity and shear distributions for flow in open channels and in particular narrow rectangular open channels. The paper is basically written based on the systematic and well thought out experimental works carried out by a number of investigators spanning over two decades on the topic at the Indian Institute of Science, Bangalore, India. Studies carried out by other investigators on the subject matter have also been incorporated in the review as and when considered necessary. Based on the review and analysis of results, a summary of important findings and suggestions of the necessity for future research have been furnished. It is believed that it would be broadly acceptable to the profession. The present review is mostly restricted to the simple case of rigid bed open channels.

1 Introduction

The flow in open channels is turbulent except in special situations. The theoretical investigations of Prandtl starting from 1914 [60] and von Karman [102], and experimental studies of Nikuradse [58] have led to rational formulas for velocity distribution and hydraulic resistance for turbulent flows over flat plates and in circular pipes. These formulas have been extended to open channel flow first by Keulegan [41] and later by many others. There are some general similarities that exist between flow through pipes and the flow through open channels. There are also certain major differences between them such as the presence of a free surface, the three dimensional nature of flow due to noncircular cross section of the flowing stream tube, the maximum velocity not always occurring at the free surface which is actually the end of a fully developed boundary layer, but at some depth below it and consequential non-uniform distribution of shear along the wetted perimeter.

However, in wide-open channels the nature of flow is two dimensional and the shear stress is constant over a greater length of the bed. Thus, the formulas for pipes can be applied to those channels by changing the constants suitably to include the free surface effects [77]. The velocity distribution in the middle region is not influenced by the side walls which are far away [77, 15] and the flow in that region can be considered as a very close approximation of the two dimensional flow.

It has become a customary practice to connect cross sectional mean velocity, V to the mean shear stress, $\bar{\tau}_0$ over the entire wetted perimeter as in the case of pipe flows by replacing the pipe diameter, d_0 with 4R in which R is the hydraulic radius of the rectangular open channels [41, 49, 15, 29, 64, 91]. However in the case of a narrow rectangular open channel neither the velocity distribution in a vertical nor along a horizontal line can be expressed in terms of the mean shear stress. The corresponding local shear stresses on the bed and wall are to be used for vertical and horizontal velocity distributions respectively.

To be able to use the relevant shear stress prior knowledge of shear stress and its distribution on the channel boundary becomes necessary. The direct measurement of shear stress at various locations of the wetted surface of the open channel is extremely difficult [3, 28] and one has to compromise and go for one or more of the indirect methods of shear measurements. One of the indirect methods commonly used is that due to Preston [61] which is based on a known velocity distribution. The velocity distribution contains some constants. If it is assumed that the values of these constants depend on the shape of the cross section of the channel in which the flow is occurring, then they have to be evaluated on the basis of the experiments. To evaluate these constants a prior knowledge of the value of local shear stress becomes necessary. Thus one gets into a vicious circle.

The problem is further complicated if the open channel is a rough one. The determination of the characteristic height of roughness elements and the location of reference datum from which distances have to be measured pose serious problems [68, 69, 70, 71].

This article presents and discusses critically the attempts made so far by different investigators in studying the velocity and shear distributions in open channels and in particular narrow rectangular open channels.

2 Various regions of flow

Several regions can be distinguished in a turbulent flow past walls [75].

1. Viscous Sublayer: In this layer the flow is laminar. The outer edge of the sublayer is approximately at a distance of $z/\delta_0 = 0.001$ in which δ_0 = the boundary layer thickness and z = the distance from the boundary.
2. Inner Region: Flow in this region is still dominated by the wall but the viscous stresses are negligible in comparison with Reynolds stresses. The layer extends up to a distance of $z/\delta_0 = 0.15$ in the case of a smooth flat plate to perhaps $z/\delta_0 = 0.27$ in the case of rough open channels. In the case of a turbulent flow past a smooth flat plate $z/\delta_0 = 0.15$ also corresponds to $zu_*/\nu = 750$ where ν is the kinematic viscosity and u_* = the shear velocity = $\sqrt{\tau_0/\rho}$ in which τ_0 is the local shear stress and ρ the mass density.

3. Outer Region: The flow in this region is influenced only by the thickness of the boundary layers. For fully developed boundary layers for pipe and channel flows, the constraints of the entire periphery of the channel geometry like the radius, depth and width replace δ_0.

3 Turbulent flow over smooth flat plate

The mixing length models form the basis for determining the velocity profiles in shear flows. Turbulent flows past walls are shear flows. The Prandtl's mixing length theory [60] results in

$$\frac{u_z}{u_*} = \frac{1}{k} \ln z + \text{Constant} \tag{1}$$

in which u_z = the velocity in the flow direction (X) at z from the wall and k = von Karman constant.

According to Batchelor [2] the mixing length expression represents the equilibrium between the production and dissipation of the turbulent kinetic energy. Ludwieg, Tillman, Hama and many others measured the mean velocity distributions across a turbulent boundary layer along a smooth flat plate with zero pressure gradient [32]. In the inner region, the data is found to fit the universal logarithmic velocity distribution given by the following equation.

$$\frac{u_z}{u_*} = A \ln\left(\frac{zu_*}{\nu}\right) + B \tag{2}$$

Clausser [16] suggested a value of 0.41 for von Karman's constant k or 2.44 for A as it is equal to 1/k and 4.9 for B. A value of A= 2.5 or k=0.4 is more widely used [30]. Nezu and Rodi [56] and Prinos and Zeris [62] obtained a value 0.412 for von Karman constant from the experiments on smooth open channels. Townsend [97] remarked that many of the observed data seem to indicate a value of B nearer to 7 than 4.9. In this article the variation of the additive constant B will be discussed at length at a later stage.

Hama [32] proposed 1/7 th power law for the velocity distribution in the inner region, for the range of zu_*/ν values from 100 to 2000 , as follows.

$$\frac{u_z}{u_*} = 8.3 \left[\frac{zu_*}{\nu}\right]^{1/7} \tag{3}$$

The value of 2000 appears to be rather a conservative estimate of the upper limit for zu_*/ν.

For the velocity distribution in the outer region Hama [32] proposed the simple empirical formula as

$$\frac{u_\delta - u_z}{u_*} = 9.6 \left(1 - \frac{z}{\delta}\right)^2 \tag{4}$$

in which u_δ is the free stream velocity.

4 Turbulent flow through a circular pipe

In the fully turbulent part of the inner region the experimental data of Nikuradse [58] indicated that the logarithmic velocity distribution given by Eq. 2 holds good with A = 2.5 and B = 5.5.

Schlichting [91] obtained 1/7 th power law with the constant 8.74 for velocity distribution in pipe flow based on the Blasius resistance law given by

$$f = \frac{0.316}{R_e^{1/4}} \tag{5}$$

in which f = Darcy- Weisbach friction factor and $R_e = d_0V/\nu = 4\,RV/\nu$. The power law fitted Nikuradse's data up to $R_e = 10^5$.

On the basis of the assumptions of a constant eddy viscosity in the outer region and linear variation of Reynolds stress a velocity distribution law for the outer region of the pipe flow similar to Eq. 4 was obtained as

$$\frac{u_m - u_z}{u_*} = 7.2 \left[1 - \frac{2z}{d_0}\right]^2 \tag{6}$$

in which u_m is the maximum velocity occurring at the center of the pipe. This fitted Lafaur's data well [52].

Zagustin and Zagustin [107] gave an analytical solution for turbulent flow in smooth pipes based on a new concept of balance of pulsation energy. The velocity distribution was obtained by them as

$$\frac{u_m - u_z}{u_*} = \frac{2}{k} \tanh^{-1}\left(\frac{d_0 - 2z}{d_0}\right)^{3/2} \tag{7}$$

5 Turbulent flow through wide-open channels

From the measurements in the middle verticals in eight different channels Bazin [41] obtained the velocity formula for wide-open channels in the form

$$\frac{u_m - u_z}{u_*} = 6.3 \quad (1 - Z)^2 \tag{8}$$

in which Z = z/d , d = the depth of flow. Eq. 8 resembles Eqs. 4 and 6 which were obtained for outer regions. It appears that velocity distribution in the outer region takes the general form

$$U_m - U_z = C \quad (1 - Z)^2 \tag{9}$$

in which $U_m = u_m/u_*$ and $U_z = u_z/u_*$. The constant C takes different values for different boundary geometries.

Keulegan [41] felt that Eq. 2, which is valid for smooth flat plates and circular pipes, could be extended to smooth wide-open channels. He used Bazin's data and verified the general logarithmic form of the velocity distribution equation to be valid.

Vanoni [99], in reference to the two dimensional channels, reported that the logarithmic velocity distribution was justified by his data up to the free surface in the middle region where the flow can be treated as being two dimensional. For aspect ratio, A_r (=width,b/ depth,d) greater than 5 there always exists a central region where the flow is two dimensional. Tracy and Lester [98] studied the velocity distribution in wide rectangular open channels and drew the following general conclusions:

1. The flow is symmetrical about the central vertical plane.
2. Each half section may be divided into two portions one in the central portion of the channel and the other in the vicinity of the wall.
3. Each velocity profile in the central region of the flow is logarithmic from the surface to a point very close to the floor. Within the limits of the experimental accuracy these curves are identical over the width of the central portion.

4. The maximum velocity in a vertical occurs at the surface for the central portion and below the free surface in the portions close to the side walls.
5. There is always a central portion for aspect ratio A_r greater than 5.

Gonchorov [23] analyzed the plane turbulent flow based on principles of perturbations. The local longitudinal velocity u_z for plane flow at any distance, z , from the bed was derived as

$$\frac{u_z}{u_d} = \frac{\ln\left(\frac{z+c}{c}\right)}{\ln\left(\frac{d+c}{c}\right)} \tag{10}$$

in which u_d is the velocity at the surface and c a constant. According to Eq. 10, $u_m = u_d$ and occurs at the free surface.

The form of the velocity distribution equation depends on the assumption made for the variation of eddy viscosity. For instance the standard logarithmic equation can be obtained by assuming the following variation for the eddy viscosity, ε.

$$\varepsilon = kzu_* \left(1-\frac{z}{d}\right)^{\frac{1}{2}} \tag{11}$$

The shear variation along the vertical is taken as linear.

Willis [103,104] developed a model for the velocity distribution for uniform flow in wide open channels by assuming an error function approximation for the distribution of eddy viscosity. The equation obtained is as follows.

$$\frac{u_m - u_z}{u_*} = \frac{-1.668}{k} \left[p\left(1-\frac{z}{d}\right) - \frac{1}{\sqrt{2\pi}} e^{-(\frac{p^2}{2})}\right] \tag{12}$$

where p is defined by Gaussian distribution function. Willis model like that of Zagustin and Zagustin [107] gives velocity distribution over the entire boundary layer. The value suggested by him for k is 0.39 which is slightly different from 0.4.

Chandrasekaran and Sakthivadivel [10] extended the use of the analytical solution for turbulent flow in pipes suggested by Zagustin and Zagustin [107] to flow in wide rivers. They verified their equation with data on rough turbulent flows in wide rivers.

Various investigators [17,56, 43, 8,85,93,105] extended the log-law for the outer region in open channels by incorporating the Coles [18] wake parameter. The

Coles wake law requires the knowledge of maximum velocity in a vertical. If the log-law fits for the entire depth the Coles wake parameter will be zero. The strength of the Coles wake function denotes the deviation of the log-law in the outer region.

6 Turbulent flow in narrow open channels

For channels of polygonal cross sections Keulegan [41] suggested an approach for obtaining the velocity distribution. He chose a trapezoidal channel cross section to illustrate the method and drew internal bisectors of the base angles. He considered that each of the three zones in which cross section of the channel was divided by the bisectors was influenced by the wall of that zone only. See Fig. 1.

The velocity distribution at any point (P or Q or R) in any particular zone (1 or 2or 3) was assumed to follow logarithmic distribution involving the shear velocity at foot of the normal drawn from the point to the corresponding wall. This method, in effect, amounts to considering that there is no

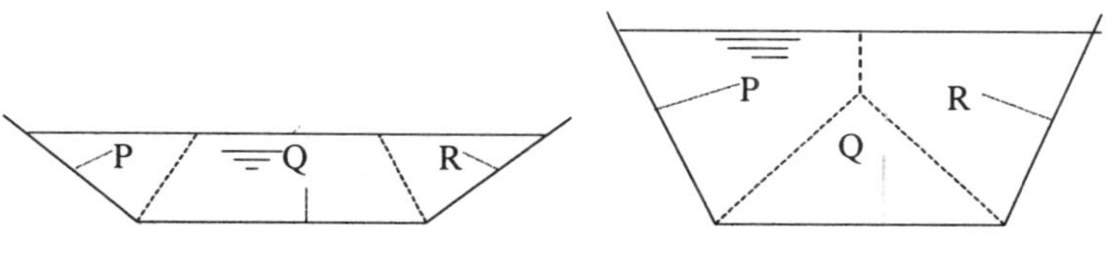

Figure 1. Division of Cross Section in to Zones of Influence of Each Wall

momentum exchange across the planes passing through the bisectors of the angles and parallel to the flow direction.

Schlichting [91] observed on the basis of Nikuradse's data that the flow at the free surface was not two dimensional for a narrow rectangular channel. The maximum velocity dipped below the free surface.

Allen [1] conducted experiments in a rectangular open channel with smooth sides and bed. He observed that for the range of measurements made in his studies, a close approximation to the mean velocity in a vertical was obtained by averaging the velocities at 0.2 and 0.8 of the depth. This implies the validity of logarithmic law up to the surface from a point very close to the bed. The measurements appear to have been made in the central portion where side wall effects were negligible and the log-law was valid.

Enger [22] conducted experiments in a trapezoidal channel having rough boundaries and concluded that the velocity distribution was logarithmic and was unaffected by any secondary currents of appreciable magnitude.

Goncharov [23] extended the application of the equation derived by him for the plane flow in a rectangular open channel. The proposed equation for velocity distribution was

$$\frac{u_{y,z}}{V} = \left[\frac{\log\left(\frac{\left(\frac{b}{2}-y\right)+C_1}{C_1}\right)}{\log\left(\frac{b}{2.7C_1}\right)}\right]\left[\frac{\log\left(\frac{z+C_2}{C_2}\right)}{\log\left(\frac{d}{5.4C_2}\right)}\right] \tag{13}$$

In which $u_{y,z}$ = the velocity at any point (y,z) in the cross section, y = the lateral distance from the side wall, b = width of the channel, and C_1 and C_2 are constants. Measured profiles, however, were not in close agreement. The measurements showed the phenomenon of 'dip' – the occurrence of maximum velocity below the free surface where as his model gives maximum velocity at the free surface.

Steffler *et al.* [94] obtained the velocity distributions in the laminar sub-layer and turbulent region for smooth channel flows by using LDA. They could successfully fit the log-law for the entire depth in the central portion of the channel. They observed the deviation from log-law near the side wall. The measurements of Rajaratnam and Muralidhar [64] on the velocity profiles, which were made over almost the entire width of the channel, suggested the validity of the logarithmic law in the vertical direction. Nearer the side walls dip could be observed. Their flows were supercritical

Chiu and his group developed the probability and entropy based velocity distribution model for flow through open channels [11] and pipes [14] , and is given as

$$\frac{u}{u_{c,m}} = \frac{1}{M}\ln\left[1+(e^M-1)\frac{\xi-\xi_0}{\xi_m-\xi_0}\right] \tag{14}$$

where u = velocity at ξ; ξ = independent variable with which u develops; ξ_0 = minimum value of ξ at which u is zero; ξ_m = maximum value of ξ at which u is maximum; $u_{c,m}$ = maximum velocity in the cross section and M = entropy parameter.

The cross sectional mean velocity, V in terms of maximum velocity is given as

$$V = \Phi(M)u_{c,m} \tag{15}$$

in which

$$\Phi(M) = \frac{e^M}{e^M - 1} - \frac{1}{M} \tag{16}$$

To use this model one should know the relation between $u_{c,m}$ and V. The occurrence of maximum velocity below the free surface was incorporated in their model.

Chow [15] divides the open channel flow into four regimes, viz., 1. Subcritical smooth turbulent flows, 2. Supercritical smooth turbulent flows, 3. Subcritical rough turbulent flows and 4. Supercritical rough turbulent flows.

At Indian Institute of Science an exhaustive experimental program was developed in 1976 to study the velocity and shear distributions for flows in narrow open channels. The study, which spanned across all the four regimes mentioned above, was carried out by Lakshminarayana [49] for Regime-1, Syamala [95] for Regime-2, Ramana Prasad [65] for Regime-4 and Sarma [79] for Regime-3. During the same period Vedula and Rao [100] proposed the binary law which connects the log-law of the inner region to the parabolic law of the outer region for the velocity distribution. This concept was later extended by Sarma and Sarma [88] and Sarma and Ramana Prasad [84]. The results of a more exhaustive study appeared in a recent publication of Sarma *et al.* [86].

The following sections discuss briefly the details of the studies. Section 7 gives the range of experiments. Section 8 gives the velocity distribution models. Section 9 deals with the local shear stress and its distribution, and mean shear stress. Section 10 deals with the side wall correction. The last section viz., Section 11 presents the resistance laws.

7 Experiments

The experiments were conducted in two flumes each exceeding 15m in length and 60cms in width and height. The artificial roughness on the channel bed and side walls was created by gluing sands of uniform size (passing through a particular size and retained on a sieve of a slightly smaller size). Velocity measurements at grid points were made with pitot tube and manometers which magnified the head differences up to 22.7 times. Fig. 2 shows the pitot tube and the manometer. Studies on subcritical flows were made in channels with horizontal floors. Ranges of experimental data are given in Table-1.

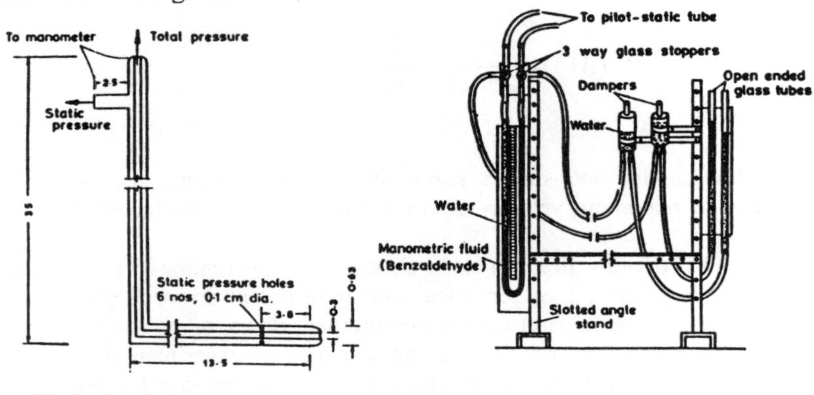

(a) Details of pitot - static tube (b) U-tube differential manometer set-up

Figure 2. Pitot static tube and manometer

8 Formulation of velocity model

The earlier studies established the following facts.

1. When the width of a rectangular channel is larger than a certain times the depth, there exists a central region where the flow can be considered as two dimensional.
2. The flow in a rectangular channel is symmetrical about the vertical passing through the mid point of the channel.
3. Close to the boundary, the velocity distribution is described by a log-law. The relevant shear velocity is the one at the foot of the normal drawn from the point to the boundary i.e., for a vertical profile, the local shear on the bed and for a horizontal profile, the local shear on the side wall which is on the same side of the mid vertical as the points are the relevant ones.
4. The log-law can be closely approximated to a power law.

Table 1. Ranges of Experimental Data

Regime	A_r Aspect Ratio	R_e (= 4 VR/ν) × 10^{-5}	F Froude number	S Slope	K_r (mm)	Length of the Channel (m)	b (m) Width
Subcritical smooth turbulent flows	1.2 to 8.13	0.47 to 2.67	0.17 to 0.69	---	---	15.25	0.61 & 0.305
Supercritica l smooth turbulent flows	2.03 to 11.73	0.67 to 5.63	1.43 to 2.65	0.007,0 .008,0. 009,0.0 11,0.01 5 & 0.02	---	15.25	0.61 & 0.305
Subcritical rough turbulent flows	1.66 to 15.10	0.11 to 6.0	0.072 to 0.743	---	0.35, 2.18 & 3.40	16.0	0.60
Supercritica l rough turbulent flows	4.11 to 15.64	1.44 to 8.3	1.364 to 1.92	0.02	0.35 & 1.85	15.25	0.61

5. The flow in a rectangular channel is symmetrical about the vertical passing through the mid point of the channel.
6. Close to the boundary, the velocity distribution is described by a log-law. The relevant shear velocity is the one at the foot of the normal drawn from the point to the boundary i.e., for a vertical profile, the local shear on the bed and for a horizontal profile, the local shear on the side wall which is on the same side of the mid vertical as the points are the relevant ones.
7. The log-law can be closely approximated to a power law.
8. There exists an outer zone away from the boundary where the velocity defect is found to follow a parabolic law.
9. The maximum velocity does not always occur at the free surface. Up to some distance from the walls it occurs below the free surface, the closer the point to the side wall, the deeper the maximum velocity point, in a vertical.

The flow in a rectangular open channel cross section can be divided up to four regions as shown in Fig.3. The definition sketch giving notation used for describing velocity field is shown in Fig. 4.

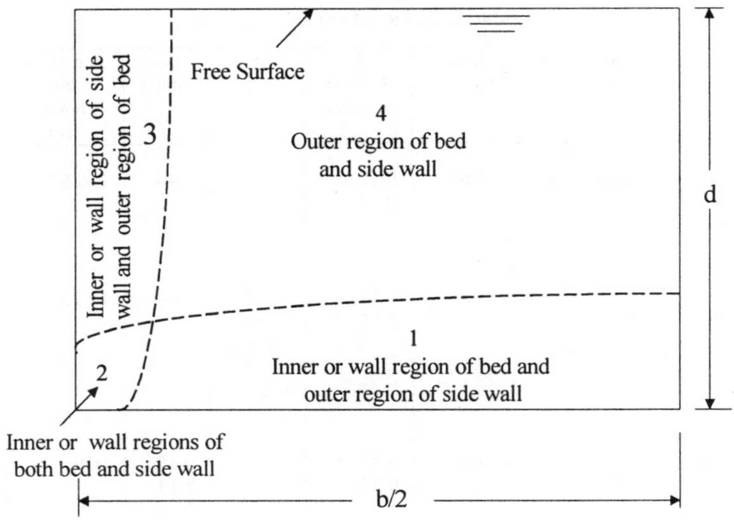

Figure 3. Various Regions in the Velocity Model

8.1 Velocity Model for smooth open channel

Region 1: The velocity distribution along the vertical in the wall region of the bed is known to follow the logarithmic law or equally well the one seventh power law. It is expressed as,

$$U_{y,z} = A \ lnZ_{*y} + B_s \tag{17}$$

or as,

$$U_{y,z} = J_s \ Z_{*y}^{1/7} \tag{18}$$

in which $U_{y,z} = u_{y,z}/u_{*y}$, $Z_{*y} = zu_{*y}/\nu$, u_{*y} = shear velocity at a distance of y from the sidewall, and B_s and J_s are constants pertaining to the smooth open channels.

The velocity distribution along the horizontal in this region is assumed to follow the parabolic law.

$$U_{c,z} - U_{z,y} = C_w(1-Y)^2 \tag{19}$$

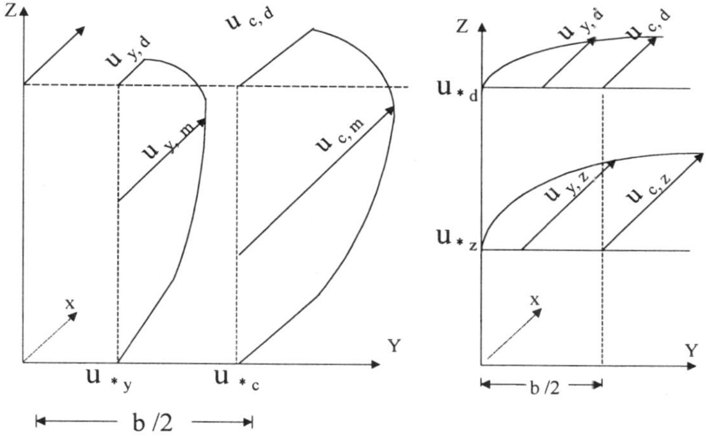

Figure 4. Definition Sketch Showing Notation for Velocity Field

in which $U_{c,z} = u_{c,z}/u_{*z}$, $u_{c,z}$ = velocity at a distance of z in a central vertical, u_{*z} = shear velocity on the side wall at a distance of z from the bed, $U_{z,y} = u_{y,z}/u_{*z}$, $Y = 2y/b$ and C_w is a constant.

Region 2: The flow in this region is influenced by both the bed and side walls. This is the corner region. The proposed model assumed that the log-law and one seventh power law are valid independently in both vertical and horizontal directions. In the vertical direction the velocity is given by either Eq.17 or 18. In the horizontal direction the velocity is given by either

$$U_{z,y} = A \ln Y_{*z} + B_s \tag{20}$$

or

$$U_{z,y} = J_s \, Y_{*z}^{1/7} \tag{21}$$

in which $U_{z,y} = u_{y,z}/u_{*z}$,u_{*z} = shear velocity at a height of z on the side wall and $Y_{*z} = yu_{*z}/\nu$.

Region 3: The velocity distribution in the vertical direction in this region requires to be modified to take into account the phenomenon of dip. The proposed modified form is

$$U_{y,m} - U_{y,z} = C_b(1 - D_y - Z)^2 \qquad (22)$$

in which $U_{y,m} = u_{y,m} / u_{*y}$, $u_{y,m}$ = maximum velocity in a vertical at a distance of y from the side wall, $Z = z/d$, D_y = the dimensionless depth below the free surface at which the maximum velocity in a vertical occurs and C_b is a constant.

The velocity distribution in the horizontal direction is given by either Eq. 20 or eq. 21.

Region 4: This region is outside inner regions of both the bed and side walls. The vertical and horizontal velocity profiles are described by Eq. 22 and Eq. 19 respectively.

When the open channel is sufficiently wide and a central region exists, the maximum velocity in a vertical occurs at the free surface. For such profiles Dy = 0.

8.2 *Velocity Model for rough open channel*

Eqs. 17 and 20 proposed for velocity distribution in the case of smooth boundaries are also known to be valid for rough boundaries provided the characteristic length used for non-dimensioning y and z is the roughness size K_r instead of ν/u_* used in the case of smooth boundaries. The equations become

$$U_{y,z} = A \ln Z_k + B_r \qquad (23)$$

and

$$U_{z,y} = A \ln Y_k + B_r \qquad (24)$$

in which $Z_k = z/K_r$ and $Y_k = y/K_r$ and B_r is the additive constant pertaining to the rough regime flow (rough channels). For the outer regions, Eqs. 19 and 22 are valid.

8.3 *Datum for rough channels*

In the case of flow over rough surface, the determination of the origin of distance coordinates viz., reference datum may be any point between the crest of the tallest roughness element and the trough of the deepest valley of the surface.

The procedure for fixing the datum for rough channels is presented in detail by Ramana Prasad and Sarma [72] earlier, which is presented briefly here. By measuring the point velocities in the flow field of wall regions at different distances from the tip of the roughness element using Clauser's technique [16] , it is possible to determine the reference datum for any roughness surface. This technique consists of shifting of origin of distance coordinate until a straight line is obtained for the logarithmic velocity distribution. By adding incremental distance (some small fractions of the size of the roughness size) to the distance measured from an arbitrary datum (the tip of the tallest roughness element) plotting $u_{y,z}$ with such z values for vertical profiles and with such y values for horizontal profiles on semi-log sheet each time fitting the best lines on the basis of correlation coefficient , the incremental distance for which the best of the best lines is obtained can be found. If the evaluation of the best fit is made on the basis of correlation coefficient for using the Clauser's technique, the reference datum would be obtained for most of the profiles as trough of the roughness element. But it is logical to expect the datum between the trough and crest of the roughness element. Table 2 gives values of datum obtained by various other investigators.

Table 2. Values of Datum Obtained by Various Investigators

Investigators	Type of Roughness	Datum below the of the element
O'Loughlin & Macdonald (1964)	Sand Roughness	0.27 K_r
Blinco Partheniades (1971)	Sand Roughness	0.27 K_r
Einstein & El-Samni (Ref. Blinco Partheniades, 1971)	Hemispheres	0.2 K_r
Goma & Gelhar (Ref. Blinco Partheniades, 1971)	Uniform Roughness	0.23 K_r
Kamphuis (1974)	Sand Roughness	0.30 K_r
Bayazit (1976)	Hemispherical elements	0.35 K_r
Cheng & Clyde (Bayazit, 1976)	Spherical elements	0.15 K_r
Grass (Bayazit, 1976)	Gravel bed	0.18 K_r
Hollingshead & Rajaratnam (1980)	Sand & Hemispheres	0.20 K_r
Zippe & Graf (1983)	Plastic grains	0.50 K_r

The distance of the reference datum from the top of the roughness element varied from 0.15 K_r to 0.5 K_r in their studies. So Clauser's method, which recommends the selection of the best of the best logarithmic lines, is perhaps not the best method. Another criterion is necessary to evaluate which of the best lines is to be taken for determining the reference datum.

This is an iterative method. By taking some incremental distance (fraction of roughness size) from the top of roughness as the arbitrary datum, the shear velocity is determined from the fitted log-law for each profile. The mean of the squares of these shear velocities are numerically integrated over the wetted perimeter to obtain the mean shear stress. This is compared with the mean shear stress determined from another independent criterion, viz., momentum principle. This criterion is described in section 9. According to this method, the datum is the distance from the top of the roughness element for which the mean shear stresses obtained from the log-law and momentum equation are close to each other. This is the method suggested by Ramana Prasad and Sarma [72]. See Fig. 5. The datum was obtained as 0.3 K_r for the experiments of Ramana Prasad [65] and Sarma [79].

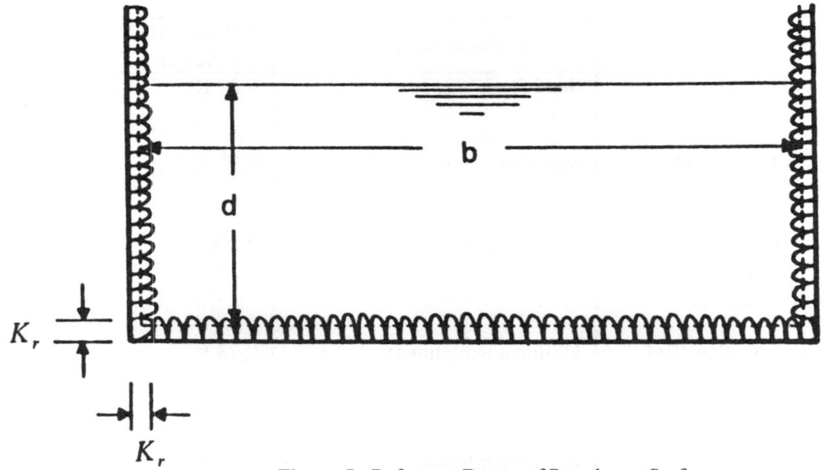

Figure 5. Reference Datum of Roughness Surface

8.4 The characteristic roughness size

In the case of an open channel artificially roughened by gluing uniform sand, the value of K_r is the same as the size of the sand used.

In the case of boundaries with non-uniform roughness consisting of roughness elements of different sizes, or in the case of artificially roughened surfaces with geometric shapes like spheres, cubes etc., where the roughness is of a particular size but the elements can be fixed in many geometrical patterns to the boundaries, describing the roughness by a single value poses another problem. The roughness in such cases is described by an equivalent sand size.

Ever since Nikuradse [58] published the results of his studies on flow through pipes artificially roughened with uniform sands, the researchers started comparing roughness surfaces used by them with those used by Nikuradse. At high Reynolds number the friction factor **f** becomes independent of R_e values
and becomes a constant. This constant value again depends on the nature of roughness of the surface and height of the characteristic roughness size relative to the flow dimension (d_0 in the case of circular pipes
and 4R in the case of open channels). The size of sand used by Nikuradse is referred as K_s and the relative roughness is then expressed as K_s/d_0. Nikuradse experimental data gives for **f**, the equation [91]

$$\frac{1}{\sqrt{f}} = 2\log\frac{d_0}{K_s} + 1.14 = 2\log\frac{R}{K_s} + 2.34 \tag{25}$$

The procedure for determining the roughness size equivalent to the sand used by Nikuradse consists of substituting the measured Reynolds number- independent **f** value and the corresponding R value in Eq. 25 and solving for K_s.

It is found that the same roughness elements arranged in different geometrical patterns give different K_s values [59]. It is also observed that fixing a uniform sand of size K_r to the boundaries by the same procedure as given by Nikuradse does not yield a K_s value which is equal to the sand size Kr [68,80]. Table 3 gives the equivalent sand sizes K_s as a multiple of the roughness size Kr.

Table 3. Nikuradse Equivalent Roughness in Sand Roughened Open Channels

Investigator	Nikuradse Equivalent Roughness , K_s
Ippen and Verma (1953)	2.28 & 2.78 K_r
O'Loughlin and Macdonald (1964)	1.5 K_r
Kamphuis (1974)	0.78 to 2.6 Kr
Hollingshead and Rajaratnam (1980)	3.32 to 3.9 Kr
Ramana Prasad (1991)	4.68 Kr
Sarma (1993)	2.64 K_r

8.5 The tangential parabola

The velocity law of the inner region and the velocity law of the outer region should both be satisfied at the common boundary of the two regions. One way is to take the common point in the vertical direction as the one at which the velocity profiles intersect. This procedure is followed by Sarma et al [82]. At the intersection point the velocity profiles did not join smoothly. See Fig. 6.

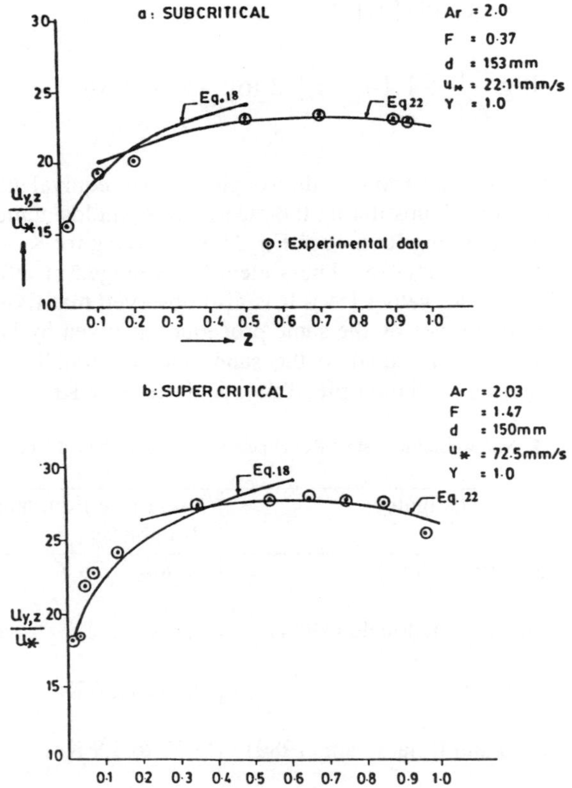

Figure 6. Power and Parabolic Law in Smooth Channels

Vedula and Rao [100] studied the conditions that are to be satisfied at the common boundary, for a wide open channel if the profiles were to join smoothly. In the central region of such a channel, the vertical velocity profiles do not show the dip and are given by Eqn. 17 and Eqn. 22 with D_y=0.

If the common boundary is at a dimensionless distance of Z_i from the bed, the velocity at the common boundary is given by both the velocity equations as

$$U_{y,z_i} = A \ln Z_i + B_s + A \ln R_{*y} \tag{26}$$

(for a smooth channel)

$$U_{y,z_i} = A \ln Z_i + B_r + A \ln D_{Kr} \tag{27}$$

(for a rough channel)

$$U_{c,d} - U_{c,z} = C_b (1 - Z_i)^2 \tag{28}$$

(for both smooth and rough channels)

in which $U_{c,d} = u_{c,d}/u_{*c}$, $u_{c,d}$ = velocity at the free surface in the central vertical, u_{*c} = shear velocity at center, $R_{*y} = du_{*y}/\nu$ and $D_{Kr} = d/K_r$.

The condition that the velocity is single valued gives

(a) for smooth channels

$$C_b(1-Z_i)^2 = A \ln Z_i + B_s + A \ln R_{*y} \tag{29}$$

(b) for rough channels

$$C_b(1-Z_i)^2 = A \ln Z_i + B_r + A \ln D_{Kr} \tag{30}$$

For the parabola of outer region to join tangentially to the logarithmic curve of the inner region at the common boundary, the first derivatives of the two curves should be single valued. For such a tangential parabola [100]

$$C_b = \frac{1.22}{Z_i(1 - Z_i)} \tag{31}$$

Table 4 gives the C_b values for some Z_i values. For narrow open channels, Sarma et al [86] showed that Eq. 31 becomes

$$C_b = \frac{1.22}{Z_i(1 - D_y - Z_i)} \qquad (32)$$

Eq. 31 or Eq. 32 is called the binary law.

Table 4. Values of the Constant of the Parabolic Law

Z_i	C_b	Remarks
0.15	9.6	Hama's Constant for smooth flat plate
0.20	7.6	
0.215	7.2	Value suggested for smooth and rough pipes of Nikuradse's data
0.25	6.5	
0.27	6.3	Bazin Constant for open channel flows

8.6 Analysis of data of IISc

The log-law is found to be valid for the entire depth at the central verticals in all the four regimes. Typical velocity distribution profiles are shown for the two regimes at the central vertical in Figs. 7.

Figs. 8. (a-e) show the variation of velocity $u_{y,z}$ with dimensionless heights Z for different dimensionless lateral distances Y in a typical run of supercritical rough turbulent flows. Similarly Figs. 9. (a-f) show the plot of $u_{z,y}$ with Y for different Z values of the same run. The similar trend in velocity profiles could be observed in the other three regimes.

For the inner region of the smooth channels both the Eqs. 17 and 18 in the vertical direction and both the Eqs. 20 and 21 in the horizontal direction are found to be valid. For the outer region Eqs. 19 & 22 are found to be valid for smooth and rough open channels. Figures of the velocity profiles in the outer region could be found in the later part of the paper.

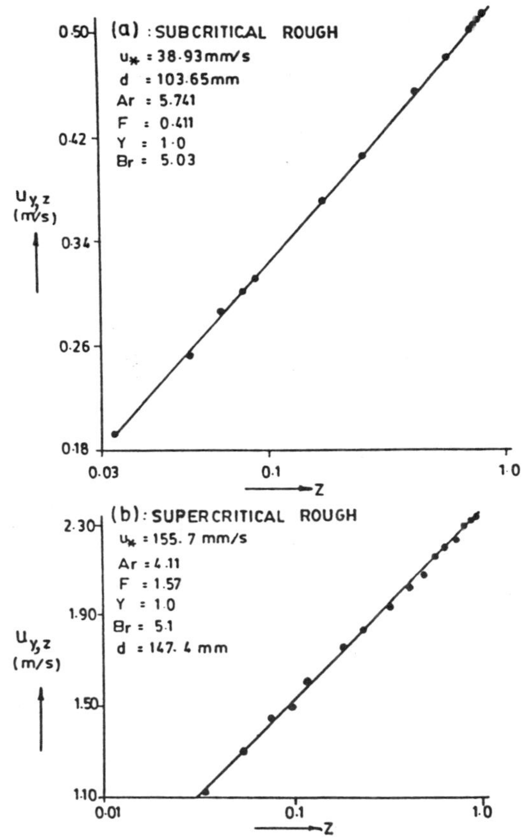

Figure 7. Extended Log-Law for the Entire Depth

Eqs. 23 and 24 describe the velocity distribution in the inner regions for rough channels in the vertical and horizontal directions respectively. See Figs. 8 & 9 for the fitness of the log-law in the inner region. The von Karman constant is assumed as 0.41 and hence the value of A is 2.44. The additive constant B_s or B_r is found varying along the perimeter. This could be attributed to the existence of non-uniform shear distribution and secondary currents. The variation of additive constant with side wall distance is also reported by Xinyu et al [105]. The values of the constants in Eqs. 17 to 22 are listed in Table 5.

8.7 Variation of Dip D_y with aspect ratio A_r and lateral distance Y

The dip, D_Y is quantified as the ratio of the depth below the free surface at which the maximum velocity occurs in the vertical section to the total depth of flow in the

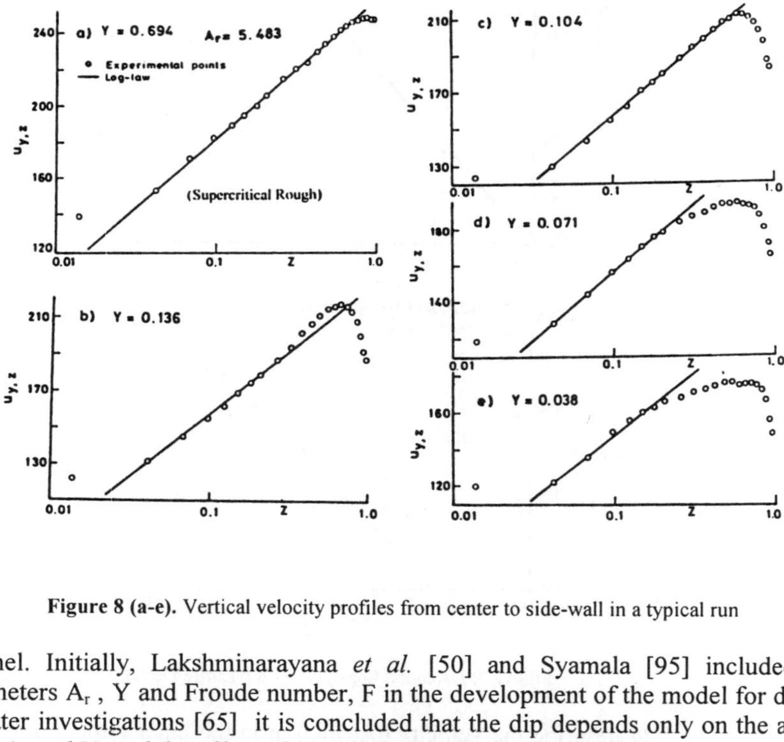

Figure 8 (a-e). Vertical velocity profiles from center to side-wall in a typical run

channel. Initially, Lakshminarayana *et al.* [50] and Syamala [95] included the parameters A_r , Y and Froude number, F in the development of the model for dip. In the later investigations [65] it is concluded that the dip depends only on the aspect ratio, A_r and Y, and the effect of Froude number probably negligible.

A simple empirical model developed from the data of supercritical rough turbulent flows [65] is given as

$$D_y = 0.45 \left(\frac{4}{A_r} - Y\right)^{0.1} \left(\frac{1}{1+0.69Y}\right) \tag{33}$$

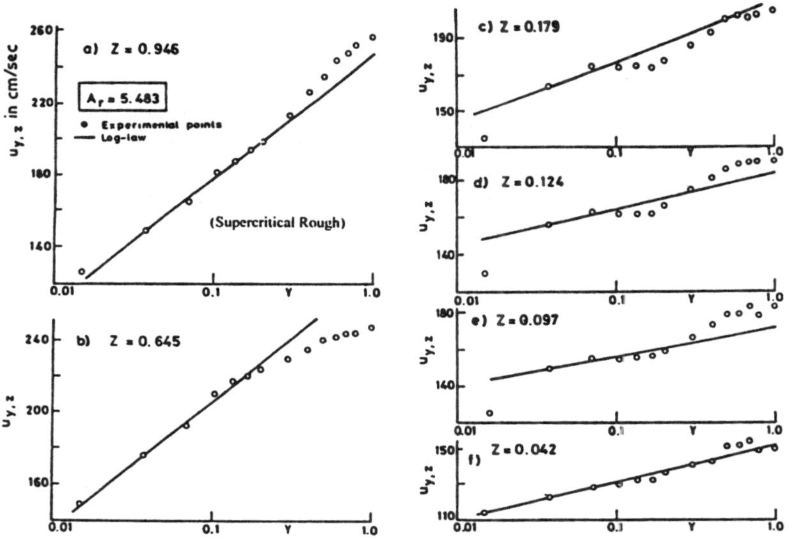

Figure 9 (a-f). Lateral velocity profiles from free surface to the bed in a typical run

Table 5. Constants of Velocity Distribution in Smooth Open Channels

Constants	Region-1	Region-2	Region-3	Region-4
J_s	8.3	8.3	8.3	--
C_w	2.4	--	--	--
C_b	--	--	9.6	9.6
A	2.44	2.44	2.44	--

This model was verified with the data of other three regimes and found good. A typical plot of validation of this model for subcritical rough turbulent flows [79] is given in Fig. 10.

8.8 Binary law

It is found that it is not always possible to fit a parabola for the velocity distribution in outer region which joins tangentially the log profile. A parabola which intersects the log profile can however be always fitted. So the study falls in to two categories, the category of vertical velocity distributions for which the binary law is applicable and the category for which it is not applicable. These two are referred as case 1 and case 2 respectively by Sarma et al [86].

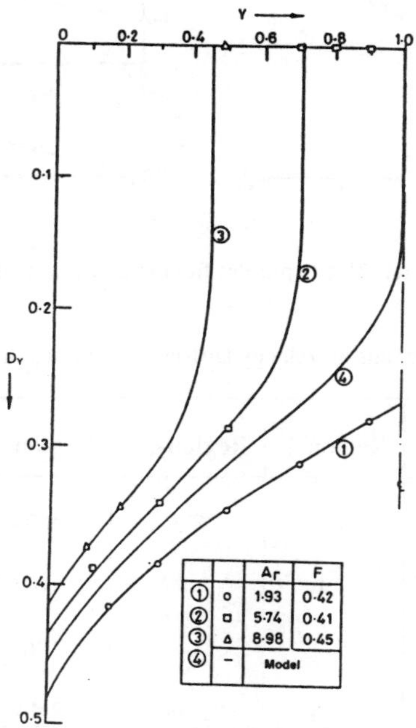

Figure 10. Comparison of the model of dip with subcritical rough data

8.8.1 Case 1:

For this case $Z_I \leq Z \leq 1.0$ and four types of tangential parabolas exist for each of the four regimes of flow mentioned earlier.
Type 1: The entire parabola curve in the range lies above the log curve for $Z > Z_i$. At $Z = Z_i$ it touches the log curve.

Type 2: The parabolic curve is above the log curve and touches it at $Z = Z_i$ and $Z =$ 1.0.
Type 3: The parabolic curve touches log curve at $Z = Z_i$ and for $Z_i < Z \leq 1$ it is partly above and partly below.
Type 4: The parabolic curve for $Z_i < Z \leq 1$ is totally below the log curve except for $Z = Z_i$ at which point they meet tangentially.

Typical curves of these four types for the case of supercritical regime are shown in Fig. 11(a-d). The similar curves could be drawn for the other three regimes.

8.8.2 Case 2:

When the entire parabolic curve is below the logarithmic curve except at $Z = Z_i$ where the two curves meet tangentially, one question arises. What is the farthest position the parabola can take below the log curve and still be tangential to it at some point Z_i in the range of 0 to 1? To answer this it should be remembered that the parabola plots as an S curve on a semi log sheet. The point of inflection is given by $Z_{in} = (1-D_y)/2$.

Noting that for values of $Z < Z_i$ the parabolic curve goes above the log-law and does not meet again, it is easy to understand that whenever it is possible to fit a tangential parabola, $0 \leq Z_i \leq Z_{in}$. That is the farthest point from the boundary up to which the common point can shift for a binary law is given by the condition $Z_i = Z_{in}$.

Therefore for the limiting parabola, not only $U_{y,z}$ and $dU_{y,z}/dZ$ should be single valued but also $d^2U_{y,z}/dZ^2$ at $Z=Z_i$ .The condition that $d^2U_{y,z}/dZ^2$ should be single valued at $Z=Z_i$, gives the following equation.

$$Z_i = \frac{1-D_y}{2} \quad ; \quad C = \frac{4.88}{(1-D_y)^2} \tag{34}$$

The maximum value Z_i can be 0.5 for a central vertical of a wide channel ($A_r > 5.0$) where D_y=0. The dimensionless maximum velocity $U_{y,m}$ in a vertical occurs at $Z_y = 1-D_y$. For limiting parabola Z_i is given by Eq. 34. Let U_{Ly} correspond to the dimensionless velocity given by log-law at y for $Z = 1-D_y$. Then $U_{Ly} - U_{y,m}$ for a tangential parabola takes a maximum value of 0.4713 when the tangential parabola is the limiting parabola. See Fig. 12 (a-b). So it is possible to fit a tangential for values of $U_{Ly} - U_{y,m} \leq 0.4713$. Beyond that only a Case-2 parabola can be fitted. See Fig. 13.

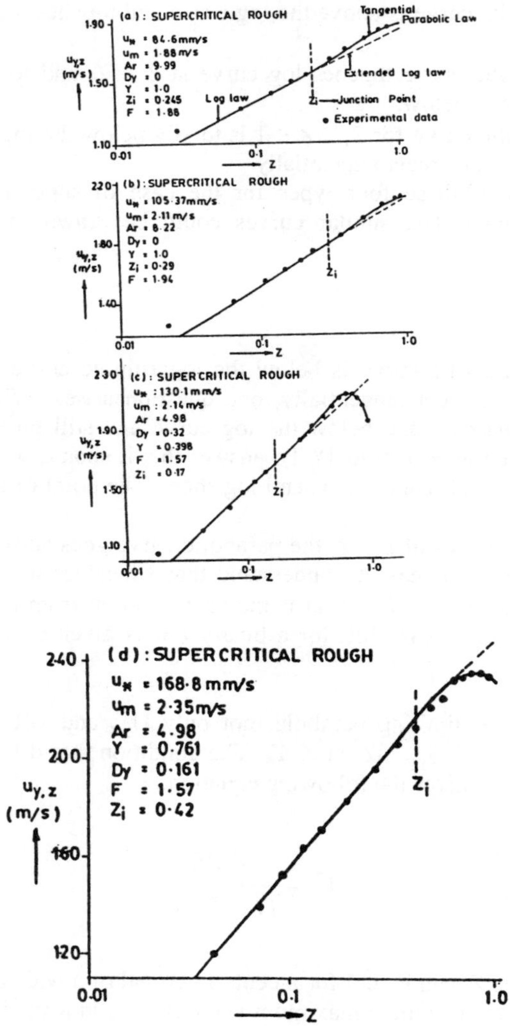

Figure 11 (a-d). Binary law in open channels (Supercritical rough)

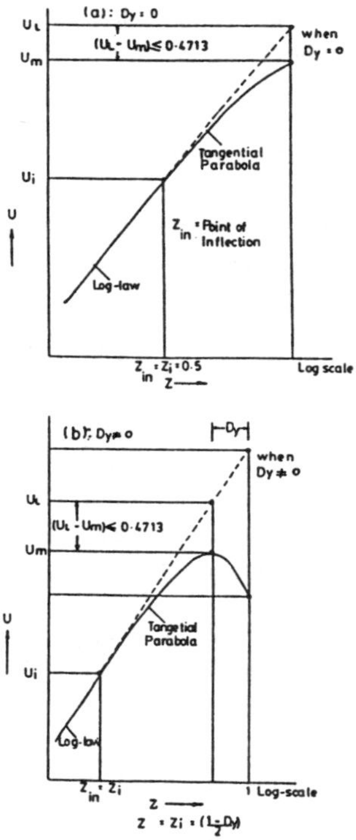

Figure 12 (a-b). Limiting parabola

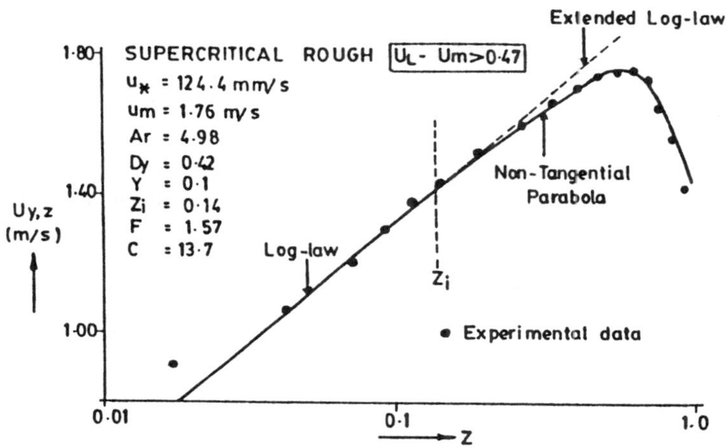

Figure 13. Non-Tangential parabola

It should be mentioned here that the log-law strictly applies only to the turbulent part of the wall (inner) region, where the shear stress is practically constant. It may no longer apply for the outer region, since the conditions on which they are based are no longer valid. From a practical engineering point of view, the log-law gives sufficiently satisfactory results even when it is applied to this outer region [30]. This approximation may be acceptable for the case of two-dimensional open channels. Cardoso et al [9] stated that the retarding flow in the near surface zone due to weak secondary currents tends to compensate the wake effect such that the log-law can approximate the entire velocity profile.

But in the case of narrow open channel flows, the maximum velocity does not occur at the free surface, but below it. It is a fact the velocity dip is caused by the secondary currents. As Nezu et al [57] noted, " Low momentum fluids are transported by the secondary motion from the near-bank to the center, and high motion fluids are moved by this motion from the free surface toward the bed". It is understood from the above discussion that the log-law will not be valid in the outer region, particularly in narrow open channels. The point of deviation from the log-law i.e., the junction point, depends on the local shear stress, dip D_y and the maximum velocity, and thus is related to width ratio and aspect ratio of the flow. It is unfortunate that the junction point Z_i or C could not be expressed in terms of width and aspect ratio because of the complexity of the problem. This leaves the scope for future research.

8.9 Frequency Power – Law for velocity distribution

The binary law proposed in terms of Z_i and D_y does not permit evaluation of the kinetic energy correction factors α and the momentum correction factor β , by direct or numerical integration as the expression for variation of Z_i possibly with Y, R_e, F and A_r is not known. So a different type of power law, which is amenable for direct integration based on the frequency distribution of velocities, is tried.

Typical isovels are plotted for a particular run of a smooth open channel flow in Fig. 14.

Let the flow area corresponding to velocities less than a certain velocity u_i be a_i. Then for $u_i = u_{c,m}$ the maximum velocity of the entire velocity field $a_i = a$, the total area of flow.

$$U_i = \frac{u_i}{u_{c,m}} \quad \& \quad P_i = \frac{a_i}{a} \quad ; \quad U = \frac{u_{y,z}}{u_{c,m}} \tag{35}$$

where P_i is the probability of $u_i < u$.

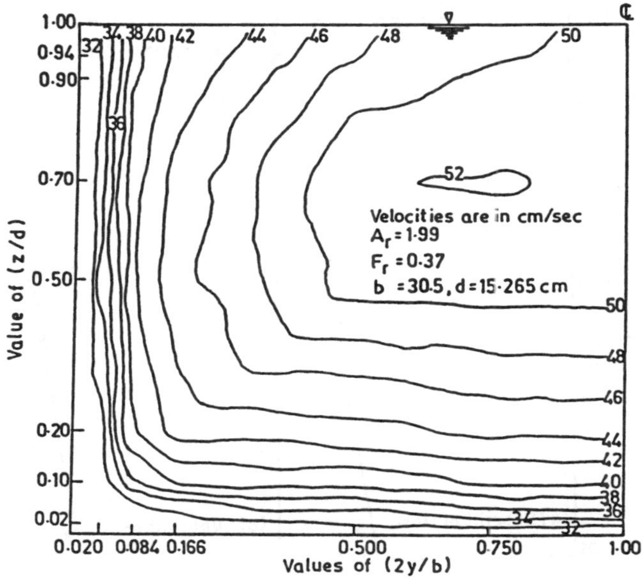

Figure 14. Isovels in a typical run (Subcritical smooth)

Thus the following power law is assumed by Sarma and Sarma [87]

$$P(U < U_i) \quad = \quad P_i = U^N \tag{36}$$

On the basis of Eq. 36, the correction factor α and β are given by

$$\alpha \quad = \quad \frac{(N+1)^3}{N^2(N+3)} \qquad \& \qquad \beta \quad = \quad \frac{(N+1)^2}{N(N+2)} \tag{37}$$

The values of U_i are plotted with measured values of P_i. A typical plot is shown in Fig. 15. Variation of N for the smooth channel data is tabulated in Table 6. The mean value $\overline{N}$ of N and the mean value of $\overline{S}$ of the standard error S are estimated as $\overline{N}$=5.9 and $\overline{S}$=0.067. For this value of $\overline{N}$, α=1.1.06 and β=1.02. Frequency power law is also applied successfully for rough open channels [81]. The corresponding values of N are tabulated in Table 7.

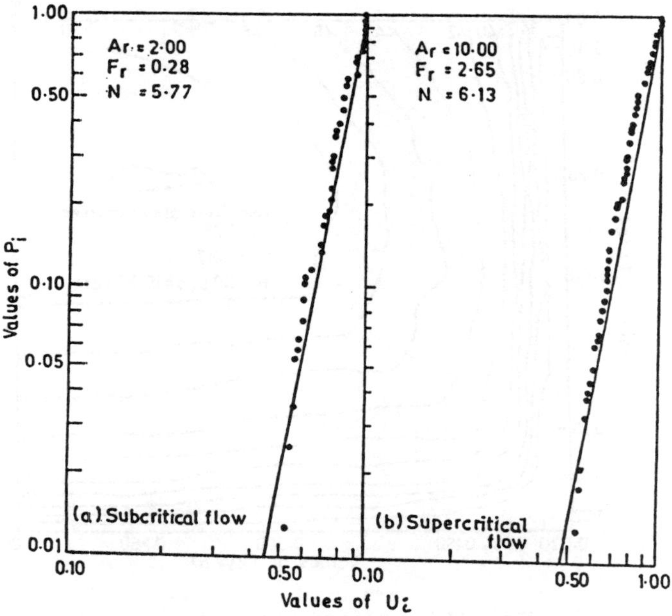

Figure 15. Frequency distribution of velocities (Smooth Channels)

Table 6. Ranges of Parameters in Frequency Power Law (Sarma & Sarma, 1990)

	Subcritical Flows	Supercritical Flows	Total Pooled Data
Number of Runs	44	30	74
Aspect Ratio , A_r	1.2 to 8.13	2.03 to 11.73	1.2 to 11.73
Froude Number, F	0.17 to 0.69	1.43 to 2.65	0.17 to 2.65
Reynolds Number, R_e	0.47 to 2.67×10^5	0.67 to 5.63×10^5	0.47 to 5.63×10^5
Value of N	4.4 to 7.8	5.03 to 7.9	4.4 to 7.9
Mean Value of N	5.8	6.1	$\bar{N} = 5.9$
Standard error of N	0.68	0.67	0.67

Table 7. Ranges of Parameters in Frequency Power Law of Rough Channels (Sarma *et al.*, 1991; Ramana Prasad, 1991)

	Aspect Ratio, A_r	Froude Number, F	Reynolds Number, R_e	Mean Value of N
Subcritical Rough Turbulent Flows	1.67 to 9.62	0.072 to 0.743	0.11×10^5 to 6.07×10^5	5.58 (For transitional turbulent flows) 4.3 (For completely rough turbulent flows)
Supercritical Rough Turbulent flows	4.11 to 15.64	1.36 to1.92	1.44×10^5 to 8.3×10^5	6.80 (For transitional turbulent flows) 4.61 (For completely rough turbulent flows)

9 Mean shear, local shear and shear distribution

The mean shear stress $\bar{\tau}_0$ over the wetted perimeter P can be computed by using the momentum equation for steady uniform flow.

$$\rho\, g \frac{d_1}{2} a_1 - \rho\, g \frac{d_2}{2} a_2 + \rho\, g \left(\frac{a_1 + a_2}{2} \right) L\, S - \overline{\tau}_0 PL \quad = \quad \rho\, Q(V_2 - V_1) \quad (38)$$

in which the subscripts denote values at sections 1 and 2 respectively by a longitudinal distance L, the slope of the bed, S, and the volume rate of flow Q.

Lakshminarayana *et al.* [51] gave the criterion that the local shear values obtained by any procedure, when summed and averaged over the entire perimeter should match with the mean values obtained from the momentum equation, viz., Eq. 38.

The model developed for shear distribution should also be consistent with the findings of the other investigators.

Enger [22] measured the shear distribution for trapezoidal channels by using two point velocity measurement method, in the inner region. This method uses the equation,

$$U_{y,z_1} - U_{y,z_2} = 2.44 \ln \frac{Z_1}{Z_2} \tag{39}$$

Here only u_{*y} is the unknown as all the other quantities are the measured ones. His measurements showed non-zero shear corner at the corner. Ippen and Drinker [33] reported that the shear distribution obtained by them was in general agreement with that of Enger. Enger further observed the u_{*y} approached the maximum values near the center of the channel. That is to say u_{*c} is the maximum of u_{*y} values. The values of τ_{0c} were 40% higher than τ_0 values. The boundary shear stress on the wall at the free surface τ_s approached small values. Most importantly he studied that the average τ_0 value computed from velocity measurements agreed with the one obtained from Eq. 38.

Smerdon and Beasley [92] found that the $\bar{\tau}_0$ in their experiments exceeded by 100% the value they obtained from the log-law velocity distribution. This shows again that log-curve should not be fitted for velocity distribution in any arbitrary way. This again stresses the importance of comparing the average of τ_0 values with the $\bar{\tau}_0$ obtained from Eq. 38 and accepting the fits that comply with the above condition only.

It may be recalled that for smooth channels the velocity in the inner region can be expressed as

$$u_{y,z} = C_n \, u_{*y}^{\frac{n+1}{n}} \tag{40}$$

and as

$$u_{z,y} = C_n \, u_{*z}^{\frac{n+1}{n}} \tag{41}$$

in which the velocity is given by one- n th power law and C_n is the associated constant. So the shear velocity is related to the power law and thus is related to the velocity at any point. Preston's technique [61] consisted of keeping the tube touching the boundary and connecting the velocity head measured in that position to the shear velocity. Davidian and Cahal [19] used the Preston tube for measuring the boundary shear stresses in smooth open channels. They observed that as the A_r increased, the flow tended to be two dimensional making the shear distribution more uniform. For higher Froude numbers, which were associated with smaller depths, also the shear distributions were more uniform.

Replogle and Chow [74] studied shear distribution in circular channels running partly full using a Preston tube. They found average shear stress estimated from log-law was 17% lower than that obtained from the momentum equation once again emphasizing the fact, that fitting the best log-curve for velocities is not the best method for estimating boundary shear stress. In the recent experimental work by Ead *et al.* [20] on corrugated circular channels, the shear stress distribution is expressed only as a function of lateral distance. In their experimental data the variation of aspect ratio is not significant and hence the influence of aspect ratio in shear distribution could not be seen in their model [67].

Rajaratnam and Muralidhar [64] while at Indian Institute of Science conducted experiments on smooth rectangular open channels in the supercritical turbulent regime. The aspect ratio varied from 0.83 to 20. They found that when A_r affects the shear distribution, the Reynolds number does not.

Gosh and Roy [28] directly measured shear stress in smooth and fully rough channels of rectangular and trapezoidal cross section. They found reasonable agreement between the directly measured values of mean shear $\bar{\tau}_0$ and those obtained from the momentum equation and velocity distributions. As the direct measurement is a very cumbersome one and the accuracy of measurement by the very nature of the measuring technique can not be very good. So it may be said within the accuracy of the experiment the agreement is good. Gosh and Jena [25] employed successfully Preston tube technique to obtain the shear stress distribution in compound smooth and rough open channels. Gosh [24] obtained the shear distribution using log-law with two point velocity measurement in rectangular channels of laterally varying wall roughness. Gosh and his team further obtained the shear distribution with laterally varying roughness in compound channels [27] and in a meandering channel with flood plain [26].

Kartha and Leutheusser [38] used a Preston tube to measure the shear stresses. They also estimated them using the slope of the entire velocity profile in the log-region. The range of aspect ratios studied was from 1 to 12.5. The two sets of shear values agreed well. The ratio of the average shear stress on the bed $\bar{\tau}_b$ to the mean shear stress over the wetted perimeter was observed by them to increase from a value less than 1.0 for A_r =1 to 1.6 for A_r =3.0. For A_r >3.0, it decreased from 1.6 to 1.0 as A_r increased.

Knight and his team [47,48,45,46] studied extensively on mean wall shear stress, mean bed shear stress and local shear stress distribution in open channels with smooth surface as well as variable roughness. Nezu *et al.* [54] evaluated shear stress distribution in smooth and rough open channels using log-law and observed the maximum shear stress at the center of the channel.

Lundgreen and Jonsson [53] developed a numerical method to determine the shear stress distribution in shallow channels but this can not be used for polygonal sections. Goncharov [23] proposed equations for distribution of boundary shear stress, both with and without secondary flows. Later Knight [44] compared Goncharov's theoretical shear stress distribution with the experimental data. Noat and Rodi [55] could come out successfully with a mathematical model for simulating secondary currents in straight open channel flows and then employed eddy viscosity relation for the shear stresses in the longitudinal momentum equation. Pizzuto [63] developed a numerical model for calculating the distributions of velocity and shear stress across irregular straight open channels of any cross sectional shape or roughness distribution. Chiu and his group [13,12] developed a mathematical model for shear stress distribution considering the effect of secondary currents. Khodashenas and Paquir [42] developed a method called Merged Perpendicular method to compute the shear stress distribution in irregular straight open channels. Ramana Prasad and Manson [66] added an analytical expression to compute percentage of total shear force carried by the walls.

It is interesting to note that Chiu's model produces maximum shear stress near the corner not at the center of the channel or at the free surface. All the other theoretical models produce the maximum bed shear at the center and the maximum wall shear stress at the free surface.

On the basis of these investigations, the following conclusions can be drawn:

1. The shear at the corner is not zero.
2. The maximum shear stress on the bed for smooth channels occurred at the center of the bed and the maximum shear stress on the wall occurs below the free surface.
3. For wider channels the shear distribution is more uniform than for the narrow ones.
4. The ratio $\bar{\tau}_b / \bar{\tau}_0$ attains a maximum value at $A_r = 3.0$
5. Reynolds number may not effect the distribution.
6. Most of the theoretical shear distributions give the maximum bed shear at the center of the channel and wall shear at the free surface.

9.1 Analysis of IISc data for shear distribution

The experimental data of shear stress distribution along the perimeter of open channels available for analysis was very limited. Chiu and Hsuing [13] made the following statement on the scarce data of shear stress distribution "There are shortages of boundary shear on laboratory flumes especially near the corners and side walls. The experimental investigations done at IISc would fill this gap.

9.1.1 Analysis of smooth channel data

Smooth Open Channel Data collected at IISc, has been analyzed by Lakshminarayana et al [51] and Sarma and Syamala [90] .

Pitot tube of radius **r** was used in a similar way as Preston tube is used to measure local shear stresses. In this method Pitot tube was kept with its bottom touching. Then u_{*y} and u_{*z} were estimated using Eqs. 42 and 43.

$$u_{*y} = \left(\frac{\upsilon}{r}\right)^{1/8}\left(\frac{u_{y,r}}{8.3}\right)^{7/8} \tag{42}$$

$$u_{*z} = \left(\frac{\upsilon}{r}\right)^{1/8}\left(\frac{u_{r,z}}{8.3}\right)^{7/8} \tag{43}$$

u_{*y} and u_{*z} were also estimated from the slope of the vertical and horizontal velocity profiles of the inner region. Table 8 shows the points of values for a few typical subcritical flows. The agreement is good.

The pairs of $\bar{\tau}_0$ values obtained from the velocity profiles and the momentum equation for supercritical flows are also in agreement.

When the Pitot tube is kept at the corner touching both the bed and side walls, it gives the shear stress at a distance of **r**, both in the horizontal and vertical direction. This common value u_{*cor}, observed to bear a constant ratio to $\sqrt{\bar{\tau}_0/\rho}$ depending on the aspect ratio. The empirical relationship is given by

$$\frac{u_{*cor}}{\sqrt{\frac{\bar{\tau}_0}{\rho}}} = \frac{0.1A_r}{1.43+0.31A_r} \tag{44}$$

9.1.2 Analysis of rough channels

In rough open channels local shear velocities were determined by fitting log-law for the inner region. While determining the shear velocity, the weightage was not given for the additive constant of log-law and von Karman constant was assumed as 0.41 [69,70,73]. The shear stress distribution along the bed and side wall for four typical runs of supercritical regime is shown in Fig. 16. The similar trend is also observed in subcritical flows. From the figures it can be seen that the maximum bed shear stress does not always occur at the center of the rectangular open channel. This is due to due to the presence of secondary currents. Tominaga et al [96] observed that the local shear is more if the secondary currents flow towards the boundary and less if they flow away from the boundary. The occurrence of a local peak near the corner for the bed shear and the maximum wall shear on the wall can be explained by the presence of secondary currents.

Table 8. Comparison of Local Shear Velocities Obtained from Surface Pitot -Static Tube and Velocity Profiles. (Lakshminarayana *et al.*, 1986)

Shear Velocities (cm/s) SP : Surface Pitot-Static Tube VP : Velocity Profile				Z/d			2y/b			
A_r	F	Method	At the corner	0.10	0.50	0.94	0.084	.25	0.5	1.0
1.0	0.3	SP	1.38	1.77	1.98	1.99	1.67	--	1.90	1.87
		VP	1.37	1.78	2.01	1.98	1.68	--	1.90	1.87
2.0	0.4	SP	1.57	1.92	2.26	2.07	1.92	--	2.11	2.22
		VP	1.57	1.91	2.25	2.06	1.88	--	2.08	2.21
3.0	0.4	SP	1.33	1.54	1.87	1.81	1.58	--	1.82	1.97
		VP	1.32	1.54	1.86	1.81	1.58	--	1.82	1.96
4.0	0.3	SP	0.93	1.04	1.24	1.17	1.12	--	1.40	1.34
		VP	0.93	1.05	1.24	1.17	1.12	--	1.40	1.33
5.0	0.2	SP	0.79	0.94	1.12	0.99	1.02	1.07	1.10	1.19
		VP	0.78	0.92	1.12	0.98	1.02	1.07	1.09	1.20
6.0	0.5	SP	1.56	1.81	2.11	2.0	1.99	2.15	2.18	2.25
		VP	1.56	1.82	2.12	1.99	2.05	2.14	2.19	2.22
7.0	0.7	SP	2.02	2.27	2.67	2.52	2.55	2.74	2.91	2.95
		VP	1.98	2.27	2.68	2.52	2.55	2.74	2.86	2.94
8.0	0.6	SP	1.54	1.73	2.10	2.11	2.02	2.22	2.35	2.42
		VP	1.55	1.73	2.10	2.11	2.02	2.20	2.34	2.41

10 Side wall correction procedure

Computation of sediment loads in an open channel involves subtracting the drag due to walls from the total drag to get the drag due to the bed. This procedure is known as the side wall correction procedure. Johnson [36] suggested this procedure first for applying wall correction in the case of smooth side walls. In this method the side wall friction should be determined from Karman-Prandtl resistance equation of smooth pipes. Einstein [21] suggested a similar method but the roughness of the side wall should be determined from the preliminary experiments. All the other assumptions are the same in both the methods. Brooks [7] made the Johnson method simpler by avoiding the trial and error procedure. Kennedy [40] modified Johnson method in such a way that friction due to smooth walls can be obtained from the figure given by them and in the case of rough side wall, they suggested the use of Nikuradse pipe friction diagram. Gosh [24] investigated the applicability of Johnson method to the channels with varying wall roughness. Knight and Macdonald [48] made an attempt to verify the assumptions of side wall correction procedure. Vittal *et al.* [101] obtained an expression for side wall friction using dimensional analysis and experimental data. Knight and his team [45,46] gave an empirical equation to compute the percentage of shear force carried by the walls in smooth and rough channels. Sarma and Ramana Prasad [83] developed a simple method for side wall correction referred to as Keulegan model which is based on Keulegan's [41] division of cross sectional flow.

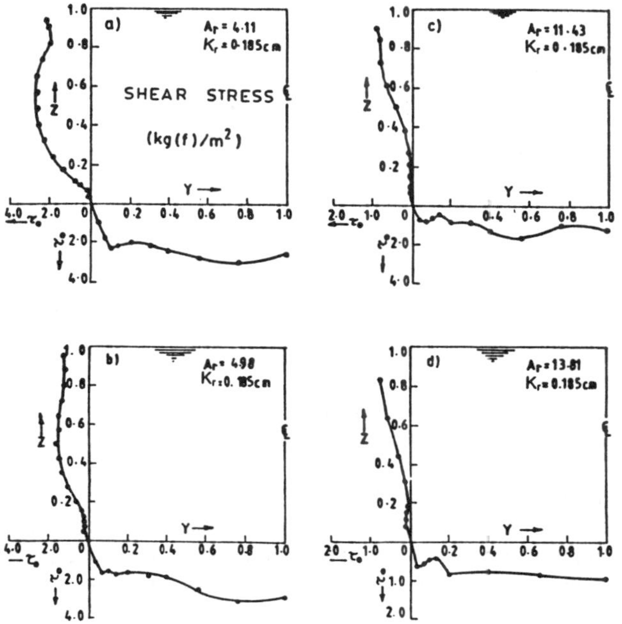

Figure 16 (a-d). Shear Distribution in Rough Open Channels (Supercritical Rough)

10.1 Keulegan Model

In this model the flow areas corresponding to the bed, A_b and side walls A_w, are delineated by drawing the bisectors of the angles at the corner. For rectangular open channels, two cases arise: one when the bisectors intersect below the free surface and the other when the bisectors intersect above the free surface. This model gives two cases.

Case 1: $A_r \geq 2$

$$\frac{\overline{\tau_{0b}}}{\overline{\tau_0}} = \frac{(A_r - 1)(A_r + 2)}{A_r^2} \tag{45}$$

where $\bar{\tau}_{0b}$ = mean bed shear stress and $\bar{\tau}$ = mean shear stress.

$$\frac{\overline{\tau_{0w}}}{\overline{\tau_0}} = \frac{A_r + 2}{2A_r} \tag{46}$$

where $\bar{\tau}_{0w}$ = mean wall shear stress.
Case 2: $A_r \leq 2$

$$\frac{\overline{\tau_{ob}}}{\overline{\tau_0}} = \frac{A_r + 2}{4} \tag{47}$$

$$\frac{\overline{\tau_{0w}}}{\overline{\tau_0}} = \frac{(4 - A_r)(A_r + 2)}{8} \tag{48}$$

This model was verified for smooth and rough open channels. The comparison of Keulegan model and Johnson model with the data of supercritical rough turbulent flows is shown in Fig. 17. The verification of Keulegan model for subcritical flows was shown in detail by Sarma and Ramana Prasad [83]. Yang and Lim [106] extended Keulegan model with more experimental data of rectangular channel and provided supporting theory for the assumptions. The extension of this model to trapezoidal and compound channels is discussed by Ramana Prasad and Manson [66].

11 Friction factor

Jayaraman [35] and Kazemipour and Apeit [39] analysis showed that friction factor f in the case of open channels is influenced by the shape of the channel. They have used A_r as the parameter representing the shape.

However, the experimental data of Tracy and Lester showed no such effect in the case of flow through rectangular open channels. For Froude number F<1.0, they obtained the friction equation as

$$\frac{1}{\sqrt{f}} = 2.03\,(\log R_e\sqrt{f} - 1) \tag{49}$$

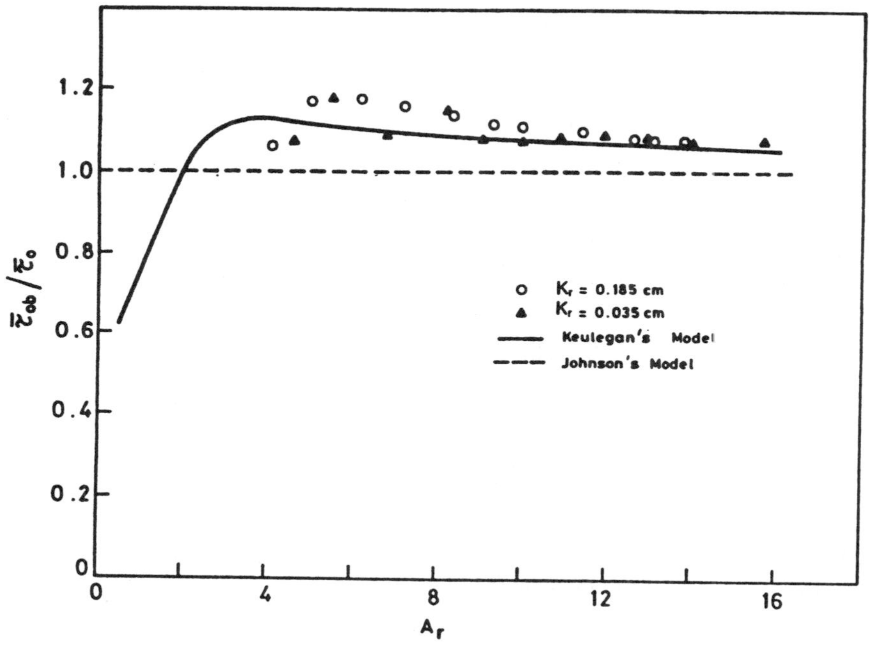

Figure 17. Keulegan model for mean bed shear stress (Supercritical rough)

Rouse *et al.* [78] observed that $F>F_s$, the Froude number at which the free surface becomes unstable, the value of **f** is affected by F. The flow, according to them has to be supercritical before it becomes unstable. For $F>F_s$, they obtained the following equation

$$\frac{1}{\sqrt{f}} = 2\log\left[\frac{R_e\sqrt{f}}{\left(\frac{F}{F_s}\right)^{\frac{2}{3}}}\right] - 0.97 \tag{50}$$

Berlamont and Vanderstappen [5] made a stability analysis and obtained the following equation for F_s in the case of smooth rectangular open channels.

$$\frac{1}{F_s^{\,2}} = \beta - 2\beta\left[1+\frac{5}{7}\left(\frac{A_r}{A_r+2}\right)\right] + \left[1+\frac{5}{7}\left(\frac{A_r}{A_r+2}\right)\right]^2 \tag{51}$$

11.1 Analysis of data of I.I.Sc. on friction

11.1.1 Smooth channels

The friction data of subcritical flow through smooth channels (regime) has been analyzed by Lakshminarayana *et al.* [51] . They obtained Eq.52. See Fig.18.

$$f = \frac{0.328}{R_e^{\,1/4}} \tag{52}$$

Sarma and Syamala [89] analyzed the friction data of supercritical flow in smooth channels (regime 2). Using the measured value of $\beta = 1.022$, they estimated F_s using Eq. 51.

So it can be stated that for $F<F_s$ the Blasius equation with a slightly different constant (Eq. 52) and for $F>F_s$ the equation of Rouse *et al.* (Eq. 50) give **f** values.

11.1.2 Rough channels

Sarma and Sarma [80] analyzed the friction data for subcritical flows through rough channels (regime 3). They obtained θ, the ratio of equivalent sand size K_s to the roughness size K_r a value of 2.64 which is within the reported values of Ippen and Drinker [33] and close to the uppermost value of Kamphuis [37].

The friction values of subcritical flows in the transition zone from smooth to rough could be described by Eq. 53. See Fig. 18.

The same data also could be described by Eq. 54.

$$f = \frac{0.328}{R_{e^{0.25}}} + 0.1545\left(\frac{K_r}{R}\right)^{0.441} \tag{53}$$

Comparing Eq. 54 with Eq.25 of Nikuradse for rough pipes K_s was obtained as $2.64K_r$

$$\frac{1}{\sqrt{f}} = 2\log\left(\frac{R}{K_r}\right) + 1.5 \tag{54}$$

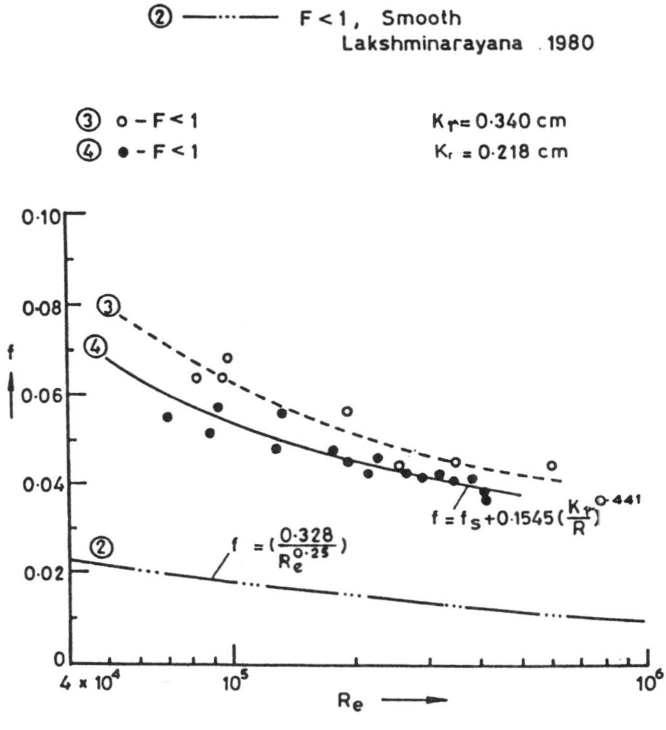

Figure 18. Friction factor in subcritical rough turbulent flows

It is a well known fact in super critical flows that the flow becomes unstable when the Froude number exceeds some critical value. Rosso *et al.* [76] expressed F_s as a function of Reynolds number, momentum coefficient, the channel roughness and channel width. Rosso *et al.* [76] analyzed the increase of friction in unstable flows by conducting experiments in rough channels. In the case of supercritical rough channels (65], the value of F_s will be greater than that of supercritical smooth channels if all the other hydraulic conditions are the same. The critical Froude number was computed on the lines given by Rosso *et al.* [76]. As the Froude numbers were less than the value of F_s the effect of F was not observed in the data of friction factor. The friction factor in the range of transition of smooth and rough, and completely rough is described by Eq. 55. See Fig. 19.

$$f = \frac{0.328}{R_e^{0.25}} + 0.239 \left(\frac{K_r}{R}\right)^{0.471} \tag{55}$$

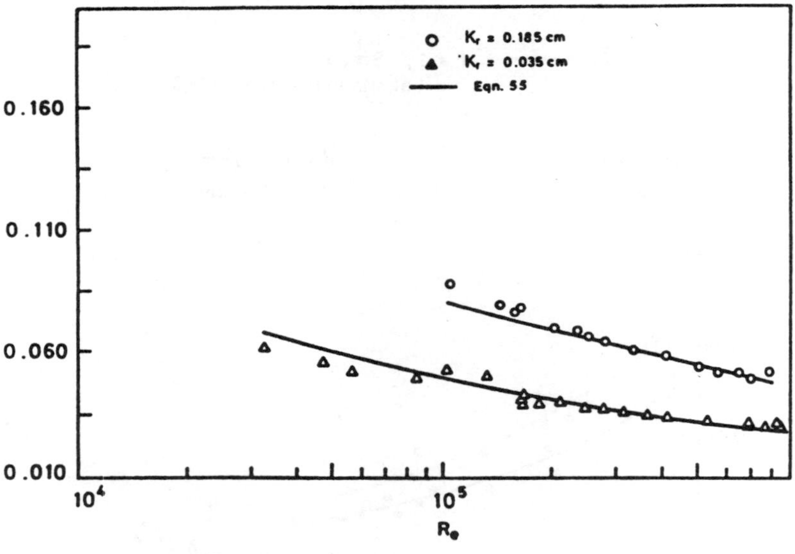

Figure 19. Friction factor in supercritical rough turbulent flows

The friction factor of completely rough turbulent flow range is also expressed as Eq. 56.

$$\frac{1}{\sqrt{f}} = 2.0 \log\left(\frac{R}{K_r}\right) + 1.0 \tag{56}$$

By comparing Eqn. 56 with Eqn.25 Nikuradse equivalent roughness is obtained as $4.68K_r$ [68]. The effect of R_e on **f** in the rough turbulent flows is rather strange. Resistance seemed to vary not as V^2 but as (V^2 + Const.$V^{1.75}$). This requires further study.

12 Conclusions

The studies on velocity distribution in flow though narrow open channels are very limited in the literature. In two dimensional (wide) open channels the velocity distributions like log-law, power law, parabolic law, Coles wake function, binary law and estimation of shear stress almost attained the level of acceptability.

The experimental data collected at Indian Institute of Science on velocity and shear distributions in narrow open channels is of great value to the profession. The entire flow region i.e. region 1,2,3 & 4, in the turbulent flow through smooth open channels is well described by the power law (Eqs.18&21) and parabolic law (Eqs.19 &22).

The validation of log-law is well established for the inner region in turbulent flows through smooth and rough open channels. However there is a strong evidence from the investigations that the additive constant is not the same along the wetted perimeter of the channel. The variation of additive constant appears to be related with the shear stress distribution and secondary currents. This leaves the scope for future research.

The binary law i.e. combination of log-law and parabolic law, is generalized for narrow open channels. The extension of Coles wake law to narrow open channels by incorporating the concept of dip may be useful and interesting topic for future research.

An alternate method to Clauser's technique is given to determine the reference datum for the flow through rough open channels. This method can also be applied to flow through pipes. But it is difficult to apply for the flow over a flat plate.

The Nikuradse equivalent roughness is found to be greater than the grain size even in uniform sand roughened open channels.

An empirical model is given for dip and it appears to be the only model by which dip can be computed with the known values of aspect ratio and lateral distance. This model is found to be valid for smooth and rough open channels.

A different type of power law called Frequency Power Law, is given for smooth and rough open channels and it permits the evaluation of the kinetic energy correction factor and momentum correction factor in narrow open channels.

Shear stress distribution in smooth rectangular open channels is described well in the form of equations 42 & 43.The shear stress distribution in rough rectangular open channels suggests that the maximum shear stress does not always occur at the center of the open channel.

The friction factor in smooth open channel flows is well explained by Blasius equation with a slightly different constant when $F < F_s$ where F_s is evaluated using

Berlamont and Vanderstappen equation. When $F > F_s$, the friction equation suggested by Hunter Rouse is used. The friction factor in rough open channels is explained by the extended version of Blasius law i.e. Eqs. 53 & 55.

A simple method for side-wall correction referred to as Keulegan model (Eqs.45, 46, 47 & 48) is given for smooth as well as rough rectangular open channels. This concept is also extended for trapezoidal and compound open channels.

Critically, it can be stated that the advanced instrumentation like Laser Doppler Anemometer could not really influence the established facts of velocity distribution but it has a great impact on the characteristics of turbulence and the pattern of secondary currents in open channels.

References

1. Allen J., Streamline and turbulent flow in open channels, *J. Sci.* **7** (1934).
2. Batchelor G. K., Note on free turbulent flows with special reference to the two-dimensional wake, *J. Aeronautical Sci.* **17** (1950).
3. Bagnold R. A., Some flume experiments on large grains but little denser than the transporting fluid and their implications, *Proc. Instn Civ. Engrs* London **4** (1955) Part-III pp. 175-205.
4. Bayazit M., Free surface flow in a channel of large relative roughness, *J. Hydr. Res.* IAHR **14** (1976) pp. 115-125.
5. Berlamont J. and Vanderstappen N., Unstable turbulent flow in open channels, *J. Hydr. Div.* **107** (1981) pp. 427-449.
6. Blinco P. H. and Partheniades E., Turbulence characteristics in free surface flows over smooth and rough boundaries, *J. Hydr. Res.* IAHR **9** (1971) pp. 43-71.
7. Brooks N. H., Mechanics of streams with movable beds of fine sand, *Trans.* **123** (1958) pp. 526-594.
8. Cardoso A. H., Graf W. H. and Gust G., Steady gradually accelerating flow in a smooth open channel, *J. Hydr. Res.* IAHR **29** (1991) pp. 525-543.
9. Cardoso A. H., Graf W. H. and Gust G., Uniform flow in a smooth open channel, *J. Hydr. Res.* IAHR **27** (1989) pp. 603-616.
10. Chandrasekaran D. and Sakthivadivel R., A note on new velocity distribution for wide rivers, *La Houille Blanche* **8** (1973) pp. 651-654.
11. Chiu C. -L., Application of entropy concept in open-channel flow study, *J. Hydr. Engg.* **117** (1991) pp. 615-628.
12. Chiu C. -L. and Chiou J. D., Structure of 3-D flow in rectangular open channels, *J. Hydr. Engg.* **112** (1986) pp.1050-1068.
13. Chiu C. -L. and Hsiung D.E., Secondary flow, shear stress and sediment transport, *J. Hydr. Div.* **107** (1981) pp. 879-898.
14. Chiu C. -L., Lin G. F. and Lu J. -M., Application of probability and entropy concepts in pipe-flow study, *J. Hydr. Engg.* **119** (1993) pp. 742-756.

15. Chow V. T., Open Channel Hydraulics (McGraw-Hill Book Co. Inc., New York, 1959).
16. Clauser F. H., The turbulent boundary layer, *Advances in Applied Mechanics* **4** (1956) pp. 1-51.
17. Coleman N. L. and Alonso C. V., Two-dimensional channel flows over rough surfaces, *J. Hydr. Engg.* **109** (1983) pp. 175-188.
18. Coles D., The law of the wake in the turbulent boundary layer, *J. Fluid Mech.* **1** (1956) pp. 191-226.
19. Davidian J. and Cahal D. I., Distribution of shear in rectangular channels, *U. S. Geological Survey Professional Paper* **475-C** (1963) pp. C206-C208.
20. Ead S. A., Rajaratnam N., Katopodis C. and Ade F., Turbulent open-channel flow in circular corrugated culverts, *J. Hydr. Engg.* **126** (2000) pp. 750-757.
21. Einstein H. A., Formulas for the transportation of bed load, *Trans.* **107** (1942) pp. 561-597.
22. Enger P. F., Tractive force fluctuations around an open channel perimeter as determined from point velocity measurements (Presented at the Convention of the ASCE, Phoenix, Arizona, USA, April 1961).
23. Goncharov V. N., Dynamics of Channel Flow (Translated from Russian; Israel Program for Scientific Translations, Jerusalem, Second Impression, 1970).
24. Gosh S. N., Boundary shear distribution in channels with varying wall roughness, *Proc. Instn Civ. Engrs* London **53** (1972) Part 2 pp. 529-544.
25. Gosh S. N. and Jena S. B., Boundary shear distribution in open channel compound, *Proc. Instn Civ. Engrs* London **49** (1971) pp. 417-430.
26. Gosh S. N. and Kar S. K., River flood plain interaction and distribution of boundary shear stress in a meander channel with flood plain, *Proc. Instn Civ. Engrs* London **59** (1975) Part 2 pp. 805-811.
27. Gosh S. N. and Mehta P. J., Boundary shear distribution in a compound channel with varying roughness distribution, *Proc. Instn Civ. Engrs* London **57** (1974) Part 2 pp. 159-163.
28. Gosh S. N. and Roy N., Boundary shear distribution in open channel flow, *J. Hydr. Div.* **96** (1970) pp. 967-944.
29. Henderson F. M., Open Channel Flow (The MacMillan Co., New York, 1966).
30. Hinze J. O., Turbulence, (McGraw-Hill Book Co., New York, 1959).
31. Hollingshead A. B. and Rajaratnam N., A calibration chart for the preston tube, *J. Hydr. Res.* IAHR **18** (1980), pp. 313-326.
32. Huffman D. C. and Bradshaw P., A note on von Karman constant in low Reynolds number turbulent flows, *J. Fluid Mech.* **53** (1972) Part-1, pp. 45-60.
33. Ippen A. T. and Drinker P. A., Boundary shear stresses in curved trapezoidal channels, *J. Hydr. Div.* **88** (1962) pp. 143-179.
34. Ippen A. T. and Verma R. P., The motion of discrete particles along the bed of a turbulent stream, *Proc. Minnesota International Hydraulics Convention* (A joint meeting of IAHR and Hydr. Div. ASCE, 1953) pp. 7-20.

35. Jayaraman V. V., Resistance studies in smooth open channels, *J. Hydr. Div.* **96** (1970) pp. 1129-1141.
36. Johnson J. W., The importance of side-wall friction in bed load investigations, *Civil Engg.* **12** (1942) pp. 329-331.
37. Kamphuis J. W., Determination of sand roughness for fixed beds, *J. Hydr. Res.* IAHR **12** (1974) pp. 193-203.
38. Kartha V. C. and Leutheusser H. J., *Distribution of tractive force in open channels, J. Hydr. Div.* **96** (1970) pp. 1469-1483.
39. Kazemipour A. K. and Apelt C. J., Shape effects on resistance to uniform flow in open channels, *J. Hydr. Res.* IAHR **17** (1979) pp. 129-147.
40. Kennedy J. F., Task committee report on sediment transportation mechanics: F. Hydraulic relations for alluvial streams, *J. Hydr. Div.* **97** (1971) pp. 101-141.
41. Keulegan G. H., Laws of turbulent flow in open channels, *J. Res. National Bureau of Standards,* U. S. Department of Commerce, Washington D.C. **21** (1938) pp. 707-741.
42. Khodashenas S. R. and Paquier A., A geometrical method for computing the distribution of boundary shear stress across irregular straight open channels, *J. Hydr. Res.* IAHR **37** (1999) pp. 381-388.
43. Kirkgoz M. S., Turbulent velocity profiles for smooth and rough channel flow, *J. Hydr. Engg.* **115** (1989) pp. 1543-1561.
44. Knight D. W., Discussion on Boundary shear distribution in channels with varying wall roughness, *Proc. Instn Civ. Engrs* London **55** (1973) Part 2 pp. 503-511.
45. Knight D. W., Boundary shear in smooth and rough channels, *J. Hydr. Div.* **107** (1981) pp. 839-851.
46. Knight D. W., Demetriou J. D. and Hamed M. E., Boundary shear in smooth rectangular channels, *J. Hydr. Engg.* **110** (1984) pp. 405-422.
47. Knight D. W. and Macdonald J. A., Hydraulic resistance of artificial strip roughness, *J. Hydr. Div.* **105** (1979) pp. 675-690.
48. Knight D. W. and Macdonald J. A., Open channel flow with varying bed roughness, *J. Hydr. Div.* **105** (1979) pp. 1167-1183.
49. Lakshminarayana P., Velocity and Shear Distributions in Smooth Rectangular Open Channels with Small Aspect Ratios (Ph.D. Thesis, Indian Institute of Science, Bangalore, India, 1980).
50. Lakshminarayana P., Sarma K. V. N. and Lakshamana Rao N. S. L., Dip in vertical velocity profiles in flows through rectangular open channels, *Proc. !3th Natl. Conf. on Fluid Mechanics and Fluid Power* (Regional Engineering College, Thiruchirapalli, India, 1984) .
51. Lakshminarayana P., Sarma K. V. N. and Lakshamana Rao N. S. L, Shear in smooth rectangular channels, *J. Instn. Engrs.* India **66** (1986) part C15 pp. 204-213.

52. Laufer J., The structure of turbulence in fully developed pipe flow, *National Advisory Committee for Aeronautics Report* **1174** (1954).
53. Lundgren H. and Jonsson I. G., Shear and velocity distribution in shallow channels, *J. Hydr. Div.* **90** (1964) HY1 pp. 1-21.
54. Nezu I., Nakagawa H. and Rodi W., Significant difference between secondary currents in closed channels and narrow open channels, *Proc. 23rd IAHR Congress,* Ottawa (1989) pp. A125-A132.
55. Naot D. and Rodi W., Calculation of secondary currents in channel flow, *J. Hydr. Div.* **108** (1982) pp. 948-968.
56. Nezu I. and Rodi W., Open-channel flow measurements with a laser doppler anemometer, *J. Hydr. Engg.* **112** (1986) pp. 335-354.
57. Nezu I., Tominaga A. and Nakagawa H., Field measurements of secondary currents in straight rivers, *J. Hydr. Engg.* **119** (1993) pp. 598-614.
58. Nikuradse J., Laws of turbulent research, *NACA Tech. Memorandum* **1292** Nov. (1950, Translated from German 1933).
59. O'Loughlin E. M. and Macdonald E. G., Some roughness-concentration effects on boundary resistance, *La Houille Blanche* **7** (1964) pp. 773-783.
60. Prandtl, Recent results of turbulent research, NACA TM **720** (Translated from German, 1933).
61. Preston J. H., The determination of turbulent skin friction by means of Pitot tubes, *J. Royal Aero. Society* **58** (1954) pp. 109-121.
62. Prinos P. and Zeris A., Uniform flow in open channels with steep slopes, *J. Hydr. Res.* IAHR **33** (1995) pp. 705-719.
63. Pizzuto J. E., A numerical model for calculating the distributions of velocity and boundary shear stress across irregular straight open channels, *Water Resour. Res.* **27** (1991) pp. 2457-2466.
64. Rajaratnam N. and Muralidhar D., Boundary shear stress distribution in rectangular open channels, *La Houille Blanche* **6** (1969) pp. 603-609.
65. Ramana Prasad B. V., Velocity, Shear and Friction Factor Studies in Rough Rectangular Open Channels for Supercritical Flows (Ph.D. Thesis, Indian Institute of Science, Bangalore, India, 1991).
66. Ramana Prasad B. V. and Manson J. R., Discussion on A geometrical method for computing the distribution of boundary shear stress across irregular straight open channels, *J. Hydr. Res.* IAHR (2001) In Print.
67. Ramana Prasad B. V. and Manson J. R., Discussion on Turbulent open-channel flow in circular corrugated culverts, *J. Hydr. Engg.* (2001) In print.
68. Ramana Prasad B. V. and Sarma K. V. N., Equivalent sand roughness in rough turbulent flows, *Procs. 8th Congr. of the Asia Pacific Div.* IAHR (C.W.P.R.S., Pune, India, Oct. 1992).
69. Ramana Prasad B. V. and Sarma K. V. N., Estimation of shear stress in rough open channels, *Proc. 11th Autralasian Fluid Mech. Conf.* (Hobart, Australia, Dec. 1992).

70. Ramana Prasad B. V. and Sarma K. V. N., Uncertainty of shear velocity, *Proc. 21st National Conf. on Fluid Mech. and Fluid Power* (Osmania University, Hyderabad, India, Dec. 1994).
71. Ramana Prasad B. V. and Sarma K. V. N., Sensitivity of wall shear stress to the reference level of the roughness elements, *Proc. 22nd National Conf. on Fluid Mech. and Fluid Power* (I.I.T., Madras, India, Dec. 1995).
72. Ramana Prasad B. V. and Sarma K. V. N., Reference datum for rough surfaces, *Procs. First International Conf. On New/Emerging Concepts for Rivers - Rivertech'96* (Chicago, USA, Sept. 1996).
73. Ramana Prasad B. V. and Sarma K. V. N., Shear distribution in rough open channels, *Natl. Conf. on Envir. Hydraulics and free surface flows* (Andhra Univ., Visakhapatnam, India, 1997).
74. Replogle J. A. and Chow V. T., Tractive-force distribution in open channels, *J. Hydr. Div.* **92** (1966) pp. 169-191.
75. Reynolds A. J., Turbulent Flows in Engineering (John Wiley and Sons, New York, 1974).
76. Rosso M., Schiara M. and Berlamont J., Flow stability and friction factor in rough channels, *J. Hydr. Engg.* **116** (1990) pp. 1109-1118.
77. Rouse H., Elementary Fluid Mechanics (Wiley Eastern Reprint, Wiley Eastern Private Ltd., New Delhi, India, 1970).
78. Rouse H., Koloseus H. J. and Davidian J., The role of Froude number in open channel resistance, *J. Hydr. Res.* IAHR **1** (1963) pp. 14-19.
79. Sarma A. K., Velocity, Shear, and Friction Studies in Subcritical Rough Turbulent Open Channel flows (Ph.D. Thesis, Indian Institute of Science, Bangalore, India, 1993).
80. Sarma A. K. and Sarma K. V. N., Rough open channel friction, *Natl. Conf. on Envir. Hydraulics and Free Surface Flows* (Andhra Univ., Visakhapatnam, India, 1997).
81. Sarma A. K., Sarma K. V. N. and Lakshamana Rao N. S., Frequency power law for the velocity distribution in smooth and rough open channels, *Proc. Seminar on Hydro Mech. and Water Resour. Engg.* (Indian Institute of Science, Bangalore, India, Dec. 1991).
82. Sarma K. V. N., Lakshminarayana P. and Lakshmana Rao N. S., Velocity distribution in smooth rectangular open channels, *J. Hydr. Engg.* **109** (1983) pp. 270-289.
83. Sarma K. V. N. and Ramana Prasad B. V., Correction for Wall Shear, Procs. *Third International Workshop on Alluvial River Problems* (Roorkee Univ., India, March 1989).
84. Sarma K. V. N. and Ramana Prasad B. V., Velocity distributions for flows in sand roughened open channels, *Proc. Seminar on Hydro Mech. and Water Resour. Engg.* (Indian Institute of Science, Bangalore, May 1990).

85. Sarma K. V. N. and Ramana Prasad B. V., Velocity-defect law in sand roughened open channel, *Proc. Seminar on Hydromech. and Water Resour. Engg.* (Indian Institute of Science, Bangalore, India, 1991).
86. Sarma K. V. N., Ramana Prasad B. V. and Sarma A. K., Detailed study of binary law for open channels, *J. Hydr. Engg.* **126** (2000) pp. 210-214.
87. Sarma K. V. N. and Sarma A. K., Velocity distribution in smooth open channels, *Proc. Instn. Engrs.* India **70** (1990) pp. 197-199.
88. Sarma K. V. N. and Sarma A. K., Velocity distributions in smooth rectangular open channels, *Proc. Seminar on Hydro Mech. and Water Resour. Engg.* (Indian Institute of Science, Bangalore, India, May 1990).
89. Sarma K.V.N. and Syamala P., Supercritical flow in smooth channels, *J. Hydr. Engg.* **117** (1991) pp. 54-63.
90. Sarma. K. V. N. and Syamala P., Shear distribution in smooth rectangular open channels for supercritical flows, *Proc. Natl. Conf. on Envir. Hydr. and Free Surface Flows* (Andhra Univ., Visakhapatnam, India, 1997).
91. Schlichting H., Boundary Layer Theory (7th Edition McGraw-Hill Classic Textbook Reissue Series, New York, 1987).
92. Smerdon E. T. and Beasley R. P., Critical tractive forces in cohesive soils, *Agricultural Engg.* (Jan. 1961) pp. 26-29.
93. Song T., Graf W. H. and Lemin U., Uniform flow in open channels with movable gravel bed, *J. Hydr. Res.* IAHR **32** (1994) pp. 861-876.
94. Steffler P. M., Rajaratnam N. and Peterson A. W., LDA measurements in open channel, *J. of Hydr. Engg.* **111** (1985) pp. 119-130.
95. Syamala P., Velocity, shear and friction factor studies in smooth rectangular channels of small aspect ratios for supercritical flows (Ph.D. Thesis, Indian Institute of Science, Bangalore, India, 1988).
96. Tominaga A., Nezu I., Ezaki K. and Nakagawa H., Three-dimensional turbulent structure in straight open channel flows, *J. Hydr. Res.* IAHR **27** (1989) pp. 149-173.
97. Townsend A. A., The Structure of Turbulent Shear Flow (Cambridge Univ. Press, New York, 1956).
98. Tracy H. J. and Lestér C. M., Resistance coefficients and velocity distribution - smooth rectangular channel, *U. S. Geological Survey Water Supply Paper* **1592-A** (1961).
99. Vanoni V. A., Velocity distribution in open channels, *Civ. Engg.* **11** (1941) pp. 356-357.
100. Vedula S. and Rao A. R., Bed shear from velocity profiles : A new approach, *J. Hydr. Engg.* **111** (1985) pp. 131-143.
101. Vittal N., Verma M. S. and Rangaraju K. G., Discussion on Open channel flow with varying bed roughness, *J. Hydr. Div.* **106** (1980) pp. 1705-1708.
102. Von Karman T., On laminar and turbulent friction, Translation by NACA TM **1092** (1942, Original Publ. in 1921).

103. Willis J. C., An error function description of the vertical suspended sediment distribution, *Water Resour. Res.* **5** (1969) pp. 1322-1329.
104. Willis J. C., A new mathematical model for the velocity distribution in turbulent shear flow, *J. Hydr. Res.* IAHR **10** (1972) pp. 205-225.
105. Xinyu L., Zengnan D. and Changzhi C., Turbulent flows in smooth-wall open channels with different slope, *J. Hydr. Res.* IAHR **33** (1995) pp. 333-347.
106. Yang S. and Lim S., Mechanism of energy transportation and turbulent flow in a 3D channel, *J. Hydr. Engg.* 123 (1997) pp. 684-692.
107. Zagustin A. and Zagustin K., Analytical solution for turbulent flow in pipes, *La Houille Blanche* **2** (1969) pp. 113-118.
108. Zippe H. J. and Graf W. H., Turbulent boundary layer flow over permeable and non-permeable rough surfaces, *J. Hydr. Res.* **21** (1983) pp. 51-65.

COMPUTATION OF OPEN-CHANNEL FLOWS WITH SHOCKS: AN OVERVIEW

S. MURTY BHALLAMUDI

Department of Civil Engineering, Indian Institute of Technology Madras, Chennai-600 036, INDIA

E-mail: bsm@civil.iitm.ernet.in

Computation of open-channel flows with shocks is required in the analysis of (1) dam-break induced flows, (2) transient flows in sewers, (3) steady flows with hydraulic jumps, (4) two-dimensional supercritical flows in non-prismatic channels etc. They are also required in the design and operation of power canals. These computations usually involve the numerical solution of either the one-dimensional or the two-dimensional shallow water equations. One of the major difficulties in the numerical solution of the shallow water flow equations has been the correct resolution of shock fronts. Classical first-order schemes introduce too much diffusion, while second-order schemes result in spurious numerical oscillations. Also, many schemes result in unphysical jumps across which there is an increase of energy. Significant contributions have been made in the last decade toward development of high-resolution schemes, which avoid these problems. The objective of this paper is to present an overview of these developments, and also bring out their relative strengths and limitations. Focus of the overview is on finite-difference and finite volume techniques for the solution of governing hyperbolic partial differential equations. Although much advancement has been made toward the front resolution, there are still some unresolved issues in the realm of shock flow computation. The paper concludes with a brief discussion of a few of these issues.

1 Introduction

Large water surface disturbances in open-channel flows are usually referred to as shocks. Unlike shocks in compressible flows, a shock in an open-channel flow is not a sharp discontinuity, and the length of the shock could be three to six times the depth of flow. Common examples of shocks in open-channel flows are hydraulic jumps, standing waves in supercritical flow (also known as oblique jumps), and sharp moving fronts in dam-break flows (also known as bores). Sharp moving fronts or shocks could also form in irrigation canals, power channels, and sewer systems due to rapid operation of control structures.

Accurate analysis of flows with shocks is very important in the design and operation of many open channels. Computation of unsteady flow in open-channels resulting from a dam-break is not only an interesting hydrodynamic problem, but also very important, as dams are potential sources of hazard. Reliable prediction of maximum flow depths and corresponding times of their occurrences is crucial for flood plain management, and for locating strategic infrastructure in river valleys. The horror of concentrated death tragedy gives special importance to analysis of dam-break floods [9].

Opening or closing of turbine gates in hydroelectric power plants and in pumped storage schemes produce steep bores in partly flowing tunnels and power canals. Bores are formed in power canals even if the gate closing is gradual, if the channel is long and friction losses are negligible. Analysis of these bores, from the point of view of bore strength and bore speed, is important for the design of power canals. Also, it is important to determine the arrival time of the reflected wave in the case of multiple machine operations [50]. Similarly, computation of storm flows in sewers involves proper resolution of location and movement of shock fronts.

Channel transitions are commonly used between canals and flumes, and between canals and tunnels to reduce energy loss [38]. They are also some times used in stilling basins, spillway chutes, and in flow measuring flumes. Supercritical flow in such conveyance structures is accompanied by oblique standing waves, and some times by hydraulic jumps. In fact, any change in the boundary alignment of an open channel produces standing waves if the flow is supercritical. Common examples are supercritical flow in a curved channel and supercritical flow through channel junctions in roof gutters and other drainage systems [72]. Consideration of shocks is very important for appropriate design and operation of all the above open-channel systems.

Computation of open-channel flows with shocks is complicated because such flows often require the solution of unsteady flow equations. Also, it may be required to consider two-dimensional flow equations in many cases. Analytical solutions are available only for some idealized cases, and one is forced to seek numerical solutions of the governing non-linear, non-homogeneous partial differential equations. In majority of the situations, complex flow domains with irregular boundaries are involved, and they introduce additional difficulties in the form of implementation of boundary conditions. All this makes shock-flow computation a fascinating area of research.

In this paper, an attempt is made to present an overview of the advances made in the computation of open-channel flows with shocks. Governing partial differential equations are presented first. Available numerical methods for solving one-dimensional shock flow problem are discussed next. This is followed by a discussion of two-dimensional shock-flow computation. The paper concludes with some suggestions for further research. The focus of this review is limited to finite-difference and finite-volume methods. Only a brief note on finite element methods is included for the sake of completeness.

2 Governing Equations

Many of the numerical models for shock-flow computation are based on the shallow water flow equations. These equations may be derived by depth averaging the general three-dimensional flow equations for conservation of mass and momentum [35]. These equations, in the *conservation form*, for a two-dimensional Cartesian co-ordinate system may be written as

$$\frac{\partial U}{\partial t} + \frac{\partial E}{\partial x} + \frac{\partial G}{\partial y} = S \tag{1}$$

in which

$$U = \begin{Bmatrix} h \\ uh \\ vh \end{Bmatrix} \quad E = \begin{Bmatrix} uh \\ u^2h + \frac{1}{2}gh^2 \\ uvh \end{Bmatrix} \quad G = \begin{Bmatrix} vh \\ uvh \\ v^2h + \frac{1}{2}gh^2 \end{Bmatrix} \quad S = \begin{Bmatrix} 0 \\ gh(S_{0x} - S_{fx}) \\ gh(S_{0y} - S_{fy}) \end{Bmatrix} \tag{2}$$

in which, h = flow depth, u = depth averaged velocity in x-direction, v = depth averaged velocity in y-direction, g = acceleration due to gravity, S_0 is the bed slope, S_f = friction slope (subscripts x and y refer to the slopes in x and y directions, respectively), t = time, and x and y are the coordinate axes. The friction slope S_f is usually calculated using the empirical steady state formula, which is similar to the Manning's equation.

$$S_{fx} = \frac{un^2\sqrt{u^2 + v^2}}{h^{1.333}} \qquad S_{fy} = \frac{vn^2\sqrt{u^2 + v^2}}{h^{1.333}} \tag{3}$$

in which n = Manning roughness coefficient. The three dependent variables h, u, and v are dependent on the three independent variables t, x, and y, and they can be determined by solving (1) - (3). Equations (1) - (3) are based on the following assumptions:

1. The pressure distribution is hydrostatic in the vertical direction.
2. The velocity distribution is uniform over the flow depth.
3. Bottom shear stress is large compared to other turbulent stresses.
4. There is no lateral inflow or outflow.
5. The bottom slope is small.

Hydrostatic pressure distribution is not valid in the vicinity of the shock, and details of flow within the shock itself are lost. However, many investigators have argued that the overall results at all other points are adequate for engineering purposes.

Liggett [99] and Jimenez and Chaudhry [89] have shown that the errors introduced by this assumption are not significant if the ratio of flow depth to width is less than 0.05, and the Froude number is not close to one.

Assumption (2) means that the momentum dispersion terms, which arise due to depth averaging of vertically non-uniform momentum terms are neglected. Again, the errors introduced by this assumption and the assumption (3) are negligible if the depth to width ratio is small [82]. Momentum dispersion terms can be viewed as stress terms, and when combined with the depth averaged turbulent stress terms, they are usually referred to as *effective stresses*. Note that circulating flows cannot be simulated without considering the effective stresses [56]. Also, simulation of flows in curved channels would require consideration of these terms because of the effect of secondary flow.

Assumption of no lateral inflow or outflow is not valid for overland flows on porous surfaces, and flows in surface-irrigation systems. Source term, S in (1) – (3) can be easily modified to take this into consideration. It is also required to supplement (1) - (3) with extra "state equations" to represent the process of lateral inflow or outflow. Readers are referred to Chow and Ben-Zvi [39], Zhang and Cundy [173] and Singh [138] for a complete discussion of these equations. It may be noted that it is possible for shocks to occur even in the case of overland flows, particularly when the infiltration characteristics are spatially varying [140, 54].

Lastly, the assumption of small bottom slope is valid in many cases of dam-break flow and supercritical flow. However, it breaks down when problems on hill-slope hydrology, and supercritical flow in steep chutes are considered. Readers are referred to Jimenez [88] and Chaudhry [35] for a complete presentation of shallow water flow equations on steep slopes.

By far the assumptions (1) and (2) are the most limiting factors in the successful application of shallow water theory in shock flow computation. However, only few studies on shock flow computation do not make these assumptions. Basco [17] and Chaudhry [35] may be referred for complete derivation of higher-order shallow water equations based on Boussinesq assumption, and their application. Readers are also referred to Steffler and Jin [143] for the derivation of vertically averaged and moment (VAM) equations, valid for moderately shallow free surface flow. These equations assume a linear longitudinal velocity profile and quadratic pressure and vertical velocity profiles.

3 One-dimensional Shock Flow

One of the classical examples of an open-channel shock flow is the dam-break flow. Studies on dam-break flow (DBF) date back to Ritter [131] who derived an analytical solution for the movement of a dam-break wave in a dry, frictionless, and horizontal channel of infinite width (Fig. 1), resulting from instantaneous and complete collapse of a dam. The most important findings of this study were: (I) the dam-site flow depth, h_d is constant and is equal to $(4/9)h_0$, and (ii) the downstream

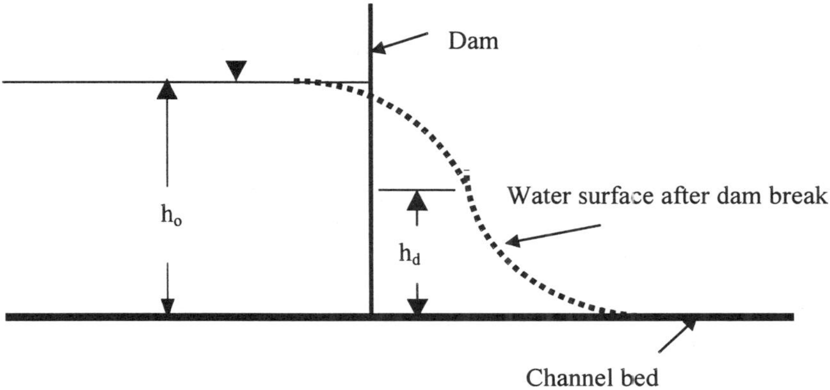

Figure 1. Definition sketch for Ritter's problem

water tip moves with a velocity equal to $2\sqrt{gh_0}$, where h_0 = flow depth upstream of the dam before the failure. Dressler [47] and Whitham [162] extended the Ritter derivation to include frictional effects, while Stoker [145] extended the derivation to include the effect of wet-bed downstream conditions. Su and Barnes [146] extended the Dressler solution to channels with different shapes, and inclined channels. Ritter's and Stoker's solutions are valid for the case of an infinite reservoir. Hunt [80, 81] removed this limitation and derived solutions for the case of finite length reservoirs. Aforementioned analytical solutions are based on the shallow water wave assumption i.e. the pressure distribution is hydrostatic in the vertical direction.

All the analytical solutions mentioned earlier are based on simplifying assumptions, and they are not valid for a dam-break wave moving in a natural river valley with varying cross sectional shape, roughness, slope, and resulting from gradual and partial dam-breaks. Therefore, numerical solutions based on shallow water wave equations are usually sought for typical engineering applications. In this section, numerical solution of governing shallow water equations for the case of one-dimensional dam-break flow is discussed. It should be mentioned here that the numerical schemes discussed in this section are equally applicable to any one-dimensional open-channel shock flow problem, such as the movement of bores in power canals, discontinuous overland flow, and transients in sewer pipes. Flows in sewer pipes could be partly full or pressurized. Traditionally the concept of Preissmann slot is used to simulate these flows using the Saint Venant equations.

3.1 One-dimensional Flow Equations

One-dimensional open-channel flow equations can be derived assuming that the lateral velocity is equal to zero, and then width averaging (1) - (2). These equations for a non-prismatic channel, and having an irregular cross-sectional shape, can be written as

$$\frac{\partial V}{\partial t} + \frac{\partial F}{\partial x} = S' \tag{4}$$

in which,

$$V = \begin{Bmatrix} A \\ Q \end{Bmatrix} \quad F = \begin{Bmatrix} Q \\ \dfrac{Q^2}{A} + gI_1 \end{Bmatrix} \quad S' = \begin{Bmatrix} 0 \\ gA(S_0 - S_f) + gI_2 \end{Bmatrix} \tag{5}$$

in which, A = Cross-sectional area, Q = flow rate, S_0 = channel slope, and S_f = friction slope. Friction slope is determined using the Manning equation. It is assumed in the above equations that the variation in channel shape with distance is smooth. I_1 and I_2 terms arise from the integration of pressure force term. These terms are determined using the following equations.

$$I_1 = \int_0^h [h - z]\, b(x,z)\, dz \tag{6}$$

$$I_2 = \int_0^h [h - z] \frac{\partial b(x,z)}{\partial x} dz \tag{7}$$

in which, h = flow depth at any x, and time t, $b(x,z)$ = flow width at any distance x and height z above the channel bottom. I_2 accounts for the effect of walls in channel transitions. Equations (4) – (7) are numerically solved to simulate one-dimensional dam-break flows, shock flows in power channels etc. These equations can be also used in a false transient approach to simulate steady shock flows such as those occur for flow over a ladder of weirs [65].

3.2 Shock Fitting Methods

It can be shown that (4) - (5) constitute a set of non-linear hyperbolic equations [43, 35]. Therefore, in normal situations, characteristic methods can be used for solving these equations. However, a dam-break wave may form a steep front while propagating along a river valley, or mathematically speaking, the governing equations (4) - (5) can admit discontinuous solutions because of the non-linearity in the flux term F. This introduces difficulties in the numerical solution of these equations. At a shock, the characteristics converge, and hence the characteristic method alone cannot be used to propagate the shock. Many alternative numerical techniques have been developed in the past [2, 43, 148, 35] to overcome this difficulty in shock-flow computation. These methods for shock-flow computation can be broadly classified into two types: (i) shock fitting methods, and (ii) shock capturing methods.

In a shock fitting method, the front is treated as an internal boundary between two regions of gradually varied flow (Fig. 2), where shallow water flow equations are valid.

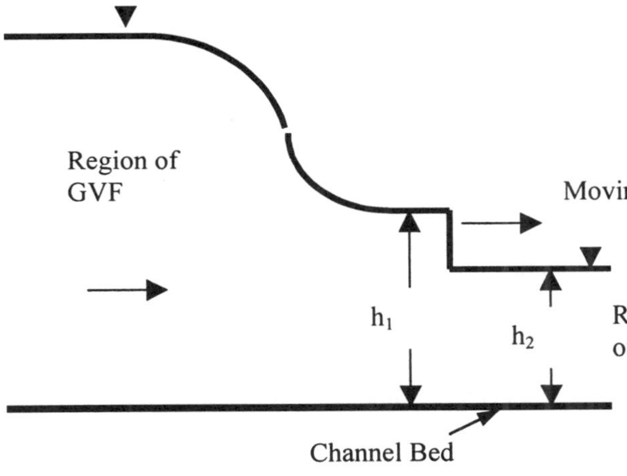

Figure 2. Schematic diagram for a shock-fitting method

The propagation of the shock is computed during each time step independently of the computations in the adjacent gradually varied flow regions. Rankine-Hugoniot relationships, which are the conservation of mass and the momentum equations for a moving hydraulic jump, are used along with the appropriate characteristic equations for this purpose [43]. Solution from the shock computation is then used as a boundary condition to compute the flow in the gradually varied flow regions to the left and to the right of the shock by any method of numerical integration of (4) and (5). Shock fitting technique is an essential feature of dam-break flow models based on the characteristic method for the solution of governing equations [105, 134, 135, 16, 37, 153, 70]. Although the shock fitting method is the theoretically correct approach for computing discontinuous flows, significant difficulties (in terms of code development) arise in tracking shocks in channel systems, and two-dimensional flows [136]. This is because irregular boundaries and channel junctions introduce a large number of shocks, and create a complicated pattern of multiple shocks due to reflections. Therefore, shock capturing or "through methods" have become more popular for computing flows with shocks.

3.3 Classical Shock Capturing Methods

Shock capturing methods involve the application of numerical techniques to the solution of basic governing equations expressed in the weak formulation [1]. In these methods, the differential equations for mass and momentum conservation are expressed in the divergent form or "conservation form". It has been shown that numerical solutions of the above differential equations tend to valid discontinuous solutions based on the integral relationships [1, 2]. The chief advantage of shock capturing methods is that they do not require any special treatment for shock propagation, and hence are easy to implement for natural and complex situations. However, one has to be cautious about the numerical diffusion, and dispersion errors introduced by shock capturing methods. These numerical errors many times may interfere with the actual solution, and may also result in run abortions.

Both finite-difference and finite-element methods have been used to solve the shallow water wave equations in the conservation form. Finite-element methods based on the classical Galerkin formulation produce very poor results when applied to shock flows. A recent collocation finite-element model for dam-break flow [6] also showed numerically generated high-frequency oscillations. Special procedures are required to introduce appropriate amount of numerical dissipation through "up-winding" to remove the numerically generated high-frequency parasitic waves. Katopodes [92] successfully developed one such dissipative Galerkin scheme for discontinuous channel flows. Molinaro [115] also reported successful application of a finite-element method to simulate dam-break flows. However, it should be mentioned here that a finite-element method is generally more complicated than a finite-difference method (in terms of code development) and uses more computational time. It does not offer any practical advantage over finite-difference methods as far as simulations of one-dimensional flows are concerned [96, 41, 95]. Therefore, finite-difference methods are more popular for solving one-dimensional shock flow problems.

Both explicit and implicit finite-difference methods have been used for dam-break flow simulations. In an explicit method, the spatial differential terms are approximated by the finite-differences using the known values of the dependent variables at time level t. Values of dependent variables at the unknown time level, t+Δt can be solved for explicitly at each node. On the other hand, in an implicit method, the spatial differential terms are approximated by the unknown values of the dependent variables at the time level t+Δt. This results in a system of non-linear equations, which needs to be solved for the values of dependent variables at time t+Δt at all the nodes simultaneously. Implicit methods involve the solution of a matrix equation at each time step. Notable among the early explicit methods are the ones developed by Liggett and Woolhiser [100], Dronkers [48], and Terzidis and Strelkoff [149]. Notable among the early implicit methods are the ones developed by Preissmann [126], Amein and Fang [10], Gunaratnam and Perkins [71], and Abbott and Ionescu [3]. Many of these methods are well documented by Mahmood and Yevjevich [103], Abbott [2], and Cunge *et al.* [43] in their classical books. It may be noted here that the shallow water flow equations are analogous to the gas dynamics equations, and numerical techniques applicable for solving the gas-dynamics equations are also applicable for shock-flows in open-channels.

Implicit methods are generally more popular than explicit methods for flood routing because the computational time step, Δt in explicit methods is usually restricted by the consideration of numerical stability or the well known Courant-Friedrichs-Lewey (CFL) condition. On the other hand, the computational time step in implicit methods is usually not restricted by the CFL condition. Also, very efficient double sweep algorithms are available to solve the block-diagonal matrix equation that results from implicit finite differencing. One of the most popular industrial codes for dam-break flow routing, National Weather Service Dam Break flow model (NWS-DAMBRK) developed by Fread [58] uses weighted four-point implicit scheme of Preissmann [126] for the solution of the governing equations. It should be mentioned here that as far as numerical stability is concerned, the computational time step in implicit methods is independent of the grid size. However, research [50] has shown that from the point of view of accuracy, the time step in implicit methods should not be very much greater than that required by the Courant stability criterion, especially in the case of shock flows. Therefore, currently, explicit finite-difference methods are more popular than implicit finite-difference methods for shock-flow computation.

Recently, Meselhe and Holly [106] have shown how one of the most popular implicit methods, the Preissmann scheme fails for transcritical flows. It has been shown that the method becomes ill conditioned if local supercritical-flow zones form next to the boundaries. Also, the scheme is only marginally stable if a critical point is encountered in the domain, and the numerical errors do not get damped. Addition of artificial viscosity to dampen the oscillations destroys the nice bi-diagonal structure of the Preissmann scheme. The above conclusion has been supported by findings of Jin and Fread [90], and Hicks *et al.* [78] also. Having found that the Preissmann scheme does not perform well for the case of transcritical flows, Jin and Fread [90] adapted a characteristic based, upwind, explicit numerical scheme, and incorporated it in the United States National Weather Service Flood

Wave (USNWS-FLDWAV) model. This method was found to give better results than Preissmann scheme for cases of large dam-break waves. On the other hand, Meselhe *et al.* [107] adapted a bi-diagonal implicit scheme developed by Casier *et al.* [32] for transonic flows, for application to transcritical open-channel flows. This scheme is second-order accurate in time and space, and requires inversion of only bi-diagonal matrices. This method is seen to fail only when the upstream to downstream depth ratio in a dam-break flow is greater than 10. It is very suitable for irrigation canal applications, where simulation times are long and one would require the use of an implicit method for computational efficiency.

Another issue of importance in the shock-flow computation is the order of accuracy of the numerical scheme. First order accurate finite-difference schemes smear the wave front, and tend to produce inaccurate results in complex flow situations (ex: surges in power canals due to load rejection followed by load acceptance). Therefore, higher-order finite-difference schemes are preferred. Also, higher-order schemes require fewer mesh points to achieve the same accuracy as a lower-order scheme. Adapting from the advances made in Computational Fluid Dynamics, Fennema and Chaudhry [50, 51] have introduced several second-order schemes for the simulation of unsteady free surface flows with shocks. Among these are the Gabutti [59], and the Lambda [120] explicit schemes, and the Beam and Warming [12] implicit scheme. These methods make use of the theory of characteristics to identify the direction of information propagation for splitting the flux vector into positive and negative parts, and for choosing appropriate finite-differencing of spatial derivative terms in the governing equations. This allows for the correct simulation of supercritical and mixed flow regimes. The Gabutti scheme is identical to the Lambda scheme in its development; however, it differs in the approximation of partial differential terms. Readers are referred to Chaudhry [35] for more details on these schemes. These three schemes use the governing equations in non-conservation form, and therefore, result in conservation errors when applied to strong shock-flow problems. Jha *et al.* [84] modified the Beam and Warming scheme by introducing a conservative splitting of flux through Roe's [132] approximate Jacobian procedure. This significantly reduced the mass balance error in the original Beam and Warming scheme at no additional cost of computation.

3.3.1 MacCormack Scheme

MacCormack scheme is one of the simplest and computationally very efficient methods for computing shock flows [61, 50, 18, 36, 23, 63, 64, 62, 4, 127, 25, 169, 170, 171, 14). Therefore, mathematical details of the method are presented briefly here. It is a variant of the second-order accurate Lax-Wendroff method, but has a much simpler programming logic. It is an explicit, two-level predictor-corrector scheme, which is second-order accurate in time and space. In this method, the finite-difference approximations of the governing equation (4) are given as below.

Predictor: Backward finite-differences are used to approximate the spatial differential terms in the predictor step.

$$V_i^* = V_i^n - \frac{\Delta t}{\Delta x}\left[F_i^n - F_{i-1}^n\right] + \Delta t \times S_i'^n \tag{8}$$

where, subscript i represents the node in the space, superscript n represents the value at the known time level, t, superscript * represents the predicted value, Δt = computational time step, and Δx = the distance step.

Corrector: Forward finite-differences using predicted values are used to approximate the spatial differential terms in the corrector step.

$$V_i^{**} = V_i^n - \frac{\Delta t}{\Delta x}\left[F_{i+1}^* - F_i^*\right] + \Delta t \times S_i'^* \tag{9}$$

where the superscript ** represents the value at the end of corrector step. Finally, the value at the end of the time step, i.e. at the unknown time level n+1 is given by

$$V_i^{n+1} = 0.5(V_i^* + V_i^{**}) \tag{10}$$

There are several variants of the above scheme. One can use forward differences in the predictor part and backward differences in the corrector part. The direction of differencing in the method can be alternated from one time step to the next. All these variations give the same level of accuracy in most of the cases. Like in many other explicit schemes, the time step in the MacCormack method is limited by the CFL condition for numerical stability i.e. the Courant number should be less than unity.

3.3.2 Artificial Viscosity

Typical first-order finite-difference methods introduce too much of numerical dissipation which smears the shock front. On the other hand, typical second-order schemes like the MacCormack scheme minimize the dissipation error, but introduce the numerical dispersion error. In these methods the shock front is not smeared but is accompanied by high frequency numerical oscillations (Fig. 3). These oscillations may mar the actual solution, and may result in run abortions. Some of the numerical schemes may also introduce shocks with no physical justification.

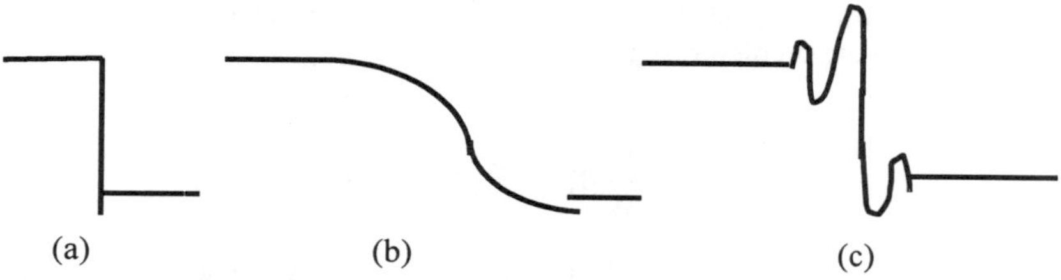

Figure 3. Numerical errors: (a) Analytical solution (b) Dissipation error (c) Dispersion error

The numerical oscillations around a discontinuity may be removed by using the "artificial viscosity" procedures. In an artificial viscosity method, artificial diffusion is introduced by adding diffusion terms to the governing equations [89]. The numerical oscillations around the discontinuity are smoothened out by the above diffusion. In an alternative way, numerical diffusion may be added by applying smoothing or filtering techniques to the computed results obtained from the application of a second-order method [50, 107]. A very effective smoothing method is the Jameson's procedure [83, 35], which introduces numerical diffusion, based on the local gradient of the parameter value. This method introduces diffusion only in regions of large gradients while leaving already smooth regions untouched.

In many artificial viscosity methods, the amount of artificial diffusion introduced depends on the grid size [97] and the numerical diffusion coefficient, appropriate value of which has to be empirically chosen based on numerical experimentation. Also explicitly introduced numerical diffusion interferes with the actual physical diffusion present in the problem in the form of bottom friction term. Therefore, although satisfactory results have been obtained by using the artificial viscosity methods [35], there has been a significant research activity in the last decade to develop numerical methods, which do not require explicit introduction of diffusion for eliminating oscillations around a discontinuity.

3.4 *Explicit Non-Oscillatory Schemes*

A new class of schemes called as "Total Variation Diminishing (TVD)" schemes originated in the field of gas dynamics in the 1980s because of the need for high-resolution solution of compressible flow equations, which admit discontinuities. These schemes are second or higher-order accurate in smooth regions, and produce high-resolution solutions near shock and contact discontinuities. The total variation of a dependent function ϕ^n (superscript n indicates the time level n) is defined as

$$TV(\phi^n) = \sum_{j=-\infty}^{\infty} \left|\phi_{j+1}^n - \phi_j^n\right| \tag{11}$$

A numerical scheme is said to be TVD if

$$TV\left(\phi^{n+1}\right) \leq TV\left(\phi^{n}\right) \tag{12}$$

This property ensures that the numerical solution around the shock is oscillation free. Harten [74] has derived the sufficient conditions that need to be satisfied for a scheme to be a TVD scheme.

Another class of schemes, known as "Essentially Non-Oscillating" (ENO) schemes have been also developed for high-resolution shock-flow computation. The main difference between TVD and ENO schemes is that certain types of ENO schemes can retain the same spatial order of accuracy even at points of extrema. On the other hand, TVD schemes reduce to spatially first-order at the points of extrema. It is to be mentioned that ENO schemes are not necessarily absolutely oscillation free, and permit oscillations across discontinuities up to the order of the truncation error. In fact, TVD schemes are a subset of ENO schemes. An excellent review of the theory and application of these schemes is available in the report prepared by Yee [166]. In the last decade many researchers have successfully adapted some of these schemes to shock flow computation in open channels. Some of the notable contributions are reviewed below.

Many of the high-resolution schemes for open-channel shock-flow computation are based on the method developed by Roe [132] for compressible flow equations. These methods essentially involve application of upwind differencing to a linearized Riemann problem and then employing sophisticated techniques for enhancing them to achieve second-order accuracy. In the modified flux approach [74] the numerical flux computations are modified to obtain second-order accuracy, while constraining the gradients in the flux function to achieve the TVD property or smooth solutions. In the MUSCL (Monotone Upwind Scheme for Conservation Laws) approach, higher-order accuracy is achieved by redefining the arguments of a numerical flux, and achieving non-oscillatory solutions by employing a slope limiter while extrapolating the dependent variables.

Glaister [69] proposed a numerical scheme based on the Roe's Riemann solver approach and applied it to wide rectangular channels. Alcrudo *et al.* [7] expanded on the idea of Roe's approximate Riemann solvers [132] and developed a first-order flux-difference-splitting technique for the case of non-prismatic channels with arbitrary cross-section. They also added a flux limiting approach [133, 147] to the standard Lax-Wendroff flux computation to construct a second-order oscillation free technique for the solution of shock-flow problems. Baines *et al.* [15] also developed a very similar technique based on Roe's upwind TVD difference scheme, and applied it to the simulation of steep waves in plant channels, in which both free-surface and pressurized flows may occur.

Savic and Holly [136] presented an integral approach based on the Godunov method for simulating dambreak flood waves in non-prismatic channels. Their work is based on the earlier work of Vanleer [158], and Colella and Woodward [40] in gas-dynamics. In this method, the piecewise-constant initial data of the Riemann problem is replaced with piecewise-linear initial data. The slope of the piecewise-linear initial data is so chosen to avoid spurious oscillations. Savic and Holly [136] have developed also another variant of the Godunov method based on the piecewise parabolic interpolation. This method is capable of simulating mixed flow regimes, and flows with strong shocks. However, in this method, the approximation of fluxes (to be used in the conservation equation) depends on whether the flow is continuous or discontinuous, and whether the flow is subcritical or supercritical. This results in algorithmic complications.

Yang *et al.* [163] developed a class of characteristic-based upwind schemes for the solution of shallow water wave equations in rectangular prismatic channels. The flux computation in their scheme follows the procedure laid out by Harten *et al.* [76] to obtain an essentially non-oscillatory solution. Yang *et al.* [163] presented second-order as well as third-order accurate non-oscillatory schemes, and in the framework of both finite-difference and Petrov-Galerkin finite-element methods. For the case of dam-break flow, the difference in results between a third-order scheme and a second-order scheme is only marginal. It is also concluded that the finite-element method gives as good results as the finite-difference method. It should be noted here that the method proposed by Yang *et al.* [163] does not avoid the occurrence of non-physical jumps in the numerical solution.

3.4.1 TVD-MacCormack Scheme

There is a different design principle for the construction of high-resolution TVD-type schemes. These schemes achieve non-oscillatory solutions by using a "smoothing procedure", based on the TVD concept, which avoids the empiricism in the artificial viscosity methods discussed earlier. These procedures are very much useful from the perspective of enhancing the capabilities of already developed numerical models, which are based on classical finite-difference schemes. For example, Garcia-Navarro *et al.* [65] presented a numerical scheme for the solution of one-dimensional shallow water equations by adding a third dissipation step to the widely used MacCormack scheme. The extra step is devised according to the theory of Total Variation Diminishing (TVD) schemes. The first two stages of this scheme i.e., the predictor and the corrector steps are same as given for the MacCormack scheme. However, the third step is modified by Garcia-Navarro *et al.* [65] as given below.

$$V_i^{n+1} = \frac{1}{2}\left[V_i^{*} + V_i^{**}\right] + \frac{\Delta t}{\Delta x}\left[D_{i+1/2}^{n} - D_{i-1/2}^{n}\right] \tag{13}$$

Addition of appropriate amount of numerical diffusion is through the terms D, which are computed based on an approximate solution of Riemann problem. The suppression of numerical oscillations in the vicinity of a shock is achieved through a limiting function in the evaluation of D. Several different limiting functions are available for this purpose [166]. Garcia-Navarro *et al.* [65] mentioned that they obtained very similar results by using different limiting functions. However, recent research has shown that the amount of numerical diffusion at the front depends on the choice of the limiting function [160]. For example, a limiting function known as the "minmod" limiter smears the shock more than the other limiters. As in the other TVD schemes discussed earlier, one main limitation of the TVD-MacCormack scheme proposed by Garcia-Navarro *et al.* [65] is that it requires field-by-field decomposition of the flux computation. Also, it uses an ad hoc procedure for avoiding the occurrence of non-physical jumps. Not withstanding these, Garcia-Navarro *et al.* [65] obtained good results for flow over a ladder of weirs, and for flow in a converging-diverging channel. It should be noted here that the TVD scheme does not involve the use of any empirically determined numerical dissipation coefficients in its formulation. Recently, this method has been used to verify the validity of shallow water flow equations for dam-break flow simulations [14], by comparing the numerical results with experimental data.

3.4.2 Local Lax-Friedrichs Scheme

Adapting from Shu and Osher [137], and Nessyahu and Tadmor [121], Rao and Latha [130], and Nujic [122] proposed different variants of an Essentially Non-Oscillating (ENO) scheme based on the Lax-Friedrichs flux computation. These schemes are known as Local Lax-Friedrichs (LLF) schemes. Unlike the earlier TVD methods, LLF methods do not require field-by-field decomposition of the flux. As far as the algorithmic complexity is concerned, these are perhaps the easiest to implement and computationally efficient non-oscillating schemes for dam-break flow simulations. One of the schemes proposed by Nujic [122] is briefly described here. It is an explicit two-step predictor-corrector scheme, which results in second-order accuracy in both space and time. In the predictor step, the finite-difference form of (4) is written as

$$V_i^* = V_i^n - \frac{\Delta t}{\Delta x}\left[F_{i+1/2}^n - F_{i-1/2}^n\right] + \Delta t\, S_i'^n \tag{14}$$

where, $F_{(i+1/2}$ represents the numerical flux through the face between cells i and i+1. This is computed using the values of V_L and V_R at the cell interface between nodes i and i+1. V_L and V_R are obtained using the information from the left-hand side, and the right-hand side of the cell face, respectively. Slope limiting procedures are applied while interpolating for V_L and V_R.

In the corrector part, the vector V at the unknown time level n+1 and at node i is computed using the following equation.

$$V_i^{n+1} = 0.5\left[V_i^n + V_i^* - \frac{\Delta t}{\Delta x}\left(F_{i+1/2}^* - F_{i-1/2}^*\right) + \Delta t\, S_i'^* \right] \tag{15}$$

Fluxes in the above equation are determined in the same way as described for the predictor step, but using the predicted values.

Nujic [122] has shown that the above method gives satisfactory results for the case of dam-break flows. However, there is more smearing of the front compared to other non-oscillating schemes. Singh and Bhallamudi [140] have adapted this scheme to simulate overland flow resulting from rainfall on a porous medium with an embedded clay lens. Their results for outflow hydrograph indicated a much better performance of local Lax-Friedrichs method as compared to the classical MacCormack scheme. It may be noted here that Yost and Rao [167] have also presented a similar scheme for the solution of Saint Venant equations. Yost and Rao [167] applied the method to simulate (1) sudden opening and closing of a sluice gate in an open-channel, (2) dam-break flow on wet bed conditions, and (3) flows with hydraulic jumps.

3.4.3 Flux-Corrected Transport Schemes

Flux-Corrected Transport (FCT) schemes, originally proposed by Boris and Book [27] have been finding application in recent years for the solution of contaminant transport equation. Yost and Rao [168] adapted this method for the solution of open-channel flow equations. This method has three essential stages: (1) transportation (advection), (2) diffusion, and (3) anti-diffusion. Any classical finite-difference scheme can be used in the first-stage. The oscillations, which result in this stage, are removed in the second stage by introducing diffusion through out the computational domain. The solution is anti-diffused in the third stage to remove diffusion where it is not required. The interaction between the diffusion and anti-diffusion stages involves a built-in flux-limiter. Recently, Srinivasan [142] has made a detailed study of three popular FCT schemes, including the one proposed by Boris and Book [27], as applied to the solution of advection-dispersion equation. His conclusions for the advection-dominated flows apply to also the solution of shallow water flow equations. It has been found [142] that not all flux-corrected transport schemes result in absolutely oscillation free, and diffusion free solutions. Also some of them may introduce unphysical jumps. Therefore, one has to be cautious while choosing a particular FCT scheme for a given application.

3.4.4 Entropy Inequality Condition

A discussion on entropy inequality condition is in order at this juncture. There are many finite-difference methods, which satisfy the Rankine-Hugoniot conditions across a jump, and hence application of these methods results in *weak solutions*, or solutions, which satisfy the integral formulation of the equations. However, some of these methods do not satisfy the entropy inequality condition. According to this condition, for the formation of a realistic jump, the characteristics must converge at a shock such that

$$\lambda(V_L) \geq \text{Shock Speed} \geq \lambda(V_R) \tag{16}$$

where, λ = characteristic speed, and subscripts L and R denote the states on the left and right sides. Shock speed can be determined using the Rankine-Hugoniot jump condition. Methods, which do not satisfy (16) result in non-physical jumps. In the case of one-dimensional dam-break flow due to instantaneous dam-breach, this may manifest as an unphysical jump at the original position of the dam (Fig. 4).

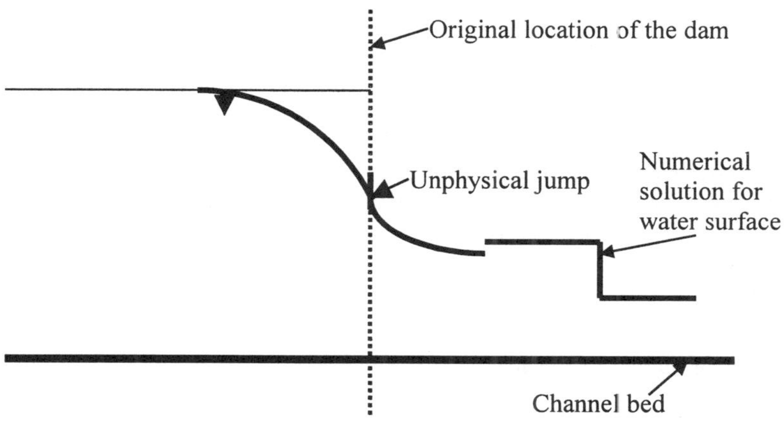

Figure 4. Schematic diagram for numerical results with non-physical jumps

Many first-order upwind schemes such as the one adapted by Jin and Fread [90] do not satisfy the entropy condition. Even some approximate Riemann solvers such as the one developed by Roe [132] do not satisfy the entropy condition. Recently, Zoppou and Roberts [177] showed that the method developed by Yang *et al.* [163] also does not satisfy the entropy inequality condition and results in unphysical jumps at sonic points i.e., where the flow changes from subcritical to supercritical conditions. As mentioned earlier, Garcia-Navarro *et al.* [65] overcame this problem in an empirical way.

Jha *et al.* [85] incorporated Harten and Hyman [75] approach for enforcing the entropy inequality condition into Roe's [132] scheme so that occurrence of unphysical jumps is avoided. This condition removes the empiricism in the method for avoiding unphysical jumps earlier adapted by Garcia-Navarro *et al.* [65]. Jha *et al.* [85] extended Roe's method to obtain second-order accuracy using (a) Lax-Wendroff numerical flux, (b) MUSCL approach, and (c) Modified flux approach. Recently, Tseng and Chu [155] extended the TVD-MacCormack method of Garcia-Navarro *et al.* [65] to include four variations of predictor-corrector steps. They also used Harten and Hyman [75] approach for enforcing the entropy inequality condition necessary for avoiding unphysical jumps. Unlike the entropy fix of Harten and Hyman [75], the scheme developed by Toro [152] implicitly includes sonic points in deriving the approximate Riemann solution.

3.4.5 Comparison of Different Approaches

Jha *et al.* [85] suggested that the Roe first-order scheme may be preferred for practical applications when CPU time, overall accuracy, and applicability are weighed. Recently, Delis *et al.* [45] evaluated four high-resolution explicit shock-capturing techniques for the simulation of one-dimensional dam-break flow. The schemes considered by them are; (i) first-order Lax-Friedrichs (LF) method, (ii) Local Lax-Friedrichs (LLF) method, (iii) Roe's approximate Riemann solver method, and (iv) Harten-Lax-Van Leer (HLLE) method. It has been shown that the Roe method introduces least amount of diffusion, while, as expected, the LF method introduces maximum diffusion. The LLF method is very effective in removing the oscillations, but is diffusive in the case of large upstream to downstream depth ratios. Therefore, care must be taken while applying this scheme to problems with strong frictional effects. However, the LLF method is very simple and takes only half the CPU time as compared to the Roe method. The HLLE method introduces slightly more diffusion than the Roe method, but is computationally more efficient.

3.5 Recent Implicit Schemes

Majority of the new methods for shallow water computation introduced in the last decade are explicit schemes. On the other hand, Garcia-Navarro *et al.* [67] introduced an implicit TVD method based on the work of Yee [165] for scalar systems. The implicit TVD scheme was found to be best suited for obtaining the continuous and discontinuous steady flows using a false transient approach. Larger time steps (not limited by the CFL condition) could be used, and second-order accuracy in space could be achieved. Also, unlike the classical implicit schemes, the above method could handle subcritical, supercritical, and mixed flow regimes without any difficulty. However, the accuracy suffered in the case of unsteady flows such as the dam-break flow if the Courant number was overly large i.e., greater than three. Non-physical spikes were also seen in their results for some cases if the Courant number was large. Also, it was advised that the implicit weighting factor be chosen equal to one, which makes the scheme only first order accurate in time for transient solutions.

Jha *et al.* [86], in continuation of their earlier work, extended the application of conservative Beam and Warming scheme to prismatic channels of arbitrary cross sections. They also incorporated the Harten and Hyman [75] approach for entropy fixing so that occurrence of non-physical jumps is avoided. It should be mentioned here that Jha *et al.* [86] presented results only for Euler implicit scheme, and not for Crank-Nicolson scheme. Therefore, their method cannot avoid diffusive errors, and the accuracy deteriorates as Courant number increases beyond two. It is to be noted that extra computational effort per time step that is needed in an implicit scheme may outweigh all the advantages of an implicit scheme, if Courant numbers much larger than one cannot be used.

Similar to the work presented by Garcia-Navarro *et al.* [67], Delis *et al.* [46] developed implicit high-resolution TVD methods for modeling one-dimensional open-channel flow. They are based on (1) Harten's modified flux method, and (2) Van Leer's MUSCL approach. It should be noted that even these methods use a value of unity for the implicit weighting factor, which means that the schemes are only first-order accurate in time. A high Courant number value of 30 to 40 could be used in a false transient approach, while simulating steady flows with shocks. However, there were slight conservation errors in the vicinity of the shock. Also, significant diffusion error was present when the methods were applied to transient dam-break flow simulation, even for Courant numbers as small as four.

It is interesting to note here that Blunt and Rubin [26] also experienced the same problem in connection with petroleum reservoir simulations. In fact, Blunt and Rubin [26] derived a limiting condition on the implicit weighting factor to retain the TVD property. This condition shows that the implicit weighting factor should approach a value of one as the Courant number increases. Recently, Verma *et al.* [159] concluded on similar lines with regards to the application of a flux-limiting scheme to implicit solution of two-dimensional contaminant transport problem. The flux-limiting procedure failed for large Courant numbers and Crank-Nicolson scheme, when the transport is highly advective. Verma *et al.* [159] suggested a value of 0.67 for the implicit weighting factor in order to obtain satisfactory results, but this introduced diffusive error.

Above discussion implies that presently available implicit TVD schemes are not as good as their explicit counterparts for a majority of strong open-channel shock-flow problems. However, they are suitable for simulating steady flows using false transient approach, and flows, which have regions of fast flow away from sharp fronts [26]. An example of such flow situation is the bore propagation through an expanding channel. An explicit scheme in such situations may require ridiculously low values of computational time step, and would be very inefficient.

Implicit schemes can be also used effectively when very high resolution of the front is not required from an engineering point of view.

3.6 Lagrangian Methods

Semi-Lagrangian methods have been popular in the area of numerical weather prediction, and are making an appearance in the current open-channel flow literature. In these methods, the advective component is treated in a Lagrangian framework. The advected quantities are determined implicitly on the grid, the location of which is determined by the particle motion. An advantage of Semi-Lagrangian methods is that the usual CFL condition for numerical stability is replaced by a weaker condition, and this permits the use of Courant numbers greatly exceeding unity. Therefore, in viscous flows, the accuracy of the numerical results away from regions of strong gradients is increased. Also, semi-Lagrangian methods are genuinely multidimensional. However, one serious disadvantage of semi-Lagrangian methods is their lack of conservation. Garcia-Navarro and Priestley [66] developed a conservative and shape preserving semi-Lagrangian method for the solution of the shallow water equations. The technique proposed by Garcia-Navarro and Priestly [66] uses Cubic Hermite polynomials for interpolation and incorporates the Flux Corrected Transport (FCT) approach to correctly reproduce the discontinuous flows. However, this method does not guarantee an unconditional shock-capturing ability. Also, the proposed technique requires linear programming procedures for the purpose of in-phase conservation recovery. This makes the technique some what complicated algorithmically.

Cui and Williams [42] presented a one-step downstream characteristic Lagrangian method for solution of transient open-channel flow equations. The main improvement is the use of a one-step moving grid process. It is claimed that this increases the accuracy, and simplifies the coding of Lagrangian process. The method presented by Cui and Williams [42] is much simpler than the one presented by Garcia-Navarro and Priestly [66]. Shocks could be handled without any special treatment. However, it introduces a limitation on the time step i.e. the time step has to satisfy the CFL condition. This takes away at least a part of the advantage of a Lagrangian method. It is also not clear how the method compares with the available TVD schemes in terms of over all accuracy and algorithmic difficulty. Recently, Wang and Shen [161] presented a pure Lagrangian method for one-dimensional dam-break flows.

3.7 Other Interesting Methods

Few techniques, which allow one to avoid the use of time-consuming Riemann solver and local characteristic decomposition, have been proposed in recent years. Such methods have an advantage that they can be extended to multi-dimensions in a simple way. One such method is the relaxation scheme proposed by Aral *et al.* [13].

This is based on an earlier work of Jin and Xin [91] for the solution of conservation equations. The key idea is to construct a semi linear hyperbolic relaxation system that approximates the original non-linear system of shallow water equations. This semi-linear relaxation system, which has linear convection and non-linear source terms, is then solved using an appropriate technique. Spatial discretization in conservation form is combined with a second-order TVD Runge-Kutta splitting scheme for time discretization. Riemann solver approach could be avoided because of the linear nature of the convection term. This method is simple to code, and preliminary results indicate solution accuracy comparable to other TVD schemes.

Chang [34] introduced a new explicit staggered grid method for solving the Euler equations based on the concept of space-time conservation element and solution element. In this method, both space and time are treated on the same footing, and flux conservation is enforced in both space and time. It uses a central finite differencing for discretization. Recently, Molls and Molls [118] have adapted the above scheme for solving the shallow water equations. Their results, without the aid of artificial viscosity, showed a significant improvement over the classical finite-difference schemes with regards to the suppression of spurious oscillations. Compared to the explicit TVD schemes, the method appears to have inherent numerical diffusion, and it is not completely oscillation free. However, Molls and Molls [118] demonstrated the simplicity of the scheme for application to two-dimensional cases.

There is also a re-thinking with regard to the artificial viscosity methods. One of the main objections to these methods has been that they involve empirical selection of artificial viscosity coefficients. Recently, Yost and Rao [170] have presented an interesting and a very simple non-linear filter method for smoothing the numerical oscillations, which result from the application of classical finite-difference methods. This non-linear filter is adapted from an earlier work of Engquist et al. [49] in the field of gas-dynamics. Advantages of this method are: (1) there is no upwind biasing in the evaluation of source terms, (2) method can be easily attached to existing finite-difference codes, (3) computational time required is much lesser compared to the TVD schemes, and (4) it does not involve any empirical coefficients. Also, extension to two-dimensional case is simple and straightforward. However, it must be noted here that Yost and Rao [170] have not demonstrated the applicability of their method for dam-break flows with very strong shocks. It is also not known how the issue of entropy inequality is addressed in their method.

4 Two-Dimensional Shock Flow

Various techniques available for computing one-dimensional dam-break flows have been discussed in the previous section. One-dimensional models usually yield satisfactory results if the channel is relatively straight and prismatic. Some studies [22, 108] have used one-dimensional models for simulating dam-break flows in curved channels by incorporating an approximate procedure for transverse water slope, to estimate the maximum and minimum water levels in a bend. Savic and Holly [136] have used their one-dimensional model to study the effect of changing width on the movement of dam-break wave, and obtained reasonably good results. Hicks *et al.* [78] also obtained satisfactory results with a one-dimensional model for dam-break flow in converging-diverging channels. Townson and Al-Salihi [153] successfully used one-dimensional equations in R-T space to compute dam-break flow in a transition. All these and other similar studies [111] indicate that one-dimensional models can be used to obtain quick and satisfactory preliminary results in some cases of two-dimensional flow. However, all these methods are either applicable to some specific cases or do not give a complete solution. For example, the model used by Miller and Chaudhry [108] predicts the super elevation in water surface satisfactorily, but does not give the complete flow information in the transverse direction. It is also obvious that supercritical dam-break flow through even a gentle transition cannot be properly simulated using a one-dimensional model because of presence of oblique jumps.

A number of models, based on the shallow water equations (1) - (3), have been developed in the last two decades for the simulation of two-dimensional dam-break and other shock flows in open channels. An over view of these models is presented in this section.

4.1 Classical Finite-Difference Methods

Many of the prior discussed classical finite-difference methods for one-dimensional flow have been extended for application to two-dimensional shock-flow problems. Katopodes and Strelkoff [94] used the method of bi-characteristics for simulating the two-dimensional dam-break flood wave. This method involves the use of a shock-fitting method for tracking the wave, and therefore, is not very convenient computationally. Garcia and Kahawita [61] have used the MacCormack shock capturing method for two-dimensional flows. Fennema and Chaudhry [53] introduced few more explicit second-order schemes (Gabutti and Lambda schemes) for the same purpose. They also introduced [52] the Beam and Warming implicit scheme for the solution of two-dimensional unsteady flows. These methods are well described in the textbook by Chaudhry [35]. Playan *et al.* [125] used a variant of Lax-Wendroff method for simulating two-dimensional flow in a basin irrigation scheme. Classical MacCormack method is still used in some, interesting two-dimensional applications. Zhang and Cundy [173], and Fiedler and Ramirez [54] have used this method for simulating discontinuous two-dimensional open-channel flow over an infiltrating surface, i.e. overland flow. Fiedler and Ramirez [54] adapted a fractional step method so that large time-steps could be used. While on the topic of classical methods, it must be mentioned that Yost and Rao [170] recently demonstrated the applicability of their non-linear filtering technique (for use with classical finite-difference schemes) to two-dimensional flows. This certainly makes the classical methods relevant even today.

4.2 High-Resolution Finite-Difference Methods

Like the classical finite-difference schemes, high-resolution finite-difference schemes also have been extended to two-dimensional flow computation. For example, Yang and Hsu [164] extended their explicit ENO scheme, based on flux vector splitting, for studying bore impinging on a circular cylinder and for bore movement through a converging-diverging channel. Governing shallow water flow equations were used in a general curvilinear coordinate system, and a Strang type dimensional splitting is used for extending the one-dimensional method to two-dimensional case. Nujic [122] has presented an application of the Local Lax-Friedrichs ENO method to the simulation of flow resulting from a partial dam-break. Fraccarollo and Toro [57], Louaked and Hanich [102], Jha *et al.* [87], and Wang *et al.* [160] used standard dimensional splitting procedure in their work on two-dimensional dam-break flow. Fraccorollo and Toro [57] used the explicit Riemann solver approach with Toro's [152]] weighted average flux method. This method has the advantage of correctly considering the movement of a dam-break wave on dry beds. Louaked and Hanich [102] used the TVD Lax-Wendroff scheme. Interesting contribution of this paper is the inclusion of an artificial compression method for obtaining high resolution near contact discontinuities, using only fewer grid points. Jha *et al.* [87] used first-order Roe's numerical flux and second-order Lax-Wendroff numerical flux procedures in their model.

Liska and Wendroff [101] have used a hybrid approach of combining the second-order Lax-Wendroff method with the first-order Lax-Friedrichs method for solving the two-dimensional shallow water equations. Wang *et al.* [160] introduced a similar hybrid TVD scheme, where in first-order upwind scheme and second-order Lax-Wendroff schemes are combined using flux limiting procedures. They used one, and two-parameter limiters in their study. Two parameter limiters gave better results than one-parameter limiters when upstream to downstream depth ratio in the dam-break flow is lower than 0.138. Also, they suggest that Van Leer's MUSCL (Monotonic Upwind Scheme for Conservation Laws) type limiter is better suited than other limiters.

4.3 Co-ordinate Transformation

Many of the two-dimensional finite-difference models, classical as well as high-resolution, use the Cartesian co-ordinate system, and represent the non-rectangular boundaries of the physical domain (Fig. 5) in an approximate way. This introduces large errors, and also may result in stability problems. In fact, this truly limits the

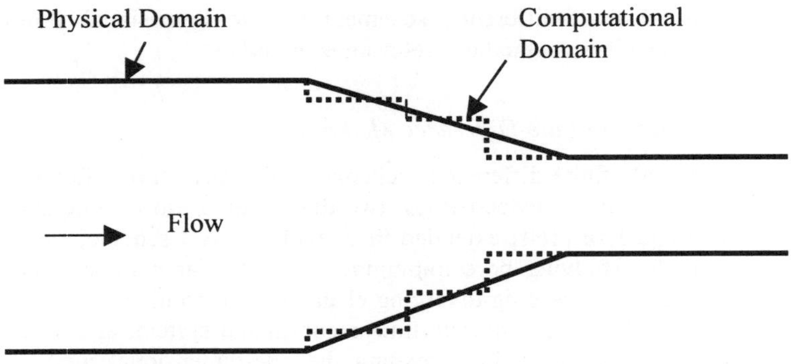

Figure 5. Approximate representation of boundaries in a finite-difference method

application of finite-difference methods to many practical problems. Grid-generation techniques [12] have helped to remedy this limitation to some extent. Dammuller *et al.* [44] applied the two-dimensional shallow water equations in a channel-fitted coordinate system to route the dam-break flow through a curved channel. Their numerical results matched satisfactorily with the experimental data obtained by Miller and Chaudhry [108] for this problem. Jimenez and Chaudhry [89] applied a very simple algebraic co-ordinate transformation technique to convert the non-rectangular physical domain of a channel transition to a rectangular computational domain. The governing equations for steady two-dimensional flow are also transformed accordingly, and are then solved using the MacCormack scheme, to simulate the steady super-critical flow in channel transitions. Hager *et al.* [73] used this model to study supercritical flow near an abrupt wall deflection. Bhallamudi and Chaudhry [24] used a similar technique for solving the unsteady flow equations to simulate flow through channel transitions. They used a false transient approach to obtain the steady flow solutions for supercritical and sub- and supercritical flows through transitions. Later, Mohapatra and Bhallamudi [111, 112] applied the same model to analyze the bed-level variations, and the dam-break flow in channel transitions. Rahman and Chaudhry [128] also used a similar approach along with an adaptive grid approach (to obtain better resolution near the shock) to simulate supercritical flow in transitions.

Algebraic transformation procedures are very simple to apply, but become problem specific. For example, simulation of supercritical flow in a combining channel junction cannot be accomplished using the model developed by Bhallamudi and Chaudhry [24]. Generalized grid-generation and coordinate transformation techniques [150] need to be adopted if one wishes to develop a two-dimensional finite-difference model for application to a complicated physical domain. Molls *et al.* [117], and Molls and Chaudhry [116] have developed one such comprehensive model. They used a generalized body-fitted coordinate system to transform the unsteady depth-averaged two-dimensional flow equations, which include a constant eddy viscosity model for turbulence. The transformed equations were then solved using (1) the ADI method on a non-staggered grid system, and (2) the MacCormack scheme. Grid packing was also included in their model to obtain high resolution in areas of interest. They obtained satisfactory results for a variety of open-channel flow cases. Recently, Molls and Zhao [119] applied this model to study the supercritical flow in a channel with wavy side walls. Younus and Chaudhry [172] used a similar approach for developing a depth-averaged k-ε turbulence model for open-channel flows with, and without shocks. They applied their model to simulate supercritical flow in transitions, and circular hydraulic jumps. Recently, Hsu *et al.* [79] adapted a similar generalized curvilinear coordinate system approach in their explicit finite-analytic model for open-channels.

On the other hand, Bellos *et al.* [19, 20] used an alternative method, which is a combination of finite-element and finite-difference methods to route dam-break flow through channels of varying width. The finite-element type approach is used to transform the locally non-rectangular physical grid to a rectangular computational grid. They used MacCormack method for solving the locally transformed equations. Soulis [141] used a very similar model to simulate steady two-dimensional supercritical flow. He used an interesting multiple grid solution procedure to obtain fast convergence to the steady state in the false transient approach. Recently, Panagiotopoulos and Soulis [123] used a similar procedure for co-ordinate transformation, but an implicit bi-diagonal MacCormack solver, to simulate steady subcritical and supercritical flows in channel transitions. The implicit solver results in a fast convergence of the solution to a steady state, when used in a false transient approach.

4.4 Finite-Element Methods

Finite-difference methods are not a natural choice for application to two-dimensional flows in complex physical domains. Although coordinate transformation techniques can be used along with these methods, the governing equations in the transformed coordinate system could become quite complicated. Finite-element methods offer a way out of the above difficulties since complex domain boundaries can be represented in a more accurate manner in these methods. Therefore, they have become very popular for simulating flows in large water bodies such as oceans, estuaries, and for simulating complex flood-plain flows, although it is more difficult to implement a finite-element method as compared to a finite-difference method. FESWMS-2DH model developed by USGS, and TABS-2 model developed by the U.S. Army Corps of Engineers, are examples of commercially available finite-element models. TELEMAC model developed by Galland *et al.* [60] is another example of a shallow water flow solver based on finite-element method. Readers are referred to Finnie [55] for a basic discussion on finite-element modeling of open-channel flows.

In the early period of the development of finite-element techniques, difficulties were encountered with regard to the computation of shocks and surges. This is because of non-dissipative nature of the classical Galerkin finite-element technique. To overcome this problem, Katopodes [93] introduced discontinuous weighting functions into a Petrov Galerkin method in order to achieve upwinding while maintaining the second-order accuracy. Readers are also referred to Malcherek and Zielke [104] for an interesting discussion on the need for, and the method of introducing upwinding in finite-element methods. Hicks and Steffler [77] compared (1) Characteristic-Dissipative Galerkin scheme (CDG), (2) Taylor-Galerkin scheme, and (3) Least-Square FEM. They concluded that all the methods required addition of artificial diffusion, but the CDG scheme required the least amount. Zienkiewiez and Ortiz [176] developed a suitable operator-splitting procedure using a characteristic based rational form of including dissipation. This method gave better results compared to Taylor-Galerkin scheme, but not fully monotonous results.

There are few other studies on shock-flow computation using finite-element models in the last decade. Akanbi and Katopodes [5] developed a finite-element model for two-dimensional wave movement on initially dry land. This is a remarkable study in the sense: (1) it uses a second-order difference scheme for time integration, (2) it is a completely non-linear implicit scheme, and (3) it uses deforming grids to account for the effects of propagating wave. Yang et al. [163], and Yang and Hsu [164] developed characteristic based high resolution non-oscillatory shock-capturing Petrov-Galerkin finite-element models. Berger and Stockstill [21] used upwind weighted test functions in a Petrov-Galerkin scheme to simulate two-dimensional high-velocity flows in channels. Their model incorporated the versatile unstructured grid procedure. This model relies on a shock detection mechanism based on elemental energy variation. Traditional finite-difference models based on coordinate transformation assume that the plan form of the domain does not change with time, i.e., the domain has vertical lateral boundaries. This limits their application. The model developed by Stockstill *et al.* [144], on the other hand, uses an implicit Petrov-Galerkin method with moving finite-element grids, and therefore, can be used when the domain is changing with time. They used this model to study two-dimensional supercritical flow in a trapezoidal curved channel. Other notable contributions in the area of finite element modeling of shock flows are by Bova and Carey [30], and by Tisdale *et al.* [151], who have presented monotone streamline upwind finite-element methods.

4.5 Finite-Volume Methods

Finite-volume methods offer a viable alternative to finite-element methods for solving flow, and transport problems [124]. These methods combine the simplicity of finite-difference methods with the flexibility of handling complex geometries, intrinsic to finite-element methods. Therefore, they have become quite popular in recent years for the solution of two-dimensional shock flow problems.

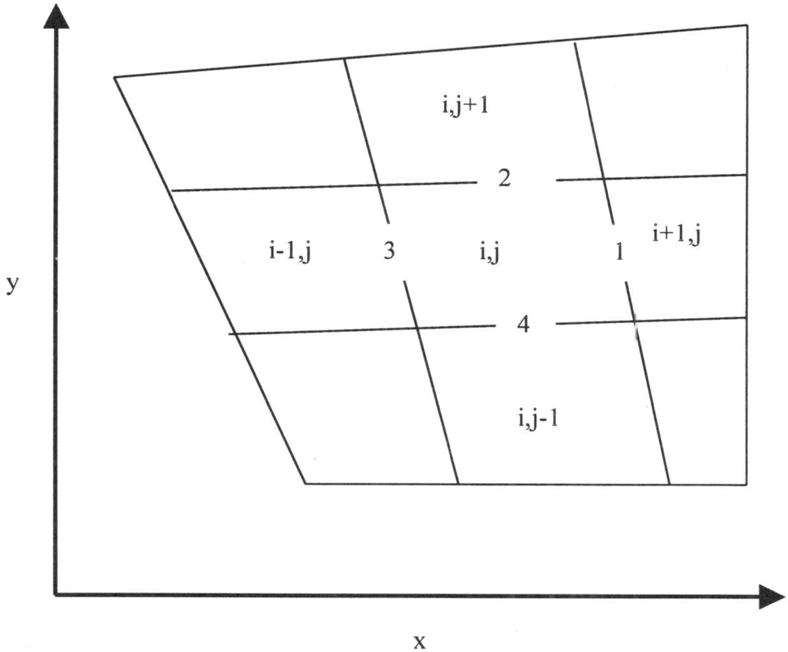

Figure 6. Definition sketch for a finite-volume method

In a finite-volume method using a structured grid, the flow domain is divided into a set of non-orthogonal cells (i, j) as shown in Fig. 6. The corresponding center point identifies each cell. The governing equations are integrated by a finite-volume technique on each of these cells covering the entire domain. Gauss divergence theorem can be applied to convert the governing equation into an integral form as given below.

$$\int_v \frac{\partial U}{\partial t} dv + \oint_s (F \bullet n) ds = \int_v S dv \tag{17}$$

where, F = flux vector at the control surface. The first term represents the integral of the time evolution of the function over the cell area. The second term represents the total normal flux through the cell boundaries.

The scalar product (F.n) in (17) can be expressed in terms of the Cartesian components as

$$F \bullet n = E n_x + G n_y \tag{18}$$

where, n_x, and n_y are the components of the unit vector n at the cell surface, in the x, and y-directions, respectively. The vector U may be assumed to be uniform over the cell, and (17) can be written as

$$\frac{\partial U}{\partial t} \Delta A + \oint_s (F \bullet n) ds = S \Delta A \tag{19}$$

where, ΔA = area of the finite-volume cell. The surface integral in (19) is determined as the sum over the four boundaries of the cell shown in Fig. 6. The numerical flux at the cell boundaries 1, 2, 3, and 4 (Fig. 6) between the node (i, j), and its four neighbors can be determined using the information available at the cell centers. Either classical or high-resolution methods may be adopted for this purpose. Either flux limiting or slope limiting procedures can be used in a high-resolution method. The time stepping can be done either explicitly or implicitly. Also, either simple Euler or predictor-corrector approaches can be employed for time stepping. Thus there are many variations in the application of finite-volume method for the solution of two-dimensional shock flow problem. A few of the recently developed methods are discussed below.

Alcrudo and Garcia-Navarro [8] developed an explicit high-resolution finite-volume scheme based on (1) construction of an approximate Jacobian, for conservative upwind discretization of equations for arbitrary cell shapes, and (2) a MUSCL variable extrapolation procedure using slope limiters. A structured grid system with quadrilateral elements is used in this model. Singh [139] has adapted the LLF method for solving the two-dimensional shallow water equations, again on a structured grid system. The numerical results for a partial dambreak flow case matched well with those obtained by Alcrudo and Garcia-Navarro [8]. However, the Local Lax Friedrichs scheme produced more smearing at the wave front, as expected. This scheme was also used successfully to simulate hydrodynamics of surface irrigation [139] in an irregular filed with high spot.

Mingham and Causon [109, 110], and Causon *et al.* [33] used an approach similar to Alcrudo and Garcia-Navarro [8] i.e., MUSCL interpolation with slope limiters, to study unsteady bore diffraction (1) on a cylindrical pier, (2) through a contraction-expansion channel, and for studying supercritical flow in spillway channels. They used a cell-centered formulation with collocated data rather than a space staggered approach. A two-stage Runge-Kutta method is used for time stepping. Tseng [154] presented a high-resolution finite-volume non-oscillatory scheme based on flux-difference splitting. A Strang type operator splitting procedure was used in order to treat the effect of source terms in an appropriate manner. Tseng and Chu [156] extended the TVD-MacCormack scheme of Garcia-Navarro *et al.* [65], for solving two-dimensional shallow water flow equations in a finite-volume framework.

Many open-channel flow systems, with complex topographical features, cannot be represented appropriately by structured grids using quadrilateral elements. Therefore, finite-volume methods with unstructured grids become necessary. Zhao *et al.* [174] have developed one such very versatile finite-volume model based on Osher's scheme for solving the Riemann problem. Their unstructured grid system uses a combination of triangular cells or quadrilateral cells. This model includes also special procedures for treating wetting and drying of elements. This model was successfully applied to simulate experiments conducted in Kissimmee River basin. Later, Zhao *et al.* [175] compared three approximate Riemann solver approaches for finite-volume method: (1) Flux-vector splitting, (2) Flux-difference splitting, and (3) Osher scheme. Based on accuracy, computational time, and numerical stability, it was concluded that all the three methods give equally good results. Anastasiou and Chan [11] developed a similar finite-volume model, based on Roe's flux function, and an unstructured mesh. Their study included both implicit and explicit formulations. Zoppou and Roberts [177] also used a similar finite-volume method based on an unstructured grid system using triangular elements. In their model, Roe's approximate Riemann solver approach was modified so as to avoid entropy-violating condition. They used Toro's method [152, 57] for this purpose. A fractional-step method was adapted for a better numerical treatment of the source terms representing the effects of topography, friction, and obstacles. An interesting feature of their model is the representation of time variant influence of obstacles using a simple empirical formulation. Zoppou and Roberts [177] have successfully applied their model to simulate flows resulting from catastrophic collapse of water supply reservoirs in urban areas. Recently, Brufau and Garcia-Navarro [29] have developed an unstructured grid solver using Delaunay triangles. The cell centered finite-volume method is based on Roe's approximate Riemann solver across the cell edges. The model was applied to simulate dam-break flow in a 45^0 channel bend.

4.6 *Boundary Conditions*

Implementation of boundary conditions is an important aspect of the numerical solutions of hyperbolic equations. The following types of boundaries usually need to be included: (i) inflow, (ii) outflow, and (iii) solid wall. Inflow and outflow boundaries are open boundaries where flow can enter or leave the computational domain. Stoker [145] has shown that the specification of boundary conditions depends on whether the flow is supercritical or subcritical. At an inflow boundary of a two-dimensional flow, three boundary conditions have to be specified if the flow is supercritical, and two boundary conditions have to be specified if the flow is subcritical. At an outflow boundary, no boundary conditions should be specified if the flow is supercritical, and one boundary condition has to be specified if the flow is subcritical. In the case of a solid wall, the no flux boundary condition has to be imposed.

Even with the specification of consistent boundary conditions, some approximations are required due to the discretization requirements of the numerical scheme, in order to calculate the remaining unknown variables at the borders. For example, a non-reflective condition may be specified at the downstream boundary. Similarly, a reflection procedure [35], or a radiation condition [9] may be adopted for solid wall boundaries. Strictly speaking, all these methods amount to over specification of boundary conditions. In order to be as correct as possible, use should be made of the information carried by the outgoing bicharacteristics, to determine the unknown variables at the boundaries [8].

5 Summary And Suggestions For Future Work

An overview of computation of open-channel flows with shocks has been presented in the preceding sections. Strengths and limitations of various numerical techniques presently available for the solution of the shallow water equations have been discussed.

It can be seen from this review that significant contributions have been made in the last decade toward the accurate solution of open-channel flows with shocks. The Total Variation Diminishing (TVD) and the Essentially Non-Oscillating (ENO) schemes can be used to obtain higher-order accurate solutions without numerical oscillations in the vicinity of shocks. Procedures have been developed for avoiding the occurrence of non-physical jumps. Recently developed high-resolution, unstructured-grid, finite-volume methods have the simplicity of finite-difference schemes (in terms of code development), while being applicable to very complex two-dimensional flow domains. However, further research needs to be carried out to address few other unresolved issues. Some of these issues are discussed in the following text.

Computationally Efficient Schemes: Explicit high-resolution schemes are limited by the CFL condition for numerical stability, which results in very small computational time steps. This is a disadvantage when one is dealing with large domains, and long period simulations. Implicit methods are a natural choice in such situations. However, as mentioned earlier, presently available high-resolution implicit schemes are not as accurate as the explicit schemes. There is a need for developing non-oscillating implicit schemes, which are (1) second-order accurate in both space and time, (2) applicable to simulations with Courant numbers much larger than one, and (3) applicable to two-dimensional flows. Alternatively, the possibility of developing hybrid explicit-implicit (explicit in the vicinity of shock and implicit in the regions away from shock) methods needs to be explored. Recently, VanderKwaak [157] has developed one such procedure for use in an integrated surface-subsurface hydrologic model. Also, further research is needed to develop computationally very efficient schemes, using the concepts such as the multiple grid algorithms [141, 169], adaptive sub-grid methods [139]etc.

Numerical Treatment of Source Terms: In the last decade, much effort has gone into developing appropriate discretization procedures for partial differential terms in the governing equations. However, flows in natural river systems are strongly affected by the bed slopes, roughness, channel-irregularity, and lateral inflow and outflow. These effects are considered in the governing equations through the source terms. Improper treatment of these source terms can result in significant numerical errors [122]. Numerical methods for accurate solution of problems with strong non-linear source terms are a comparatively new area of research [177], and this is very much relevant to the field of computational hydraulics.

Non-Hydrostatic Pressure Distribution: There has been a concern in the literature regarding the assumption of hydrostatic pressure distribution in the vicinity of a shock. For example, comparison of numerical and experimental results, for supercritical flows in transitions, channel junctions etc., has indicated that shallow water flow equations break down when depth to width ratio is large. Recent studies by Mohapatra et al. [113] using a two-dimensional model in vertical plane has shown that the assumption of hydrostatic pressure distribution gives erroneous results for the movement of dam-break flood wave in a relatively dry channel, although this effect is not significant for wet downstream conditions. More significantly, Ramsden [129] has shown that the above assumption is completely invalid in the case of a bore impacting on walls. Boussinesq equations (also termed as Serre equations by some researchers), which take into account the non-hydrostatic effect, could be solved numerically [68, 31] for this purpose. Similarly, the more advanced vertically averaged and moment (VAM) model proposed by Steffler and Jin [143] could be used. However, both Boussinesq model and the advanced VAM model are based on certain simplifying assumptions for the vertical distribution of velocity and pressure. Alternatively, Navier-Stokes-type solvers with procedures for including moving boundaries can be used [98, 28, 113, 114] to obtain a detailed solution without having to make any assumption regarding either pressure or velocity distribution in the vertical direction. One of the main concerns in this regard is the development of robust, accurate, and computationally efficient algorithms for tracking the rapidly changing free-surface profile. Another important issue is the suppression of spurious high-frequency numerical oscillations without affecting the high-frequency physical oscillations. Further research needs to be carried out in this direction.

Complex Internal Conditions: Successful application of advanced numerical models to practical cases of flows in complex river systems, and flood plain flows very much depends on the numerical treatment of internal conditions. For example, routing of dam-break flows through complex river networks and simulations of open-channel flows with shocks in canal networks must incorporate appropriate internal conditions for the junctions. Efforts by Fread and his associates resulted in a very versatile one-dimensional NWS-DAMBRK [58], which can be used in many practical cases. Garcia-Navarro and Saviron [64], Garcia-Navarro [62] discussed in detail the importance of junction conditions, and presented appropriate procedures for the same. Although Zhao et al. [174], and Zoppou and Roberts [177] paid a significant attention to the incorporation of time varying internal conditions (Ex: building failures due to wave impact, breaching of embankments etc.) in two-dimensional models, it is felt that further research needs to be done in this direction. Only then realistic application of the advanced numerical models to practical problems becomes common.

References

1. Abbott M. B., Weak-solutions of the equations of open-channel flow. In K. MAHMOOD and V. YEVEJEVICH (eds.) *Unsteady Flow in Open Channels* (Water Resources Publications, Fort Collins, 1975).
2. Abbott M. B., *Computational Hydraulics: Elements of Free Surface Flow* (Pitman Publishing Ltd., London, 1979).
3. Abbott M. B. and Ionescu F., On the numerical computation of nearly horizontal flows, *J. of Hydraulic Research* **5** (1967) pp. 97-117.
4. Aguirre-Pe J., Placho F. P. and Quisca S., Tests and numerical one-dimensional modeling of high-viscosity fluid dam-break wave, *J. of Hydraulic Research*, **33** (1995) pp. 17-26.
5. Akanbi A. A. and Katopodes N. D., Model for flood propagation on initially dry land, *J. Hydraulic Engineering* **114** (1988) pp. 689-706.
6. Alam M. M. and Bhuiyan M. A., Collocation finite-element simulation of dam-break flows, *J. of Hydraulic Engineering* **121** (1995) pp. 118-128.
7. Alcrudo F., Garcia-Navarro P. and Saviron J. M., Flux difference splitting for 1D open channel flow equations, *Int. J. for Numerical Methods in Fluids* **14** (1992) pp. 1009-1018.
8. Alcrudo F. and Garcia-Navarro P., A high resolution Godunov type scheme in finite volumes for the 2D shallow water equations, *Int. J. Numerical Methods in Fluids* **16** (1993) pp. 489-505.
9. Almeida A. B. and Franco A. B., Modeling of dam-break flow, In M. HANIF CHAUDHRY and L. W. MAYS (eds.) *Computer modeling of free-surface and pressurized flows* (Kluwer,Dordrecht, The Netherlands, 1994) pp. 343-373.
10. Amein M. and Fang C. S., Implicit flood routing in natural channels, *J. of Hydraulics Division* **96** (1970) pp. 2481-2500.

11. Anastasiou K. and Chan C. T., Solution of 2D shallow water equations using finite-volume method on unstructured triangular meshes, *Int. J. for Numerical Methods in Fluids* **24** (1997) pp. 1225-1245.
12. Anderson D. A., Tannehill J. C. and Pletcher R. H., *Computational Fluid Mechanics and Heat Transfer* (McGraw-Hill, New York, 1984).
13. Aral M. M., Zhang Y. and Jin S., Application of relaxation scheme to wave propagation simulation in open-channel networks, *J. of Hydraulic Engineering* **124** (1998) pp. 1125-1133.
14. Aureli F., Mignosa P. and Tomirotti M., Numerical simulation and experimental verification of dam-break flows with shocks, *J. of Hydraulic Research* **38** (2000) pp. 197-206.
15. Baines M. J., Maffio A. and Filippo A. D., Unsteady 1-D flows with steep waves in plant channels: the use of Roe's upwind TVD difference scheme, *Advances in Water Resources* **15** (1992) pp. 89-94.
16. Balloffet A. *et al.*, Dam collapse wave in a river, *J. of the Hydraulics Division* **100** (1974) pp. 645-665.
17. Basco D. R., *Computation of rapidly varied, unsteady free-surface flow* (Report No. WRI 83-4284, U. S. Geological Survey, 1983).
18. Bellos C. V. and Sakkas J. G., 1-D dam-break flood wave propagation on dry bed, *J. of Hydraulic Engineering* **113** (1987) pp. 1510-1524.
19. Bellos C. V., Soulis J. V. and Sakkas J. G., Computation of two-dimensional dam-break induced flows, *Advances in Water Resources* **14** (1991) pp. 31-41.
20. Bellos C. V., Soulis J. V. and Sakkas J. G., Experimental investigation of two-dimensional dam-break induced flows, *J. Hydraulic Research*, **30** (1992) pp. 47-63.
21. Berger R. C. and Stockstill R. L., Finite-element model for high-velocity channels, *J. of Hydraulic Engineering* **121** (1995) pp. 710-716.
22. Bhallamudi S. M., Miller S. and Chaudhry M. H., One-dimensional modeling of dam-break flows, In S.R. ABT and J. Gessler (eds.) *Proc. ASCE 1988 National Conf.* Colorado-Springs (1988) pp. 576-581.
23. Bhallamudi S. M. and Chaudhry M. H., Numerical modeling of aggradation and degradation in alluvial channels, *J. of Hydraulic Engineering* **117** (1991) pp. 1145-1164.
24. Bhallamudi S. M. and Chaudhry M. H., Computation of flows in open channel transitions, *J. Hydraulic Research* **30** (1992) pp. 77-93.
25. Bhallamudi S. M. and Eswaran V., A shock-capturing scheme for avalanche modeling. In K. C. AGRAWAL (ed.) *Proc. Of SNOWSYMP 94* (SASE, Manali, India, 1994) pp. 264-268.
26. Blunt M. and Rubin B., Implicit flux-limiting scheme for petroleum reservoir simulation, *J. Computational Physics* **102** (1992) pp. 192-210.
27. Boris J. P. and Book D. L., Flux-corrected transport I: SHASTA, a fluid transport algorithm that works, *J. Computational Physics* **11** (1973) pp. 38-69.

28. Bradford S. F. and Katopodes N. D., Non-hydrostatic model for surface irrigation, *J. Irrigation and Drainage Engineering* **124** (1998) pp. 200-212.
29. Brufau P. and Garcia-Navarro P., Two-dimensional dam-break flow simulation, *Int. J. for Numerical Methods in Fluids* **33** (2000) pp. 35-58.
30. Bova S. W. and Carey G. F., An entropy variable formulation and application for the two-dimensional shallow water equations, *Int. J. for Numerical Methods in Fluids* **22** (1996) pp. 429-444.
31. Carmo J. S. A., Santos F. J. S. and Almeida A.B., Numerical solution of the generalized Serre equations with the MacCormack finite-difference scheme, *Int. J. for Numerical Methods in Fluids* **16** (1993) pp. 725-738.
32. Casier F., Deconnick H. and Hirsch C., A class of bidiagonal schemes for solving the Euler equations, *AIAA Jl.* **22** (1984) pp. 1556-1563.
33. Causon D. M., Mingham C. G. and Ingram D. M., Advances in calculation of methods for supercritical flow in spillway channels, *J. of Hydraulic Engineering* **125** (1999) pp. 1039-1050.
34. Chang S. C., The method of space-time conservation element and solution element: A new approach for solving the Navier-Stokes and Euler equations, *J. Computational Physics*, **119** (1995) pp. 295-324.
35. Chaudhry M. H., *Open-Channel Flow* (Prentice-Hall, Englewood Cliffs, N.J., 1993).
36. Chaudhry M. H., Bhallamudi S. M., Martin C. S. and Naghash M., Analysis of transient pressures in bubbly, homogeneous, gas-liquid mixtures, *J. of Fluids Engineering* **112** (1990) pp. 225-231.
37. Chen C., Laboratory verification of a dam-break flood model, *J. of Hydraulics Division* **106** (1980) pp. 535-556.
38. Chow V. T., *Open-Channel Hydraulics* (McGraw-Hill Book Co., N.Y., 1959).
39. Chow V. T. and Ben-Zvi A., Hydraulic modeling of two-dimensional watershed flow, *J. of Hydraulic Division* **99** (1973) pp. 2023-2040.
40. Colella P. and Woodward P. R., The piecewise parabolic method (PPM) for gas-dynamical simulations, *J. Computational Physics* **54** (1984) pp. 174-201.
41. Cooley R. L. and Moin S. A., Finite element solution of Saint-Venant equations, *J. of Hydraulics Division* **102** (1976) pp. 759-775.
42. Cui G. and Williams B., Downstream Characteristic Lagrangian hybrid method for flows in open channels, *J. of Hydraulic Research* **36** (1998) pp. 379-396.
43. Cunge J. A., Holly F. M. and Verwey A., *Practical Aspects of Computational Hydraulics* (Pitman Advanced Publishing Program, London, 1980).
44. Dammuller D. C., Bhallamudi S. M. and Chaudhry M. H., Modeling of unsteady flow in curved channel, *J. Hydraulic Engineering* **115** (1989) pp. 1479-1495.
45. Delis A. I., Skeels C. P. and Ryrie S. C., Evaluation of some approximate Riemann solvers for transient open channel equations, *J. of Hydraulic Research* **38** (2000) pp. 217-231.

46. Delis A. I., Skeels C. P. and Ryrie S. C., Implicit high-resolution methods for modeling one-dimensional open-channel flow, *J. of Hydraulic Research* **38** (2000) pp. 369-382.
47. Dressler R. F., Hydraulic resistance effect upon the dam-break functions, *J. of Research* NBS **49** (1952) pp. 217-225.
48. Dronkers J. J., Tidal computations for rivers, coastal areas and seas, *J. of the Hydraulics Division* **95** (1969) pp. 29-77.
49. Engquist B., Lotstedt P. and Sjogreen B., Nonlinear filters for efficient shock computation, *Math. and Computation*, **52** (1989) pp. 509-537.
50. Fennema R. J. and Chaudhry M. H., Explicit numerical methods for unsteady free surface flows with shocks, *Water Resources Research* **22** (1986) pp. 1923-1930.
51. Fennema R. J. and Chaudhry M. H., Simulation of one-dimensional dam-break flows, *J. of Hydraulic Research* **25** (1987) pp. 41-51.
52. Fennema R. J. and Chaudhry M. H., Implicit methods for two-dimensional unsteady free-surface flows, *J. Hydraulic Research* **27** (1989) pp. 321-332.
53. Fennema R. J. and Chaudhry M. H., Numerical solution of two-dimensional transient free surface flows, *J. Hydraulic Engineering* **116** (1990) pp. 1013-1034.
54. Fiedler F. R. and Ramirez J. A., A numerical method for simulating discontinuous shallow flow over an infiltrating surface, *Int. J. of Numerical Methods in Fluids* **32** (2000) pp. 219-240.
55. Finnie J. I., Finite-element methods for free-surface flows, In M. H. CHAUDHRY and L. W. MAYS (eds.) *Computer Modeling of Free-Surface and Pressurized Flows* (Kluwer, The Netherlands, 1994).
56. Flokstra C., The closure problem for depth-averaged two-dimensional flow, *!7th Congress of IAHR* (Baden-Baden, Vol.2, 1977) pp. 247-256.
57. Fraccarollo L. and Toro E. F., Experimental and numerical assessment of the shallow water model for two-dimensional dam-break type problems, *J. of Hydraulic Research* **33** (1995) pp. 843-864.
58. Fread D. L., *DAMBRK: the NWS dam-break flood forecasting model* (Office of Hydrology, National Weather Service, Silver Spring, Maryland, 1979).
59. Gabutti B., On two upwind finite-difference schemes for hyperbolic equations in non-conservative form, *Computers and Fluids* **11** (1983) pp. 207-230.
60. Galland J-C., Goutal N. and Hervouet J-C., TELEMAC: A new numerical model for solving shallow water equations, *Advances in Water Resources* **14** (1991) pp. 138-148.
61. Garcia R. and Kahawita R. A., Numerical solution of the St. Venant equations with the MacCormack finite-difference scheme, *Int. J. for the Numerical Methods in Fluids* **6** (1986) pp. 259-274.
62. Garcia-Navarro P., Surges through an open-channel junction, *J. of Hydraulic Research* **31** (1993) pp. 79-87.

63. Garcia-Navarro P. and Saviron J. M., McCormack's method for the numerical simulation of one-dimensional discontinuous unsteady open channel flow, *J. of Hydraulic Research* **30** (1992) pp. 95-92.
64. Garcia-Navarro P. and Saviron J. M., Numerical simulation of unsteady flow at open channel junctions, *J. of Hydraulic Research* **30** (1992) pp. 595-609.
65. Garcia-Navarro P., Alcrudo F. and Saviron J. M., 1-D open-channel flow simulation using TVD-MacCormack scheme, *J. of Hydraulic Engineering* **118** (1992) pp. 1359-1372.
66. Garcia-Navarro P. and Priestley A., A conservative shape preserving semi-Lagrangian method for the solution of the shallow water equations, *Int. J. for Numerical Methods in Fluids* **18** (1994) pp. 273-294.
67. Garcia-Navarro P., Priestley A. and Alcrudo F., An implicit method for water flow modeling in channels and pipes, *J. of Hydraulic Research* **32** (1994) pp. 721-742.
68. Gharangik A. M. and Chaudhry M. H., Numerical simulation of hydraulic jump, *J. Hydraulic Engineering* **117** (1991) pp. 1195-1211.
69. Glaister P., Approximate Riemann solutions of the shallow water equations, *J. Hydraulic Research* **26** (1988) pp. 293-306.
70. Gozali S. and Hunt B., Dam-break solutions for a partial breach, *J. of Hydraulic Research* **31** (1993) pp. 205-214.
71. Gunaratnam D. and Perkins F. E., *Numerical solution of unsteady flows in open channels* (Report 127, Massachusetts Institute of Technology, Cambridge, MA, 1970).
72. Hager W., Supercritical flow in open channel junctions, *J. of Hydraulic Engineering* **115** (1989) pp. 595-616.
73. Hager W., Schwalt M., Jimenez O. F. and Chaudhry M. H., Supercritical flow near an abrupt wall deflection, *J. of Hydraulic Research* **32** (1994) pp. 103-118.
74. Harten A., High resolution schemes for hyperbolic conservation laws, *J. Computational Physics* **49** (1983) pp. 357-393.
75. Harten A. and Hyman J. M., Self adjusting grid methods for one-dimensional hyperbolic conservation laws, *J. Computational Physics* **50** (1983) pp. 235-269.
76. Harten A., Engquist, Osher S. and Chakravarthy S., Uniformly high order accurate non-oscillatory schemes III, *J. Computational Physics* **83** (1989) pp. 148.
77. Hicks F. E. and Steffler P. M., Comparison of finite-element methods for the St. Venant equations, *Int. J. for Numerical Methods in Fluids* **20** (1995) pp. 99-114.
78. Hicks F. E., Steffler P. M. and Yasmin N., One-dimensional dam-break solutions for variable width channels, *J. of Hydraulic Engineering* **123** (1997) pp. 464-468.
79. Hsu C-T., Yeh K-C. and Yang J-C., Depth-averaged two-dimensional curvilinear explicit finite analytic model for open-channel flows, *Int. J. for Numerical Methods in Fluids* **33** (2000) pp. 175-202.

80. Hunt B., Asymptotic solution for dam-break problem, *J. of the Hydraulics Division* **108** (1982) pp. 115-125.
81. Hunt B., A perturbation solution of the flood routing problem, *J. of Hydraulic Research* **25** (1987) pp. 215-234.
82. Ippen A. T. and Harleman D. R. F., Studies on the validity of the hydraulic analogy to supersonic flow: Parts I and II (USAF Technical report, No. 5985, 1950).
83. Jameson A., Schmidt W. and Turkel E., Numerical solutions of the Euler equations by finite volume methods using Runge-Kutta time stepping scheme, *AIAA 14th Fluid and Plasma Dynamics Conference* (Palo-Alto, Ca, AIAA-81-1259, 1981).
84. Jha A. K., Akiyama J. and Masaru U., Modeling of unsteady open-channel flows - Modifications to Beam and Warming scheme, *J. of Hydraulic Engineering* **120** (1994) pp. 461-476.
85. Jha A. K., Akiyama J. and Masaru U., First- and second-order flux difference splitting schemes for dam-break problem, *J. of Hydraulic Engineering* **121** (1995) pp. 877-884.
86. Jha A. K., Akiyama J. and Masaru U., A fully conservative Beam and Warming scheme for transient open-channel flows, *J. of Hydraulic Research* **34** (1996) pp. 605-622.
87. Jha A. K., Akiyama J. and Masaru U., Flux-difference splitting schemes for 2D flood flows, *J. of Hydraulic Engineering* **126** (2000) pp. 33-42.
88. Jimenez O. F., *Computation of supercritical flow in open-channels* (M. S. thesis, Department of Civil Engineering, Washington State University, Pullman, WA, 1986).
89. Jimenez O. F. and Chaudhry M. H., Computation of supercritical free surface flows, *J. of Hydraulic Engineering* **114** (1988) pp. 377-395.
90. Jin M. and Fread D. L., Dynamic flood routing with explicit and implicit numerical solution schemes, *J. of Hydraulic Engineering* **123** (1997) pp. 166-173.
91. Jin S. and Xin Z., The relaxation schemes for systems of conservation laws in arbitrary space dimensions, *Communications in Pure and Applied Mathematics* **XLVII** (1995) pp. 235-276.
92. Katopodes N. D., A dissipative Galerkin scheme for open-channel flow *J. of Hydraulics Division* **110** (1984) pp. 450-466.
93. Katopodes N. D., Two-dimensional surges and shocks in open channels, *J. of Hydraulics Division* **110** (1984) pp. 794-812.
94. Katopodes N. D. and Strelkoff T., Computing two-dimensional dam-break flood waves, *J. Hydraulics Division* **104** (1978) pp. 1269-1288.
95. Keuning D. H., Application of finite element method to open channel flow, *J. of Hydraulics Division* **102** (1976) pp. 459-468.

96. King I. P., Finite element models for unsteady flow routing through irregular channels, In C. A. BREBBIA *et al.* (eds.) *Finite Elements in Water Resources* (Pentech Press, London, 1976) pp. 4.165-4.184.
97. Kroll N. and Jain R. R., Solution of two-dimensional Euler equations: Experience with a finite volume code, *DFVLR-FB87-41*, DFVLR (Institut fur Enwurfsaerodynamik, Braunschweng, 1987).
98. Lemos C. M. and Martins B. D., A numerical study of the impact of solitary waves on obstacles of simple geometric configurations. In M. RAHAMAN and C. A. BREBBIA (eds.) *Ist Int. Conf. On Advances in Fluid Mechanics* **9** (1996) pp. 113-123.
99. Liggett J. A., Basic equations of unsteady flow. In K. MAHMOOD and V. YEVJEVICH (eds.) *Unsteady Flow in Open Channels* (Water Res. Publications, Fort Collins, Colorado, 1975) pp. 29-62.
100. Liggett J. A. and Woolhiser D. A., Difference solutions of the shallow water equations, *J. of the Engineering Mech. Division* **93** (1967) pp. 39-71.
101. Liska R. and Wendroff B., Two-dimensional shallow water equations by composite schemes, *Int. J. for Numerical Methods in Fluids* **30** (1999) pp. 461-479.
102. Louaked M. and Hanich L., TVD scheme for the shallow water equations, *J. of Hydraulic Research* **36** (1998) pp. 363-378.
103. Mahmood K. and Yevjevich V(eds.) *Unsteady Flow in Open Channels* (Water Resources Publications, Fort Collins, Colorado, 1975).
104. Malcherek A. and Zielke W., Upwinding and characteristics in FD and FE methods. In M. H. CHAUDHRY and L. W. MAYS (eds.) *Computer Modeling of Free-Surface and Pressurized Flows* (Kluwer, The Netherlands, 1994).
105. Martin C. and Defazio F. G., Open-channel surge simulation by digital computer, *J. of Hydraulics Division* **95** (1969) pp. 2049-2070.
106. Meselhe E. A. and Holly F. M. Jr., Invalidity of Preissmann scheme for transcritical flow, *J. of Hydraulic Engineering* **123** (1997) pp. 652-655.
107. Meselhe E. A., Sotiropoulos F. and Holly F. M. Jr., Numerical simulation of transcritical flow in open-channels, *J. of Hydraulic Engineering* **123** (1997) pp. 774-783.
108. Miller S. and Chaudhry M. H., Dam-break flows in curved channels, *J. Hydraulic Engineering* **115** (1989) pp. 1465-1478.
109. Mingham C. G. and Causon D. A., High resolution finite-volume method for shallow water flows, *J. of Hydraulic Engineering* **124** (1998) pp. 605-614.
110. Mingham C. G. and Causon D. M., Calculation of unsteady bore diffraction using a high resolution finite volume method, *J. of Hydraulic Research* **38** (2000) pp. 49-56.
111. Mohapatra P. K. and Bhallamudi S. M., Bed-level variation in channel expansions with movable beds, *J. Irrigation and Drainage Engineering* **120** (1994) pp. 1114-1121.

112. Mohapatra P. K. and Bhallamudi S. M., Computation of a dam-break flood wave in channel transitions, *Advances in Water Resources* **19** (1996) pp. 181-187.
113. Mohapatra P. K., Eswaran V. and Bhallamudi S. M., Two-dimensional analysis of dam-break flow in vertical plane, *J. Hydraulic Engineering* **125** (1999) pp. 183-192.
114. Mohapatra P. K., Bhallamudi S. M. and Eswaran V., Numerical simulation of impact of bores against inclined walls, *J. Hydraulic Engineering* **126** (2000) pp. 942-945.
115. Molinaro P., Dam-break wave analysis: A state of the art. In D. BENSARI, C. A. BREBBIA and D. OVAZAR (eds.) *Computational Water Resources* (Computational Mechanics Publications, London, 1991) pp. 105-124.
116. Molls T. and Chaudhry M. H., Depth-averaged open channel flow model, *J. of Hydraulic Engineering* **121** (1995) pp. 453-465.
117. Molls T., Chaudhry M. H. and Khan A. W., Numerical simulation of two-dimensional flow near a spur-dike, *Advances in Water Resources* **18** (1995) pp. 227-236.
118. Molls T. and Molls F., Space-time conservation method applied to Saint Venant equations, *J. of Hydraulic Engineering* **124** (1998) pp. 501-508.
119. Molls T. and Zhao G., Depth-averaged simulations of supercritical flow in channel with wavy side wall, *J. of Hydraulic Engineering* **126** (2000) pp. 437-446.
120. Moretti G., The λ - scheme, *Computers and Fluids* **7** (1979) pp. 191-205.
121. Nessyahu H. and Tadmor E., Non-oscillatory central-differencing for the hyperbolic conservation laws, *J. of Computational Physics* **87** (1990) pp. 408-463.
122. Nujic M., Efficient implementation of non-oscillatory schemes for the computation of free-surface flows, *J. of Hydraulic Research* **33** (1995) pp. 101-112.
123. Panagiotopoulos A. G. and Soulis J. V., Implicit bi-diagonal scheme for depth-averaged free surface flow equations, *J. of Hydraulic Engineering* **126** (2000) pp. 425-436.
124. Peyret R. and Taylor T. D., *Computational Methods for Fluid Flow* (Springer, New York, 1983).
125. Playan E., Walker W. R. and Merkley G. P., Two-dimensional simulation of basin irrigation. I: Theory, *J. Irrigation and Drainage Engineering* **120** (1994) pp. 837-855.
126. Preissmann A., Propagation of translotaory waves in channels and rivers. In *Proc. Of First Congress of French Association for Computation* (Grenoble, September 1961) pp. 433-442.
127. Rahman M. and Chaudhry M. H., Simulation of hydraulic jump with grid-adaptation, *J. of Hydraulic Research* **33** (1995) pp. 555-569.

128. Rahman M. and Chaudhry M. H., Computation of flow in open-channel transitions, *J. of Hydraulic Research* **35** (1997) pp. 243-256.
129. Ramsden J. D., Forces on a vertical wall due to long waves, bores and dry-bed surges, *J. Waterways Ports Coastal and Ocean Engineering* **122** (1996) pp. 134-141.
130. Rao V. S. and Latha G., A slope modification method for shallow water equations, *Int. J. for Numerical Methods in Fluids*, **14** (1992) pp. 189-196.
131. Ritter A., The propagation of water waves, *Ver Deutsch Ingenieur Zeitschr*, Berlin, **Vol. 36,** part 3, No. 33 (1892) pp. 947-954.
132. Roe P. L., Approximate Riemann solvers, parameter vectors and difference schemes, *J. Computational Physics* **43** (1981) pp. 357-372.
133. Roe P. L., Efficient construction and utilization of approximate Riemann solutions. In R. GLOWINSKI (ed.) *Computing Methods in Applied Sciences and Engineering* (North Holland, Amsterdam, 1984) pp. 499-518.
134. Sakkas J. G. and Strelkoff T., Dam-break flood in a prismatic dry channel, *J. of Hydraulics Division* **99** (1973) pp. 2195-2216.
135. Sakkas J. G. and Srelkoff T., Dimensionless solution of the dam-break waves, *J. Hydraulics Division* **102** (1976) pp. 171-184.
136. Savic L. J. and Holly F. M. Jr., Dambreak flood waves computed by modified Godunov method, *J. Hydraulic Research* **31** (1993) pp. 187-204.
137. Shu C. W. and Osher S., Efficient implementation of essentially non-oscillatory shock capturing schemes, *J. Computational Physics* **77** (1988) pp. 439-471.
138. Singh V., *Computation of shallow water flow over a porous medium*, (Ph. D. thesis, Department of Civil Engineering, Indian Institute of Technology, Kanpur, India, 1996).
139. Singh V. and Bhallamudi S. M., Hydrodynamic Modeling of Basin Irrigation, *J. Irrigation and Drainage Engineering* **123** (1997) pp. 407-414.
140. Singh V. and Bhallamudi S. M., Conjunctive surface-subsurface modeling of overland flow, *Advances in Water Resources* **21** (1998) pp. 567-579.
141. Soulis J. V., Multiple grid solution of the open-channel flow equations using a marching finite-volume method, *Advances in Water Resources* **14** (1991) pp. 203-214.
142. Srinivasan C., *Analysis of solute transport in porous media for nonreactive and sorbing solutes using hybrid FCT model* (M.S. thesis, Department of Civil Engineering, Indian Institute of Science, Bangalore, India, 2000).
143. Steffler P. M. and Jin Y., Depth averaged and moment equations for moderately shallow free surface flow, *J. Hydraulic Research* **31** (1993) pp. 5-17.
144. Stockstill R. L., Berger R. C. and Nece R. E., Two-dimensional flow model for trapezoidal high-velocity channels, *J. Hydraulic Engineering* **123** (1997) pp. 844-852.
145. Stoker J. J., *Water Waves* (Interscience Publishers 1957).

146. Su S. T. and Barnes A. H., Geometric and frictional effects on sudden releases, *J. Hydraulics Division* **96** (1970) pp. 2185-2200.
147. Sweby P. K., High resolution schemes using flux limiters for hyperbolic conservation laws, *SIAM J. Numer. Anal.* **21** (1984) pp. 995-1011.
148. Tan W., *Shallow Water Hydrodynamics* (Elsewier, Amsterdam. 1992).
149. Terzidis G. and Strelkoff T., Computation of open channel surges and shocks, *J. Hydraulics Division* **96** (1970) pp. 2581-2610.
150. Thompson J., Thames F. and Mastin W., Automatic numerical generation of body-fitted curvilinear coordinate system for field containing any number of arbitrary two-dimensional bodies, *J. Computational Physics* **15** (1974) pp. 299-319.
151. Tisdale T. S., Scarlatos P. D. and Hamric J. M., Streamline upwind finite-element method for overland flow, *J. Hydraulic Engineering* **124** (1998) pp. 350-357.
152. Toro E. F., A weighted average flux method for hyperbolic conservation laws, *Proc. Royal Society Series A* **423** (1989) pp. 401-418.
153. Townson J. M. and Al-Salihi A. H., Models of dam-break flow in R-T space, *J. Hydraulic Engineering* **115** (1989) pp. 561-575.
154. Tseng M. H., Explicit finite volume non-oscillatory schemes for 2D transient free-surface flows, *Int. J. for Numerical Methods in Fluids* **30** (1999) pp. 831-843.
155. Tseng M. H. and Chu C. R., The simulation of dam-break flows by an improved predictor-corrector TVD scheme, *Advances in Water Resources* **23** (2000) pp. 637-643.
156. Tseng M. H. and Chu C. R., Two-dimensional shallow water flows simulation using TVD-MacCormack scheme, *J. Hydraulic Research* **38** (2000) pp. 123-131.
157. Vanderkwaak J. E., *Numerical simulation of flow and chemical transport in integrated surface-subsurface hydrologic systems* (Ph. D. thesis, Department of Earth Sciences, University of Waterloo, Ontario, 1999).
158. Vanleer B., Towards the ultimate conservative difference scheme V: A second order sequel to Godunov's method, *J. Computational Physics* **32** (1979) pp. 101-136.
159. Verma A. K., Bhallamudi S. M. and Eswaran V., Overlapping control volume method for solute transport, *J. Hydrologic Engineering* **5** (2000) pp. 308-316.
160. Wang J. S., Ni H. G. and He Y. S., Finite-difference TVD scheme for computation of dam-break problem, *J. Hydraulic Engineering* **126** (2000) pp. 253-262.
161. Wang Z. and Shen H. T., Lagrangian simulation of one-dimensional dam-break flow, *J. Hydraulic Engineering* **125** (1999) pp. 1217-1220.
162. Whitham G. B., The effect of hydraulic resistance in the dam-break problem, *Proc. Royal Society of London Series-A* (1955) pp. 226-227.

163. Yang J. Y., Hsu C. H. and Chang S. H., Computation of free surface flows, Part-1: One-dimensional dam-break flow, *J. Hydraulic Research* **31** (1993) pp. 19-34.
164. Yang J. Y. and Hsu C. A., Computation of free surface flows, Part-2: Two-dimensional unsteady bore diffraction, *J. Hydraulic Research* **31** (1993) pp. 403-413.
165. Yee H. C., Construction of explicit and implicit symmetric TVD schemes and their applications, *J. Computational Physics* **68** (1987) pp. 151-179.
166. Yee H. C., *A class of high resolution explicit and implicit shock-capturing methods* (NASA Technical Memorandum 101088, NASA Ames Research Center, California, 1989).
167. Yost S. and Rao M. S. V. P., A non-oscillatory scheme for open-channel flows, *Advances in Water Resources* **22** (1998) pp. 133-143.
168. Yost S. and Rao M. S. V. P., Flux-corrected transport technique for open-channel flow, *Int. J. for Numerical Methods in Fluids* **29** (1999) pp. 951-973.
169. Yost S. and Rao P., A multiple grid algorithm for one-dimensional transient open-channel flows, *Advances in Water Resources* **23** (2000) pp. 645-651.
170. Yost S. and Rao P., A non-linear filter for one and two-dimensional open channel flows with shocks, *Advances in Water Resources* **24** (2000) pp. 187-193.
171. Yost S., Rao P. and Brown R. M., Absorbing boundary technique for open channel flows, *Int. J. for Numerical Methods in Fluids* **33** (2000) pp. 641-656.
172. Younus M. and Chaudhry M. H., A depth-averaged k-ε turbulence model for the computation of free-surface flow, *J. Hydraulic Research* **32** (1994) pp. 415-444.
173. Zhang W. and Cundy T. W., Modeling of two-dimensional overland flow, *Water Resources Research* **25** (1989) pp. 2019-2035.
174. Zhao D. H., Shen H. W., Tabios III G. Q., Lai J. S. and Tan W. Y., Finite-volume two-dimensional unsteady-flow model for river basins, *J. Hydraulic Engineering* **120** (1994) pp. 863-883.
175. Zhao D. H., Shen H. W., Lai J. S. and Tabios III G. Q., Approximate Riemann solvers in FVM for 2D hydraulic shock wave modeling, *J. Hydraulic Engineering* **122** (1996) pp. 692-702.
176. Zienkiewicz O. C. and Ortiz P., A split-characteristic based finite element model for the shallow water equations, *Int. J. for Numerical Methods in Fluids* **20** (1995) pp. 1061-1080.
177. Zoppou C. and Roberts S., Catastrophic collapse of water supply reservoirs in urban areas, *J. Hydraulic Engineering* **125** (1999) pp. 686-695.

SCOURING HORSESHOE VORTEX

T. GANGADHARAIAH
Civil Engg., I.I.T. Kanpur – 208016,
INDIA
E-mail: tganga@iitk.ac.in

MOHAMMED MUZZAMMIL
Civil Engg., Aligarh Muslim University, Aligarh, U.P.
INDIA
E-mail: muzzammil.786@rediffmail.com

BALDEV SETIA
Civil Engg., Regional Engg. College, Kurukshetra, Haryana
INDIA
E-mail: setia_b@rediffmail.com

A. K. GUPTA
Aerospace Engg., I.I.T. Kanpur – 208016,
INDIA
E-mail: akgupta@iitk.ac.in

Knowledge on the mechanism of local scour around a bridge pier is of importance in predicting not only the maximum depth of foundation for piers and abutments but also to protect them from scouring phenomena. Horseshoe vortex formed at the junction of the bridge pier and riverbed and its interaction with the sediment is the cause of the scouring process. In a series of investigations carried out in the Hydraulics laboratory of the Dept. of Civil Engineering at I.I.T. Kanpur and spread over a decade, special aides, devices and techniques have been developed to visualize and measure the size, rotational speed and location of the horseshoe vortex before and during the scouring process. It is observed that the horseshoe vortex characteristics remain fairly constant when it is formed on the rigid bed. However, on a mobile bed as the scouring process is initiated, the horseshoe vortex size increases and vortex rotational speed decreases as the scour depth increases. The computed horseshoe vortex strength initially increases, reaches a maximum value and then starts decreasing as the scour hole deepens.

1 Introduction

An obstruction such as a bridge pier placed on a riverbed induces an adverse pressure gradient in the oncoming flow resulting in three-dimensional boundary layer separation on the bed in front of the pier. This causes the system of vortices, which wraps around the pier in the shape of the horseshoe vortex. Such a vortex system causes high local shear stresses on the riverbed resulting in the scouring of sediment in the immediate vicinity of the pier. Due to scouring action by the horseshoe vortex, the scouring hole deepens leading to sinking of the horseshoe vortex. The size and rotational velocity of the horseshoe vortex change as the scour hole deepens. At some stage of the scouring process the strength of the vortex gets adjusted such that the deepening of the scour hole is slowly stopped and an equilibrium state is reached whereby the incoming sediment will be equal to the scouring and transport of the sediment out of the scour hole. This horseshoe vortex associated with scouring action is called scouring horseshoe vortex. Knowledge on the variation of the size, rotational velocity and strength of the horseshoe vortex during scouring will help in arriving at the maximum depth of scour hole, an information needed in the design of foundation depth of the pier.

A large body of literature on the topic of local scour around bridge piers is available. A few of the relevant and recent papers are by Hjorth [9], Ahmed and Rajaratnam [1], Graf and Yulistiyanto [7]. It is important in the design of scour protection works to reduce the depth of scour or to create no scouring condition around the pier. This can be achieved by knowing the behavior of horseshoe vortex before and during scouring action. This paper focuses on the behavioral details of horseshoe vortex before initiation of scour and during scouring of sediment around a pier or any hydraulic structure built on a mobile bed in clear water scour conditions. Methodology adopted is first to visualize the horseshoe vortex and then to measure the vortex rotational speed, size and location of its centre of rotation before and during the scouring process.

2 Visualization of Horseshoe Vortex (HSV)

In three-dimensional separation of flow on the bed at the junction corner of the pier, the oncoming flow wraps around the pier in the shape of horseshoe vortex. The sizes of separation zone and the vortex are mainly functions of the pier geometry and flow Reynolds number. It is important to make this vortex visible for the measurement of its characteristics during scouring process. At first, many fluid flow visualizers like ink, lime water, potassium permanganate solution and other coloured dye were injected through a small tube near the corner of the pier model and bed on the line of symmetry to make the vortex visible. It was observed that these dyes got rapidly diffused in the fluid resulting in blurring of the visualization of HSV. In order to overcome the rapid diffusion, mud suspension water was injected into the region of horseshoe vortex. With proper lighting arrangement it turned out to be a great success in visualizing the vortex.

This method of visualization essentially makes use of a transparent glass cylinder to serve as a model bridge pier. But for a small (approximately 2 mm) longitudinal slit running through the length of the cylinder, the rest of the cylinder was painted black or wrapped from inside with a thick black paper. A bulb as a source of light was lowered into the hollow glass cylinder pier when in position. The light from the bulb passed through the thin slit and illuminated a vertical plane. A turbid suspension of mud in water was used as the visualization media. Mud suspension in water had an advantage over other media like colour dyes, as it did not diffuse rapidly in turbulent flow field and hence the vortex was clearly visible under this lighting arrangement. The mud suspension from a small external reservoir (wash bottle) was injected in the junction region of the pier through a nozzle. A metallic tube 3 mm in diameter having a nozzle at its end was bent in semicircular shape to embrace the cylindrical pier. The nozzle tube being kept flush with the bed caused minimum interference to the vortex under consideration. Fig. 1 and Fig. 2 show the vortex formed before the initiation of scour and during the scour on the line of symmetry in front of cylindrical pier model.

Figure 1. Horseshoe vortex on rigid bed in flow from right to left in front of pier on the line of symmetry.

Figure 2. Horseshoe vortex in scour hole in flow from right to left in front of pier on the line of symmetry.

3 Horseshoe Vortex Characteristics on Rigid Bed

Flow characteristics of horseshoe vortex on rigid bed represent the conditions before initiation of scouring action. Fig. 1 shows the presence of primary vortex along with the corner vortex on a rigid bed.

Pressure on the surface plate, on which the cylindrical pier was mounted, was measured and plotted as pressure co-efficient contours C_p, where $C_p = \dfrac{P - P_o}{\frac{1}{2}\rho U_0^{\ 2}}$ and P is the pressure at the point of interest, P_o is the pressure of the upstream undisturbed flow, ρ is water density and U_0 is the maximum velocity of upstream undisturbed flow. Horseshoe vortex imprints on the surface plate were obtained by wet paint technique. This method consists of painting the plate around the cylinder and keeping this in specified flow conditions till the full flow patterns has developed due to erosion of paint. Fig. 3 shows one such sample photo taken after removing the plate and the model from the flow and drying the plate. The traces of a similar photo are shown in Fig. 4. Paint impression lines indicate line of separation, primary vortex, secondary vortex surrounded by corner vortex in the corner of the pier, and counter vortex in between the primary and secondary vortices. These vortices wrap around the pier as parts of the total horseshoe vortex system. The presence of vortex causes a depression of the pressure [4] as indicated in Fig. 5, obtained from data of pressure measurements on the plate. Fig. 6 shows the contours of pressure coefficients C_p on one half portion of the plate and scour depth contours for the other half portion of the bed, the direction of flow being top to bottom.

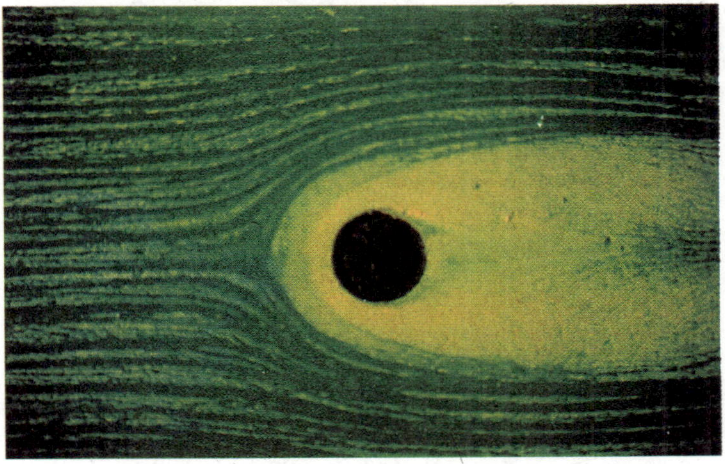

Figure 3. Imprints of vortex and separation on a plate around a cylindrical pier using wet paint technique (flow from left to right).

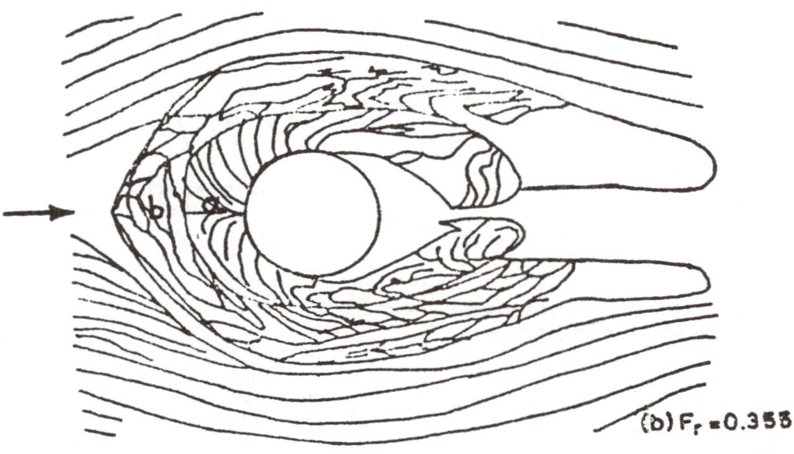

Figure 4. Sketch showing the vortices taken from similar wet paint imprints.

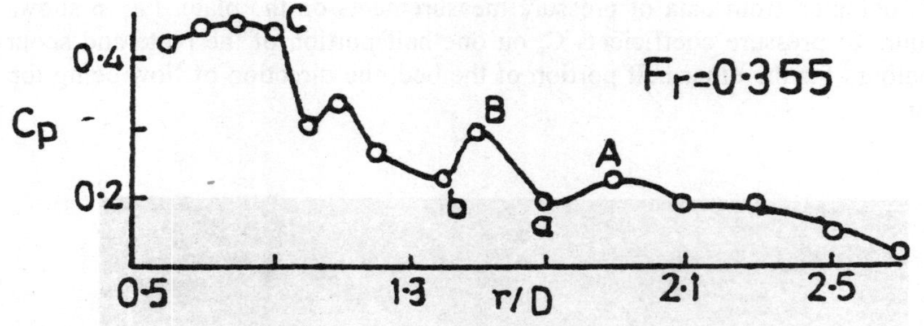

Figure 5. Upstream pressure distribution on the line of symmetry in vortex zone.

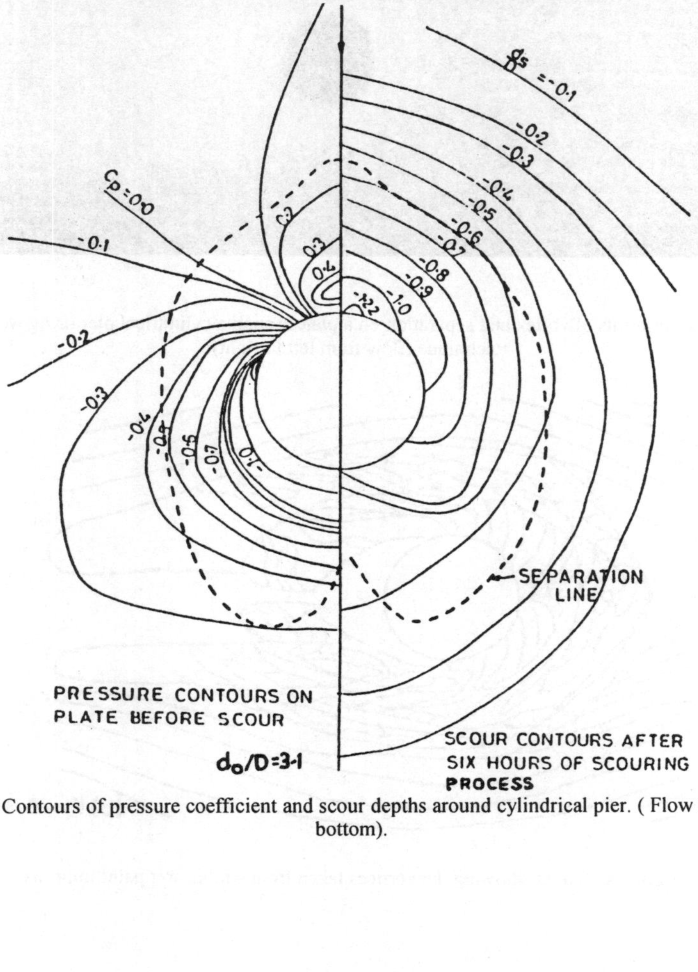

Figure 6. Contours of pressure coefficient and scour depths around cylindrical pier. (Flow from top to bottom).

Muzzammil [12] used the measured velocity contours of Mellivile and Raudkivi [10] to show that these vortices possessed characteristics of a forced vortex before and during scouring process. The strength of the horseshoe vortex on the line of symmetry can be computed using forced vortex theory as follows:

$$\frac{\Gamma}{\pi D U_o} = \left(\frac{2r_*}{D}\right)\left(\Delta C_p\right)^{1/2} \tag{1}$$

where Γ is the vortex strength, D is diameter of the pier, U_0 is maximum velocity of flow, 2 r_* is the width of depression of pressure measured on line of symmetry and ΔC_p is the difference in pressure co-efficient between the crest and trough of the pressure depression. This was related to the upstream flow parameter $\frac{U_0 D}{V_* \delta}$, where V_* is bed shear velocity and δ is boundary layer thickness at the position of cylindrical pier in its absence. Fig. 7 shows the relationship between these two parameters made dimensionless with respect to the vorticity of the overcoming flow represented by $U_0 D$. It is observed that as $\frac{U_0 D}{V_* \delta}$ increases the vortex strength increases. In the case of fully developed flows δ corresponds to the depth of flow.

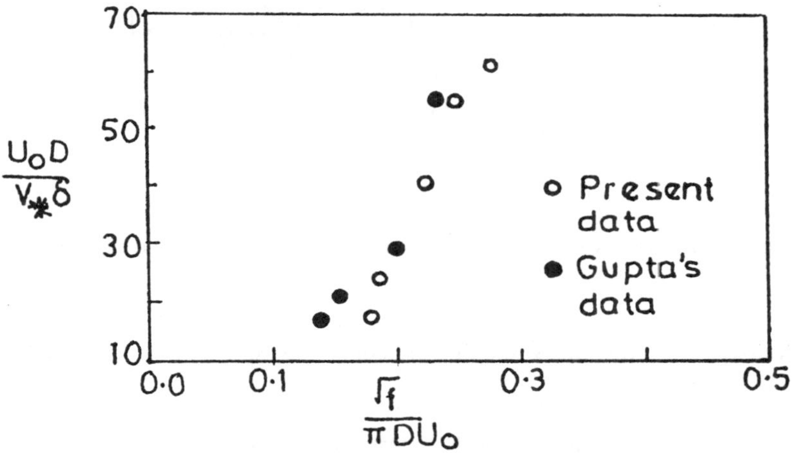

Figure 7. Vortex strength as a function of flow parameters.

Vortex dimensions and vortex rotations were measured using specially designed vortex gauge and vortex meter respectively. Vortex gauge consists of a thin wire supported U-shaped fork such that its location near the vortex should not disturb the oncoming vortex. Similarly for vortex meter, a metal foil on the shaft in place of wire was used for measuring rotational speed of the vortex when it was placed in the centre of the vortex. Special counting device similar to velocity current meter was installed on the vortex probe. The vortex dimensions D_v and rotational speed 'N' measured are plotted in Fig. 8 and Fig. 9, respectively. Fig. 8 shows that the vortex dimensions $\frac{D_v}{D}$ are fairly constant over a large range of pier Reynolds number $R_{eD} = \frac{U_0 D}{\nu}$ with a value of $\frac{D_v}{D} = 0.2$. Fig. 9 shows the vortex rotational speed 'N'' increasing with pier Reynolds number.

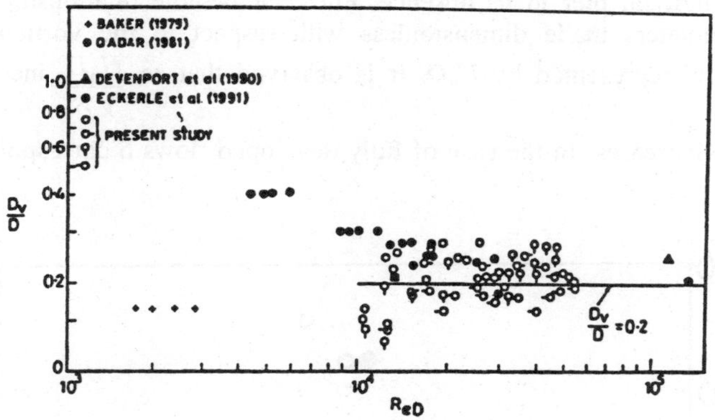

Figure 8. Horseshoe vortex size variation with pier Reynolds number on the rigid bed.

Vortex velocity V_θ and vortex strength Γ are computed as,

$$V_\theta = \pi D_v N \text{ and } \Gamma = \pi D_v V_\theta \tag{2}$$

Where D_v is measured in meters and N in terms of number of rotations per second. After non-dimensionalising these values with flow velocity U_0 and pier diameter D, the results are plotted in Fig. 10 and Fig. 11. It is observed that there is a mild

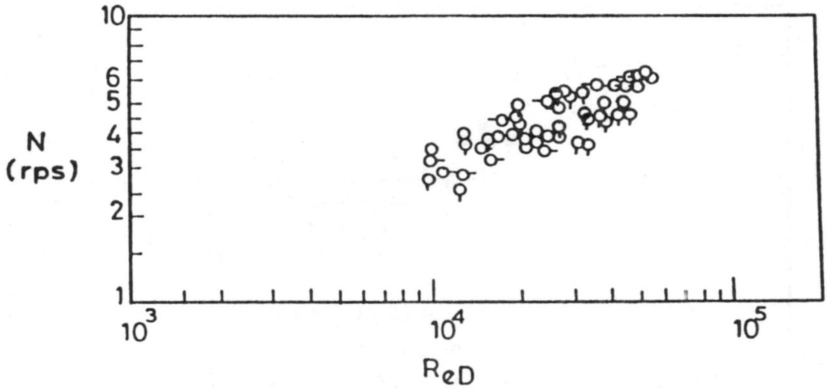

Figure 9. Rotational speed of horseshoe vortex as a function of pier Reynolds number on Rigid bed.

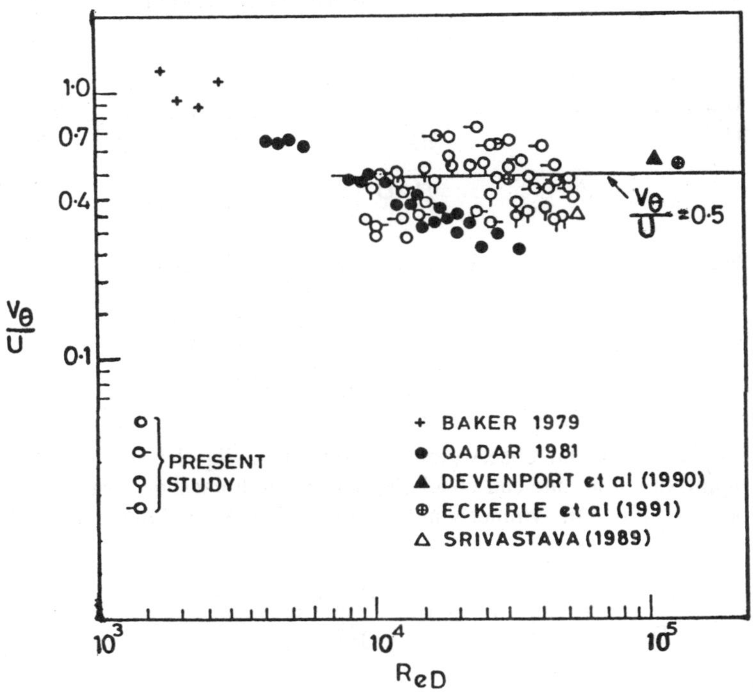

Figure 10. Vortex velocity as a function of pier Reynolds number on rigid bed.

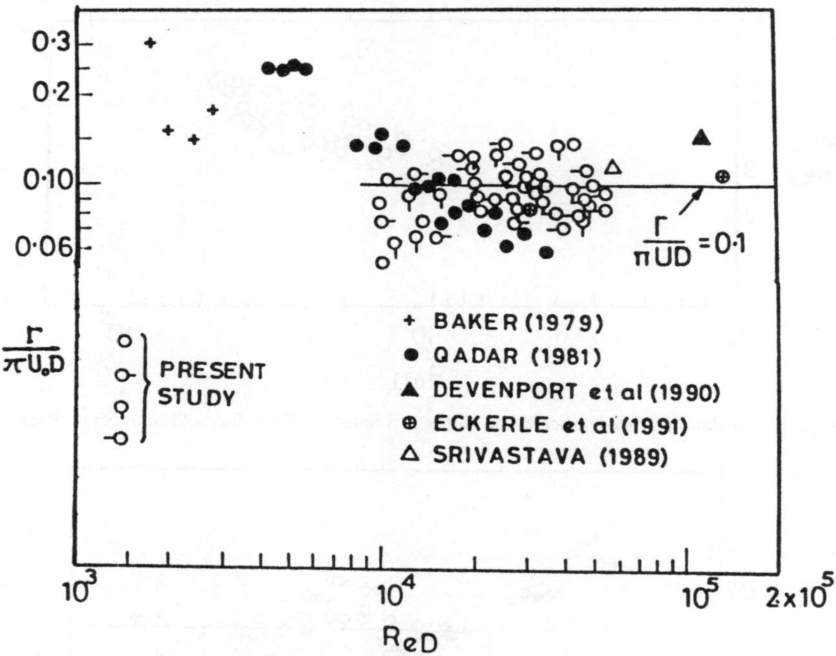

Figure 11. Vortex strength as a function of pier Reynolds number on rigid bed.

reduction in $\frac{V_\theta}{U_0}$ and $\frac{\Gamma}{\pi DU_0}$ with increase in Reynolds number. However, large scatter in the experimental observations is noticed in Fig. 10 and Fig. 11. With judicious approximation, one can conclude that for the horseshoe vortex formed on the rigid bed on the line of symmetry in front of pier model, following values can be obtained.

$$\frac{D_v}{D} = 0.2; \ \frac{V_\theta}{U_0} = 0.5 \text{ and } \frac{\Gamma}{\pi DU_0} = 0.1 \tag{3}$$

5 Characteristics of Scouring Horseshoe Vortex

The rigid bed in the flume was replaced by mobile beds using alluvial sediments with median diameter d_{50} = 0.16 mm and standard deviation σ_g = 1.38 collected from river Ganga, and coarse sand with d_{50} = 0.60 mm and σ_g = 1.72 from river Yamuna near Chambal. A glass cylinder as a pier model is employed for scouring studies. Experiments on scour for duration of 10 hours of continuous flow were carried out in the laboratory flumes. Scour depth H_s, horseshoe vortex size and rotation speed were measured during the scouring process using vortex size D_v, and vortex rotation speed N, the vortex velocity V_θ and vortex strength Γ were computed and non-dimensionalised. Fig. 12 shows the vortex size variation $Y_1 = D_v/h$ plotted against $X_1 = \left((H_s + 0.1D)/h\right)\left(\frac{D}{h}\right)^{1.43}$. Here the origin is located at the vortex centre ie.0.1D above the bed level before scour. From the statistical analysis of the data, a relation between these variables is written as:

$$\frac{D_v}{h} = 0.54\left[\left(\frac{H_s + 0.1D}{h}\right)\left(\frac{h}{D}\right)^{1.43}\right]^{0.41} \tag{4}$$

$$\text{or} \qquad Y_1 = 0.54\ X_1^{0.41} \tag{5}$$

$$X_1 = \left(\frac{H_s + 0.1D}{h}\right)\left(\frac{D}{h}\right)^{1.43} \tag{6}$$

The vortex rotation N, non-dimensionlised with D and U_o is plotted in Fig. 13. large scatter in Fig. 12 and Fig. 13 can be attributed to the unsteady nature of turbulent horseshoe vortices. From statistical analysis, the following expressions were determined.

$$Y_2 = \frac{\left[\dfrac{ND}{U_o}\right]}{\left[R_{eD}/\left(\dfrac{h}{D}\right)^{0.16} . F_D^{1.6}\right]^{0.94}} \tag{7}$$

$$\text{and} \qquad X_2 = \left[\frac{H_s + 0.1D}{h}\right] / \left[F_D^{1.15}\left(\frac{h}{D}\right)^{0.77}\right] \tag{8}$$

which are shown plotted in Fig. 13. The relation between them using regression analysis is obtained as

$$Y_2 = e^{-0.44\,X_2}\,(8.23 \times 10^{-6}) \tag{9}$$

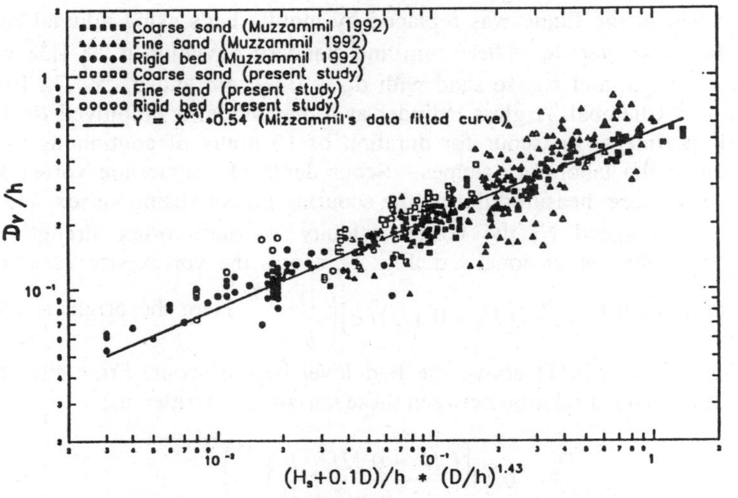

Figure 12. Variation of characteristic size of vortex with scour depth. (combined data for rigid and mobile bed).

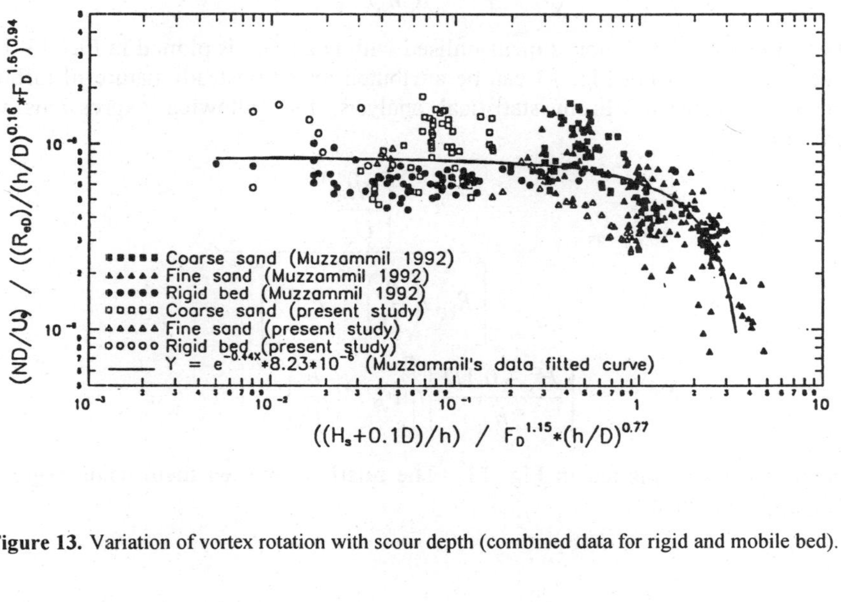

Figure 13. Variation of vortex rotation with scour depth (combined data for rigid and mobile bed).

It is observed that value of rotational speed decreases with increasing scour depth. Fig. 14 shows the computed values of vortex velocity non-dimensionlised as $\frac{V_\theta}{q}$ where q is flow intensity plotted $\left(\frac{D}{h}\right)^{1.43}\left(\frac{H_s+0.1D}{h}\right)$. The relation between vortex velocity and scour depth is obtained by regression analysis and is shown in Fig. 14.

$$Y_3 = \left[\frac{V_\theta D}{q}\right]\left[\frac{(h/D)^{0.16} F_D{}^{0.16}}{R_{eD}}\right]^{0.94} \tag{10}$$

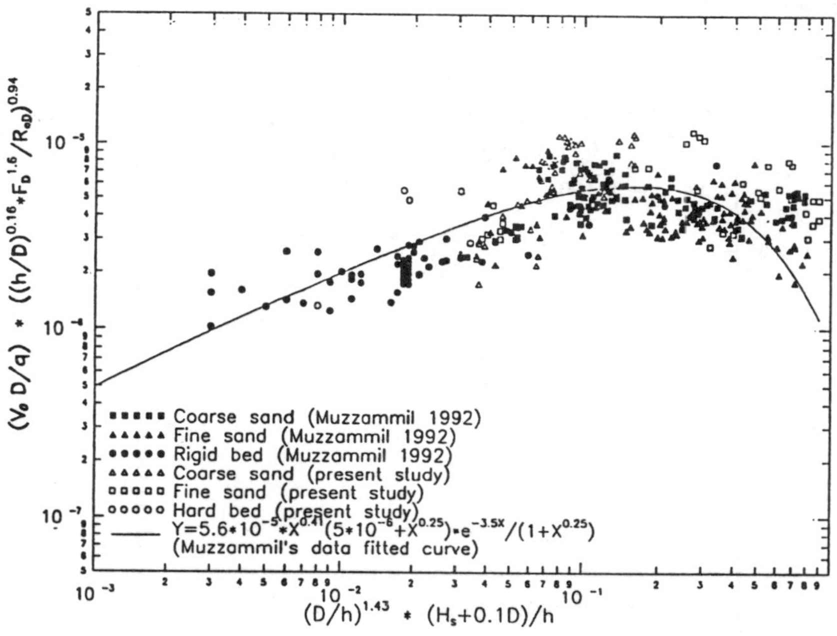

Figure 14. Variation of vortex velocity with scour depth (combined data for Rigid and mobile bed).

and
$$X_3 = \left(\frac{h}{D}\right)^{1.43}\left[\frac{H_s+0.1D}{h}\right] \tag{11}$$

$$Y_3 = \frac{\left(5.6*10^{-5}\right) X_3^{0.41} \left(5*10^{-6} + X_3^{0.25}\right) e^{-3.5X_3}}{\left(1 + X_3^{0.25}\right)} \tag{12}$$

It is observed that non-dimensional vortex velocity increases up to $X_3 = 0.2$ and with further increase in X_3, Y_3 decreases as can be seen in Fig. 14.

Vortex strength variation with scour depth for a particular flow is shown in Fig. 15. It is observed that the vortex strength increases steeply from its value before initiation of scour as soon as scouring process gets initiated, reaches a maximum and decreases with further increase in scour depth. This trend in the variation was observed for most of the flows whose initial value was known. Using this, the variation of vortex strength is studied by non-dimensionalising it with its maximum value and its position at which maximum value occurs. This variation is shown in Fig. 16.

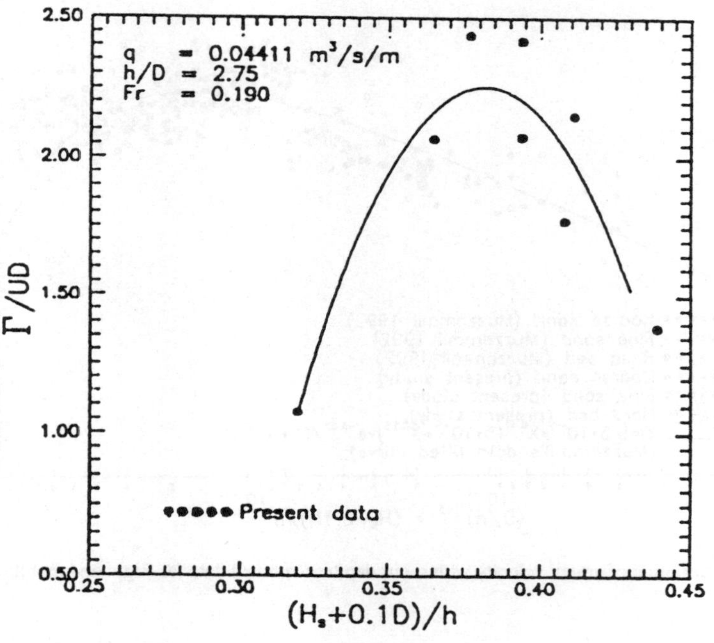

Figure 15. Variation of vortex strength with scour depth (mobile bed).

The maximum value of vortex strength Γ_{max} is plotted against depth of flow accounting for Reynolds number and Pier Froude F_0 number as shown in Fig. 17. Position at which this maximum value Γ_{max} occurs is shown in Fig. 18.

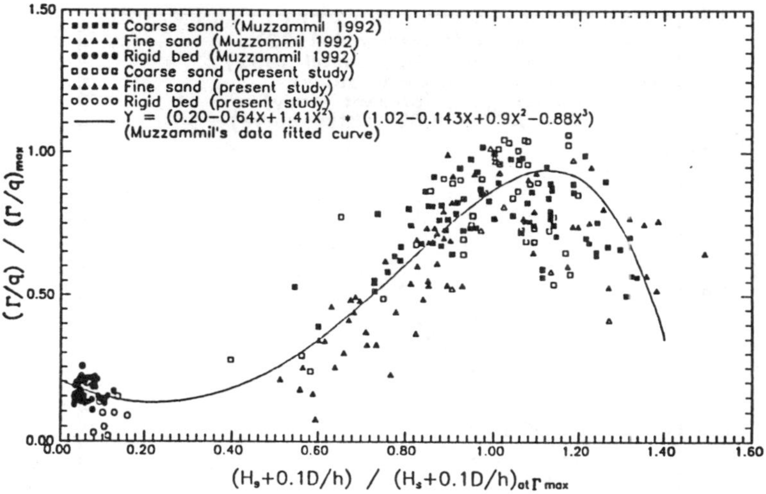

Figure 16. Variation of vortex strength with scour depth (Rigid bed and mobile bed data).

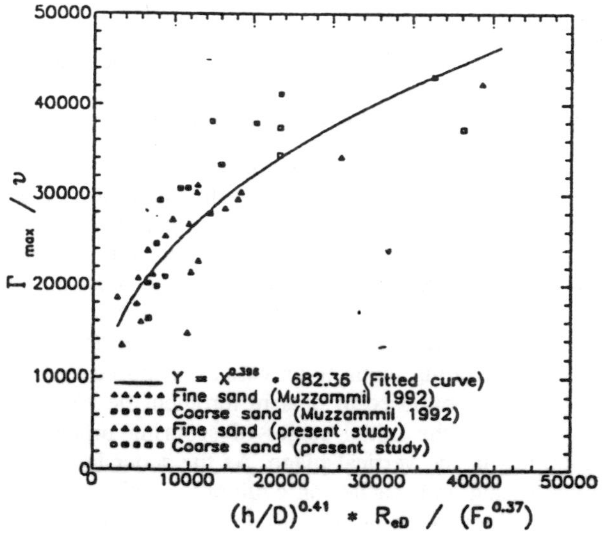

Figure 17. Variation of vortex strength scale Γ_{max} with hydraulic parameter.

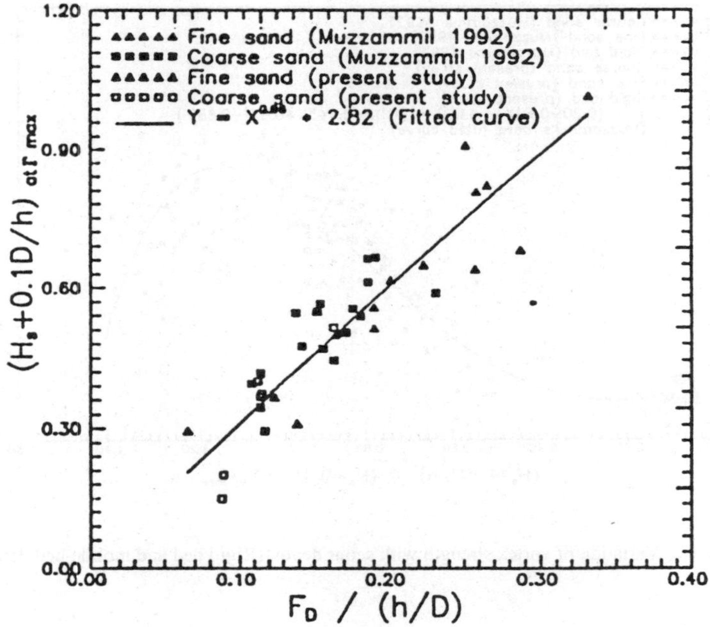

Figure 18. Variation of scour depth scale with hydraulic parameters.

6 Observations on Horseshoe Vortex Computations

The size and rotational speed of the vortex formed in the scour hole during scouring process were measured by making the vortex visible and locating a vortex probe in the centre of the vortex for its rotation and using a vortex gauge for its size measurements. The vortex velocity and vortex strength were computed based on the observed nature of the vortex as a forced vortex. The present study confirms the experimental observation of Melville [10] that the vortex strength increased during scouring process. However, the present investigation contradicts the assumption of constant vortex strength during scouring process as proposed by Baker [2, 3] for the laminar flow condition as opposed to present turbulent flow condition. Increase in vortex strength occurs in the initial stages of scouring processes due to inward curving of streamlines in the vertical plane. As the scour hole deepens the vortex strength starts decreasing and reaches an equilibrium value. Equilibrium condition is reached when the vortex velocity will not be able to lift and transport the sediment particle from its deepest portion, and the rate of erosion in the scour hole due to vortex interaction is equal to the rate of deposition.

The scatter in the results may be explained on the basis of the highly unstable nature of the vortex. The vortex probe position may be slightly different from vortex core when the vortex sways back and forth. Also, though it is assumed that vortex rotation N is an instantaneous value corresponding to that scour depth, actually it is an averaged value at a location spanned over a finite time period. The characteristic size of the vortex is based on observable physical position of the rim formed in the scour hole at the junction of primary and secondary vortex and also by making the vortex visible clearly, using mud visualization technique. Hence it would not introduce much variation.

7 Vortex Concept for Prediction of Scour Depth

Vortex velocity starts decreasing when the inward curving of streamlines remains constant i.e. the quantity of water flowing into the scour remains constant. This nature of variation can be seen in Fig. 14. The vortex velocity data collected during this time onwards were re-analyzed keeping in view that at the equilibrium stage of the scouring process the vortex velocity reached a limiting value which may be proportional to the critical velocity of sediment particle lying in the deepest portions of the scour hole. With this in view, V_θ is plotted as a function of scour depth and properties of the sediment as shown in Fig. 19. Fig. 19 indicates that the vortex velocity decreases with increase in scour depth having the following relation,

$$\frac{V_\theta D}{q} = \left\{ \frac{H_s}{D} \cdot \left(\frac{h}{D}\right)^{2.18} \left(\frac{V_{*c}}{\sqrt{\left(\frac{\Delta\rho}{\rho} g d_{50}\right)}} \right)^{8.25} \right\}^{-0.46} * 4.4 * 10^{-4} \quad (13)$$

At equilibrium stage V_θ attains a value V_{*c} , which is analogous to the stage of flow at which sediment in the scour hole is about to be dislodged from the bed. If V_θ is related to the threshold velocity U_c of the sediment bed upstream of the pier, a relationship for equilibrium scour depth may be obtained in terms of V_* which will be easy to use in practice. The magnitude of V_{*c} was obtained from Shield's criteria of initiation of sediment motion. V_{*c} represents the shear velocity at which sediment is about to move, and therefore represents the resisting tendency of the flow on the sediment particle in the scour hole. Thus it is reasonable to equate V_θ to V_{*c} at equilibrium stage.

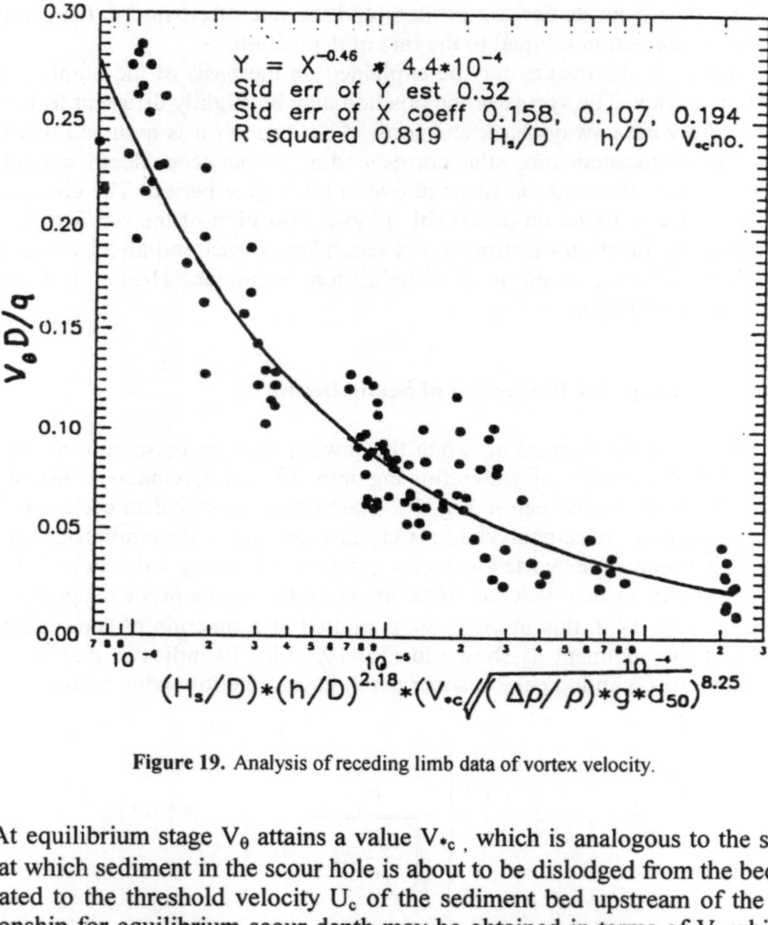

Figure 19. Analysis of receding limb data of vortex velocity.

At equilibrium stage V_θ attains a value V_{*c} , which is analogous to the stage of flow at which sediment in the scour hole is about to be dislodged from the bed. If V_θ is related to the threshold velocity U_c of the sediment bed upstream of the pier, a relationship for equilibrium scour depth may be obtained in terms of V_* which will be easy to use in practice. The magnitude of V_{*c} was obtained from Shield's criteria of initiation of sediment motion. V_{*c} represents the shear velocity at which sediment is about to move, and therefore represents the resisting tendency of the flow on the sediment particle in the scour hole. Thus it is reasonable to equate V_θ to V_{*c} at equilibrium stage.

8 Conclusions

Horseshoe vortex formed at the junction corner of a pier with rigid and mobile bed is studied. To make this horseshoe vortex visible for measurements a special technique called mud flow visualization technique was developed and used successfully. Vortex rotation and its size were measured using specially designed and fabricated vortex meter and vortex gauge, respectively. The imprints of flow separation and horseshoe vortex on the rigid bed were obtained using a wet paint technique developed indigenously. Using the above measurement techniques, horseshoe vortex size, rotation and its location were measured on rigid beds and on mobile bed during the scouring process. It has been observed that the size, velocity and strength of horseshoe vortex on rigid bed remain fairly constant when normalized in terms of pier diameter D, and upstream flow velocity U_0. In contrast the horseshoe vortex size increases with increase in scour depth for mobile bed conditions. Horseshoe vortex rotation speed 'N' decreases with increase in scour depth. Vortex strength increases initially, reaches a maximum value and decreases further with increase in scour depth. These results on horseshoe vortex behavior during scouring process will help in the better understanding of the scouring process.

9 Acknowledgements

Author's wishes to record their sincere thanks to INCH (Indian National Committee for Hydraulics) for providing financial assistance in the form of a research project on local scour protection around bridge piers and abutments.

References

1. Ahmed I. and Rajaratnam N., Flow around bridge piers, *J. of Hydraulic Engg.*, **124** (1998) pp. 288-299.
2. Baker C. J., Laminar horseshoe vortex, *J. of Fluid Mechanics,* **95** (1979) pp. 347-367.
3. Baker C. J., Theoretical approach to prediction of local scour around bridge piers, *J. of Hydraulic Research*, **18** (1980) pp. 1-12.
4. Belik L., Secondary flow about circular cylinders mounted normal to a flat plate, *Aeronautical Quaterly*, **24** (1973) pp. 211-252.
5. Devenport W. J. and Simpson R. L., Time dependent and time averaged turbulence structure near the nose of wing-body junction, *J. of Fluid Mech.*, **210** (1990) pp. 23-55.
6. EcKerle W. A. and Longston L. S., Horseshoe vortex formation around a cylinder, *J. of Turbomachinery,* **109** (1987) pp. 278-285.
7. Graf W. H. and Yulistiyanto., Experiments on flow around a cylinder, the velocity and velocity fields, *J. of Hydraulic Research*, IAHR, **36** (1998) pp. 637-653.
8. Gupta A. K., Experimental investigations of boundary layer flow past a circular cylinder mounted on a flat plate (M. Tech. thesis, Department of Civil Engg., I.I.T. Kanpur, India, 1984).

9. Hjorth P., Studies on the nature of local scour (Dept. Of Water Resources Engg., University of Lund, Bulletin No.**6**, 1975).
10. Melville B. W. and Raudkivi A. J., Flow characteristics in local scour at bridge piers, *J. of Hydraulic Research,* IAHR, **15** (1977) pp. 373-380.
11. Muzzammil M., Characteristics of horseshoe vortex at cylindrical bridge pier models (Ph.D. thesis, Department of Civil Engg., I.I.T. Kanpur, India, 1992).
12. Muzzammil M., Experimental investigation: Open channel flow past a circular cylinder mounted on rigid bed and on mobile bed (M.Tech. thesis, Dept. of Civil Engg., I.I.T. Kanpur, India, 1985).
13. Qadar A., The vortex scour mechanism at bridge piers, *Proc. Inst. of Civil Engg.*, **71** (1981) pp. 739-757.
14. Setia B., Scour protection around bridge piers and abutments (Ph.D. thesis, Dept. of Civil Engg., I.I.T. Kanpur, India, 1998).
15. Srivastava P., Strength of the horseshoe vortex (M.Tech. Thesis, Department of Civil Engineering, I.I.T. Kanpur, India, 1989).

WASTEWATER RENOVATION USING SOIL-AQUIFER TREATMENT SYSTEM

C. S. P. OJHA
Civil Engineering Department, University of Roorkee, Uttranchal - 247 667, INDIA
E-mail: cojhafce@rurkiu.ernet.in

The soil aquifer treatment (SAT) concept combines a series of processes, viz. physical, chemical, and biological for the removal of organic matter present in the wastewater. It involves use of the soil and the aquifer as treatment media with controlled passage of wastewater through the unsaturated (vadose) and saturated zones and its subsequent recovery from the aquifer. Most of the treatment is achieved in the vadose zone. The process has received wide recognition in developed countries. However, its use in India is not yet established. Present study compiles experiences gained with SAT system in various countries. Based on laboratory and field studies, it is inferred that SAT is an effective alternative wastewater treatment process and produces an effluent better than produced by a secondary (biological) process. SAT systems should be considered appropriate where subsoil and ground water conditions are favorable.

1 Introduction

With the growing quantum of wastewater and increasing stress on the available water resources in any region, there is a growing consensus on the uses and recycle of treated wastewater. Many of the wastewater treatment units such as Activated Sludge Process, Upflow Anaerobic Sludge Blanket Reactor Systems and other anaerobic digestion units have been in use but these are not cost effective. At the same time, these require specific maintenance. Probably, for these reasons, use of soil aquifer treatment system has been exploited in many developed countries including USA, Israel, to name a few.

Fig. 1 shows a schematic diagram of SAT system. It basically consists of an infiltration basin. The waste water is stored in the infiltration basin where a part of the impurities are trapped and lead to the development of a biofilm on the bed of the infiltration basin. The biofilm may further arrest the incoming impurities and develop with time. A certain part of the impurities may be arrested in the upper layers of the bed of the infiltration basin. It is also possible that many impurities including microbial population may travel deep into the underlying aquifer. Thus, in SAT system, it is important to know the penetration of different types of impurities. In fact, such a penetration depth will depend on the type of bed material including its properties, viz. conductivity, adsorption capacity. Also, the removal of different types of impurities, the allowable loading rate and the sequence of operation of such

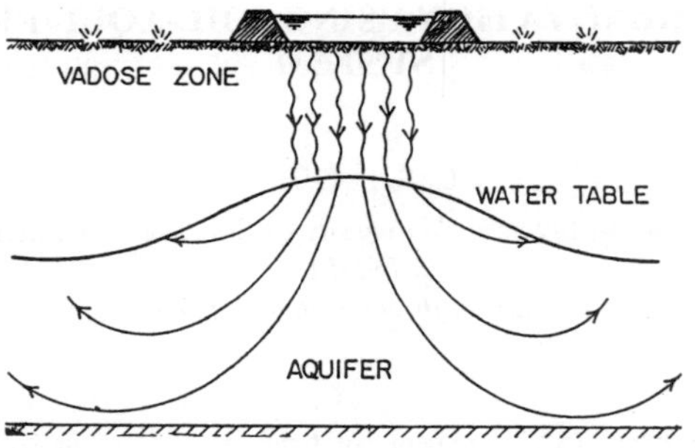

Figure 1. Schematic diagram of an SAT system

Units may be of relevance to the users of such systems. Some of these issues have been the subject of several investigations in the past. The objective of this article is to look into these investigations and to assess the potential of waste water renovation using SAT system for a variety of effluent waters, viz. primary, secondary and tertiary effluents or their combinations. The classification of effluents into primary, secondary and tertiary basically represents the level of treatment of a given wastewater. Generally, primary treatment is considered to be physical in nature and thus, a primary treated wastewater is having higher pollution potential compared to secondary or tertiary treated effluents. In terms of quality, secondary effluent lies in between the primary and tertiary treated effluents and normally results after a combination of physical and chemical/biological treatment. The source of such effluents may be municipal or industrial wastewater.

SAT system was initially used for the treatment of secondary effluents [4, 5]. Later, the process was extended to treat primary, secondary as well as tertiary effluent waters or their combinations. Thus, to present work done on SAT system, the following sequence is preferred: (i) treatment of secondary effluents, (ii) treatment of primary/secondary effluents, (iii) treatment of secondary/tertiary effluents, (iv) treatment of municipal waste waters, (v) removal of harmful organic pollutants, and (vi) use of SAT as a recharge system.

2 Treatment of Secondary Effluents

Bouwer *et al.* [4, 5] reported results of a 5-year pilot study on high rate land treatment of conventional secondary effluent for unrestricted irrigation and recreation at Phoenix, Arizona. In this study, it was observed that: (i) infiltration rates decreased linearly during flooding and were restored during drying of the basin; and (ii) the hydraulic conductivity of the aquifer was not measurably affected by groundwater recharge with sewage effluent.

The filtration of the secondary effluent through the sand and gravel resulted in essentially complete removal of SS, BOD and fecal coliform bacteria. Almost all the fecal coliform bacteria were removed in the first 0.6 m (2 ft.) of the soil but further penetration was observed for the first ten days of a new flooding period following an extended dry period. No fecal coliforms were encountered after 91 m (300 ft.) of horizontal travel of the renovated wastewater. Sequence of short, frequent flooding, and drying periods of several days each, yielded essentially complete conversion of the nitrogen in the influent (mostly ammonia) to nitrate in the renovated water but no removal of nitrogen altogether. With flooding and drying periods of 2 weeks each, ammonium was adsorbed in the soil during flooding and nitrified and partially denitrified during drying. This yielded renovated wastewater with alternating low nitrogen levels and nitrate peaks and a net nitrogen removal of about 30 percent. Laboratory studies indicated that this performance could be doubled or almost tripled by adding organic carbon to the effluent by recycling a portion of the renovated water containing the nitrate peak through the soil or by reducing the hydraulic loading rate. Care should be taken that no more ammonium is adsorbed in the soil during flooding that can be nitrified during drying otherwise the ammonia content of the renovated wastewater would increase.

Phosphate removal was about 50% after 9.1 m (30 ft.) of underground travel. A 100 m lateral movement resulted in more than 90% removal of phosphate. Phosphate removal was generally due to gradual precipitation in the sand and gravel, probably as calcium phosphate.

Amongst heavy metals, copper and zinc concentrations were reduced by about 80% whereas cadmium and lead concentration remained about the same as the wastewater moved through the sand and gravel. Mercury levels were reduced from about 2 to 1.2 mg/L.

Total dissolved solids in the renovated wastewater increased up to 2 percent than that of the influent secondary treated wastewater. The pH of the influent and renovated wastewater was around 7 and 8 respectively. In summary, the study demonstrated that a high-quality renovated wastewater suitable for unrestricted irrigation and recreation could be obtained with a rapid infiltration system in the riverbed.

It is well understood that the efficacy of SAT may decrease with longer runs. With this in view, Bouwer *et al.* [6] reported results of a 10 years field study at the experimental rapid-infiltration site (Flushing Meadow Project in the Salt River-Bed at Phoenix) as well as laboratory columns which received secondary treated wastewater for renovation. The maximum hydraulic loading was obtained with a water depth of 0.3m and flooding and drying cycle duration of 2-3 weeks and 10-20 days respectively. The soil aquifer underlying the riverbed did not show up any sign of clogging even after 10 years of operation, as no reduction in hydraulic loading was visible.

Nitrogen removal obtained varied from 30% to 65% depending on the hydraulic loading whereas phosphate removal increased with increasing distance of underground movement of wastewater and ranged between 40-80% at different hydraulic loading. Additional lateral movement of about 100m through aquifer increased the removal of phosphate to 94% (influent concentration, C_o = 9.1 mg/L). Like nitrogen, no decrease in phosphate removal efficiency of the soil aquifer system was seen after 10 years of operation.

The BOD and COD reduction were also significant. BOD was reduced virtually to zero in the renovated wastewater (C_o = 10-20 mg/L). Influent and effluent COD concentration were 30-60 mg/L and 10-20 mg/L respectively. Average TOC contents in the influent and effluent were 19.2 mg/L and 3.3-5.2 mg/L respectively indicating that the total removal of TOC was not achievable.

Viruses and fecal coliforms removal was almost total after varying vertical and lateral travel of wastewater through the soil aquifer. Simulated laboratory studies however showed that the rain water could remobilize previously adsorbed viruses in the soil during the first 24 hours of a drying period but the addition of $CaCl_2$ or drying for more than 24 hours prevented most of this desorption.

Bouwer and Rice [3] reported results of their study on wastewater renovation through SAT at the 23rd Avenue project. Good removals of nitrogen, phosphorous as well as bacteria and viruses were obtained and the SAT renovated wastewater conformed to regulatory standards for unrestricted irrigation and recreation. The removal of algae cells in the secondary effluent increased the hydraulic loading almost 5 folds.

3 Treatment of Secondary/Primary Effluents

Considering the potential of SAT for the treatment of secondary effluents, its utility was also examined for the treatment of combined waste waters, viz. secondary and primary effluents, secondary and tertiary effluents. The work in respect of two studies is presented here. Rice *et al.* [13] observed that higher nitrogen removal percentage could be achieved from primary effluent with a higher C/N ratio. Phosphate (PO_4-P) removal however, was almost same with primary effluent and secondary effluent both although their concentration differed. It was indicated the PO_4-P increased with the decrease in infiltration rate. Also, organic content of the primary effluent did not show any effect on PO_4-P removal indicating that phosphorus removal was probably a result of adsorption and chemical precipitation and the process of biodegradation had little influence on it.

Organic carbon removal occurred from the wastewater as it moved through the soil column. However, the organic carbon in the effluent from the columns was same for primary as well as secondary effluent.

The fecal coliforms (FC) concentration of the primary effluent decreased from 0.65×10^6 – 23×10^6 FC/100 ml to 0-186 FC/100 ml as it moved through the soil. Virus adsorption for primary effluent resembled with that of secondary effluent.

Rice and Bouwer [18] demonstrated the wastewater renovating efficiency of soil aquifer treatment using primary treated effluent and compared the findings with secondary effluent. It was shown that the quality of renovated water with primary treated effluent was equal or even better than that of secondary treated wastewater as influent. It was concluded that soil system could effect removal of high organic load, and bacteria and viruses in the primary treated effluent. The infiltration capacity of the system depended on the hydraulic conductivity of the soil and the concentration of organic and inorganic fractions of the suspended solids in the infiltration wastewater. Superficial clogging of the soil layer due to high organics could be restored after drying or by scarifying the surface by raking or harrowing while the in-depth clogging caused by the fine inorganic solids could need annual deep cultivation. A hydraulic loading of 27 to 51 m/year was obtained with alternating flooding and drying duration of 1 week each in summer and 1 week wet followed by 2 to 3 weeks dry in winter. The hydraulic loading was 5 to 8% of the infiltration capacity of the soil with pure water and about 50% of the infiltration capacity of the soil capacity with the secondary effluent. Using primary effluent bypasses the expensive and energy-intensive secondary treatment process and provides renovated water at a lower cost, provided that the land is available at a reasonable price.

4 Treatment of Combined Secondary and Tertiary Waste Waters

Considering the use of SAT for the treatment of combination of secondary and primary effluents, studies related to the treatment of combinations of secondary and tertiary effluents also took place. As the tertiary treatment basically aims at the removal of microorganisms including viruses, many of the observations focus the removal of fecal coliforms, viruses, etc. Carlson *et al.* [8] made an assessment of the relative efficacy of soil aquifer treatment to provide tertiary or a combined secondary and tertiary treatment. A comparative study of the effect of mass loading and hydraulic loading on the performance of the system indicated that the hydraulic loading was more important of these two loading parameters. The observation was in agreement with the findings reported by Thomas and Bendixen [20] regarding role of hydraulic loading on effluent quality.

Powelson *et al.* [17] assessed the efficacy of soil-aquifer treatment in removal of viruses from secondary and tertiary treated sewage effluents. Two small basins were constructed at a recharge site. Infiltration rates ranged from 0.2 to 16.8 m/d and samples of renovated wastewater were drawn from depths of 0.3-6.08 m from the unsaturated zone. 37-99.7% of the viruses was removed at depth of 4.3 m. The study concluded that a greater removal could be expected for human enteric viruses due to their tendency to adsorb more to soil under conditions similar to the study.

Powelson and Gerba [16] carried out soil column batch studies using secondary sewage effluent to investigate the effect of virus type, effluent type and water saturation on virus transport. It was found that: (1) the transport of polio virus was significantly retarded compared to other viruses; (2) virus transport was not affected by effluent type; (3) virus removal was about three times more in unsaturated zone that saturated zone and column retardation of virus transport were only 0.8-8% of that predicted by adsorption coefficients determined from the batch studies.

5 Municipal Wastewater Treatment

The objective of presenting this section is to emphasize the use of SAT for the treatment of domestic/municipal wastewater. It is noted that the use of SAT has not been confined to the treatment of only municipal wastewater [21]. Municipal wastewater can be treated using SAT provided the wastewater is given a physical treatment in the form of sedimentation. Anderson *et al.* [1] demonstrated the ability of a soil-turf filter in removal of nitrogen from municipal wastewater. Removals of the order of 33% and 17% were achieved by soil processes and turf growth respectively. Nitrogen concentration of the wastewater influenced the application rate. The study concluded that a soil with higher clay content would considerably increase the degree of purification of the wastewater but decrease the volume of renovated water. The system was recommended for areas of water deficit where wastewater is disposed off into the river.

In the second part of the study [1], the efficiency of the soil-turf-filter in removing P, B, Na and Cl from wastewater was assessed. It was observed that chlorine is removed from wastewater by light-induced photodegradation. Boron is not retained by the soils; Sodium concentration as observed by low SAR leachate values was not hazardous. Phosphorous was removed by chemical precipitation and was found to be a function of loading rate as well as precipitative capacity of the soil. The wastewater reclamation system was recommended for the water deficit areas.

George *et al.* [9] studied the short-term effects of the application of municipal wastewater on crops, soil and ground water. Authors reported an overall improvement in the ground water quality during monitoring period.

Suzuki *et al.* [19] studied the behavior of nitrogen removal in the land application of wastewater including its mass balance. Clarified domestic wastewater was applied to three lysimeters. The results of the study showed the removal efficiency of NH_4-N and organic nitrogen more than 90% in soil water against the influent wastewater. However, the removal of total nitrogen was less than 50%. The concentration of NO_3-N in the soil water increased significantly.

6 Removal of Harmful Organic Pollutants

McCarty *et al.* [12] have shown that the principal processes that affect the movement and fate of organic compounds in infiltration system are volatilization, sorption and chemical or biological transformation. The effect of these processes on a specific compound depends on two factors namely: (1) physical and chemical properties of the compound, and (2) sub-surface environment.

Bouwer *et al.* [7] studied the efficiency of soil-aquifer treatment system for removal of potentially harmful organic pollutants which are present in trace quantities in the wastewater. It was shown that the volatilization was important for low-molecular weight volatile compounds as water moves through the infiltration basins and a decrease in concentration from 30 to 70% was found between the basin inflow and outflow.

The role of biodegradation was seen in reduced concentrations of TOC, dissolved oxygen and many other micropollutants in the percolated water from the basin. The biological process involved both oxic and anoxic conditions. The contribution of the adsorption process could be assessed through nonhalogenated aliphatic and aromatic hydrocarbons and the pollutants like ethyl benzene, napthalene, phenanthrene, etc. which exhibited a concentration reduction between 50-99% in the percolated water. Halogenated organic compounds and organic substances represented by TOX showed smaller reduction through soil passage.

The study suggested that although by treatment of wastewater through rapid infiltration, reduction in organic micropollutants by sorption and biodegradation take place, the possibility of some organic contaminants reaching the groundwater could not be ruled out. Adequate precaution is, therefore, necessary to meet such situation and system should be designed to localize contamination of the aquifer.

The travel of organic pollutants from the ground surface through the unsaturated zone to the groundwater table at a citrus grove irrigated by sewage effluent has been reported by Muszkat *et al.* [13]. It was observed that the pollutant concentration increased with depth at the irrigated site. On the other hand, pollutants concentration decreased with depth through unsaturated zone at the control sites which were irrigated by groundwater or received only precipitation. The downward mobility of the organics in increasing concentration was attributed to sewage itself. The study emphasized the need for due care while considering sewage as a readily available source of irrigation water especially in the water scarcity areas due to inherent risks associated with its use.

7 Sat as a Recharge Basin

Ho *et al.* [10] conducted a study at a ground water recharge site in Western Australia with the primary objective to study the improvement in the removal efficiency of fecal coliforms and nutrients (N and P) by amending the existing sand (Bassendean type) of the recharge basin with gypsum-neutralized red mud. Earlier studies with the Bassendean sand alone had shown that it was not efficient in removal of enteric bacteria, viruses, nutrients and heavy metals.

The results of the study showed that the bacteria and viruses removal was better in the amended sand, which was attributable to increased die-off rate due to lower infiltration rates. Simple filtration, i.e. straining and adsorption also contributed to the removal mechanism. Phosphorus removal was more than 80%. Adsorption was considered to be the predominating mechanism in phosphate-P removal.

The nitrogen removal was the maximum (~ 45%) when the primary effluent was used with the basin operating in flooding and drying mode. A higher degree of nitrogen removal appeared possible by optimizing the lengths of the flooding and drying periods. Adopting lower infiltration rate could also result in greater denitrification. An increase in total nitrogen removal from 30 to 70% by decreasing the infiltration rate from 120 to 70 m/year has been reported by Bouwer [2].

As mentioned earlier, the primary effluent alone instead of a mixture of primary and secondary effluent gave a greater nitrogen removal. Some breakthrough of fecal coliforms in the groundwater was observed initially but after about two weeks of operation, the removal attained consistency and the count never exceeded 200/100 ml. The study [2] further suggested that the primary effluent was preferable to secondary effluent for artificial recharge, which was in agreement with findings of the other workers [8, 18].

Kamarek *et al.* [11] reported data from the Dan Region wastewater reclamation project in Israel which has been in operation since 20 years and highlighted the performance in terms of removal of various pollutants including bacteria, viruses and heavy metals.

The SAT system consisted of recharge basin covering a total area of 42 ha and subdivided into sub-basins of 1.5-2.0 ha each. The basins were operated in alternating flooding (~ 1 day) and drying (~ 2-3 days) mode in order to ensure that the aerobic conditions prevailed in unsaturated zone and the underlying aquifer. The performance of the SAT system was monitored in terms of removal of organic pollutants, nutrients as well as bacteria and viruses. Irrigation water quality parameters were also determined simultaneously. The quality data showed that the SAT renovated wastewater was of high quality which was suitable for unrestricted irrigation, industrial uses and other non-municipal applications, viz. horticulture, landscaping, toilet flushing and recreational purposes. The role of SAT system as a seasonal and multiannual underground storage was also acknowledged by Ojha [15]. In a study done at IISc campus, Ojha [15] assessed the recharge potential of wastewater, which was treated primarily by an SAT system.

8 General Observations about Sat System

Despite limited studies on SAT, it is possible to make certain general observations as follows:

- The infiltration capacity of the system depends on the hydraulic conductivity of the soil and the concentration of organic and inorganic fractions of the suspended solids in the infiltration wastewater.
- A soil with higher clay content would considerably increase the degree of purification of the wastewater but decrease the volume of renovated water.
- Superficial clogging of the soil layer due to high organics can be restored after drying or by scraping the surface by raking or harrowing while the in-depth clogging caused by the fine inorganic solids could need annual deep cultivation.
- With flooding and drying periods of 2 weeks each, ammonium is found to be adsorbed in the soil during flooding and nitrified and partially denitrified during drying.
- Adopting lower infiltration rate results in greater denitrification.
- Primary effluent alone instead of a mixture of primary and secondary effluent gives a greater nitrogen removal.
- Adsorption is considered to be the predominating mechanism in phosphate-P removal.
- Use of primary effluent avoids the expensive and energy-intensive secondary treatment process and provides renovated water at a lower cost, provided that the land is available at a reasonable price.
- Primary effluent is preferable to secondary effluent for artificial recharge.

9 Facts for Future Investigations

From the reported review in this study, some of the observations, which are not fully resolved at this stage, can be described as follows:

- The removal of algae cells in the secondary effluent can increase the hydraulic loading almost 5 folds
- Virus adsorption for primary effluent resembles with that of secondary effluent. Virus removal is not dependent on the effluent type. One needs to quantify the minimum depth of penetration so that virus does not appear in the treated water.
- Organic carbon removal occurs from the wastewater as it moves through the soil column. However, the organic carbon in the effluent from the columns remains same for primary as well as secondary effluent.

- A comparative study of the effect of mass loading and hydraulic loading on the performance of the system indicates that the hydraulic loading is more important of these two loading parameters. This conclusion of [8] needs reappraisal.
- SAT system can work effectively for many years. However, this will require a certain permissible loading of suspended solids. This needs to be quantified.
- Planning of wetting and drying schedules of SAT systems remains flexible and needs to be studied considering the role of bio-film developing on the bed of infiltration basin.
- Extrapolation of lysimeter experiments to forecast the SAT performance is poorly understood. A theoretical approach based on filtration theory is needed to study the clogging of the soil media.
- In Indian context, the treatment using this process is yet to gain momentum [13]. Economic feasibility is needed to justify the use of SAT system at a larger scale.

10 Summary

This review has systematically presented the various studies on SAT system in other countries. It is apparent that SAT system has been used for the treatment of primary, secondary, and tertiary effluents as well as their combinations. However, in many instances the conclusions appear to be qualitative. The relevant statistics of various SAT systems is reviewed and summarized in Table 1. From the review of work done in this area, it is apparent that there have not been many studies to help the operation of SAT system. Although Table 1 presented here can be helpful for planning SAT studies in certain similar situations, it is not comprehensive enough to be used in a wide range of operating conditions. For the soil conditions different from those reported here, the use of lysimeter tests is a must. As the lysimeter tests are to be run with a higher loading rates in order to plan and execute a SAT system in minimum possible time, one may need to undertake such studies in future to develop lysimeter experiments and the associated hydraulic loading rates which could be used to forecast the performance of SAT systems for a pre-assigned duration of treatment. Much of the experimental findings remain qualitative and modeling of removal of various pollutants need to be quantified through SAT systems along with their impact on underlying ground water reservoir. With such studies in future, SAT is bound to have a great potential in India.

Table 1. Review of SAT Projects in Other Countries

Project - 1

Name of country	USA
Name of project	Flushing Meadows Project Arizona (1976)
Site	Salt Riverbed
Size of basins including sub-basin if any	6.1 x 213 m : 6 Nos. in parallel, Spacing 6.1 m
Type of effluent	Secondary treated municipal wastewater
Hydraulic loading rate	121 m/year (0.33 m/day)average; 151 m/year (0.41 m/d) maximum
Hydraulic conductivity of soil	1.21 m/d (445 m/year)
Duration of experiment	10 years
Wetting-drying cycle	Flooding 2-3 weeks : drying 10-20 days (winter/summer)
Infiltration rate	0.3 – 0.1 m/d
Location of GWT	3 m
Soil characteristics	1 m fine loamy sand + coarse + gravel layers up to 0.75 m followed by clay layer
Total infiltration	754 m
Removal of pollutants	
TDS	2% increase (influent TDS 1000-1200 mg/L)
BOD	~ 100%
SS	~ 100%
Org. Carbon	~ 65-100%
COD	~ 65%
Total N	~ 30-65%
Phosphorous (PO_4 – P)	~ 40-94%

Project – 1 *(contd.)*

Metals	Zn, Cu, Cd and Hg concentration in influent was quite low below permissible limits in drinking water and permanent irrigation except of lead which was 0.082 mg/L in influent and 0.066 mg/L in renovated waste-water (~ 20% R). Drinking water limit ~ 0.05mg/L and the limit of permanent irrigation 5 mg/L
Bacteria	C_o (FC) = $10^5 - 10^6$ ml/100 ml. During first 5 years of the project, C_e = 10-500 / 100 ml C_e = 0-1/100 ml with continued flooding
PH	Influent: 8; Effluent: 6.7 – 7

Remarks:

1. No decrease in removal efficiency after 10 years
2. Accumulation of organics and SS only in top 5 cm of soil below the infiltration basin after 10 years of operation.
3. Modeling of SAT system not attempted.
4. Column lysimeter studies not conducted.

Project - II

Name of Country	USA
Name of Project	Mesa Arizona (1977)
Site	Mesa wastewater treatment plant
Size of basins including sub-basin if any	3 x 9 m; 4 nos. in parallel spacing 6.1 m.
Type of effluent	Primary treated wastewater
Hydraulic loading rate	27-51 m/year
Hydraulic conductivity of Soil	---
Duration of experiment	2 years
Wetting-drying cycle	1 week – 1 week (summer) 1 week – 2 week (winter)
Infiltration rate	0.1 – 0.7 m/d
Location of GWT	---
Soil characteristics	Loamy sand to sandy loam
Total infiltration	---
Depth of liquid in infiltration basin	0.25 m

Remarks:

1. Modeling of SAT system not attempted
2. Column lysimeter studies not conducted.

Project - III

Name of Country	USA
Name of Project	23rd Avenue Rapid Infiltration Project (1975)
Site	Phoenix
	89 x 465 m (4 ha) 4 basins
Type of effluent	ASP treated secondary effluent (chlorination was added later)
Hydraulic loading rate	16-31 m/year (average 21.2 m/year)
Soil characteristics	Loamy sand/sand and gravel
Depth of GWT	3-20 m (average 15 m)
Hydraulic conductivity of soil	1.21 m/d (445 m/year)
Wetting-drying cycle	2 weeks – 2 weeks initially ; 1 week/4 week later
Depth of water in the basin	0.2 m
Removal of pollutants	
Total N	~33%
NH_4-N	~33%
NO_3-N	0.1-0.2 mg/l (increase)
Org. N	2-3 mg/l (increase)
Phosphate (O)	C_o : 4-10 mg/l, C_e : 0.2-0.4 mg/l
Fecal coliforms	$C_o = 1.8 \times 10^6$/100 ml, C_e = 22/100 ml (av.)
Viruses	C_e = 0.15 – 2 PFU/100 l
Organic Carbon	C_o = 8-15 mg/l, C_e = 1.9 mg/l

Remarks:

1. Modeling of SAT system not attempted
2. Column lysimeter studies not conducted

Project – IV

Name of Country	USA	
Name of Project	DAN Region project (1977)	
Site	Rolling sand dunes	
Size of basins including Sub-basin, if any	Total area: 29 ha., 4 basins. Out of the four basins one was used as test basin and subdivided into 5 basins.	
Type of effluent	Tertiary treated wastewater	
Hydraulic loading rate	200-250 m/year	
Duration of experiment	2.5 years	
Removal of pollutants		
	C_o(mg/l)	C_e(mg/l)
SS	15	0
BOD	8	<1
COD	60	10
$KMnO_4$ consumption (as O_2)	12	2
DOC	18	3.2
Amm. N	15	3
Total N	7-10	8
Phosphorous	1	0.02
PH	9.5	10.8

Remarks :

1. Modeling of SAT system not attempted
2. Column lysimeter studies not conducted

Acknowledgements

Author would like to thank his former Ph.D. student P. Nema, for his kind permission to reproduce certain material from his Ph.D. thesis. Author would also like to thank Prof. A. Kumar of University of Roorkee, and Prof. P. Khanna, former director, NEERI, for their useful input in the present study and his respected teachers and friends at IISc Bangalore for their encouragement to prepare this article.

References

1. Anderson E. L., Pepper I. L., Kneebone W. R. and Darke R. J., Reclamation of wastewater with a soil-turf-filter-II: Removal of phosphorus, boron, sodium and chlorine. *JWPCF* **53**(1981) pp. 1408-1412.
2. Bouwer H., Role of groundwater recharge in treatment and storage of wastewater for reuse. *Water Sci. Technol.* **24**(1991) pp. 295-302.
3. Bouwer H. and Rice R., Renovation of wastewater at the 23rd Avenue rapid infiltration project, *JWPCF* **36**(1984) pp. 76-83.
4. Bouwer H., Rice R. C. and Escaraga E. D., High-rate land treatment I: infiltration and hydraulic aspects of the Flushing Meadows Project, *JWPCF* **46**(1974a) pp. 834-843.
5. Bouwer H., Lance J. C. and Riggs M. S. (1974 b)., High-rate land treatment II : water quality and economic aspects of the Flushing Meadows Project, *JWPCF* **46** (1974b) pp. 844-859.
6. Bouwer H., Rice R. C., Lauce J. C. and Gilbert R. G., Rapid infiltration research at flushing meadows project Arizona, *JWPCF* **52**(1980) pp. 2457-2470.
7. Bouwer E. J., McCarty P. L., Bouwer H., and Rice R. C., Organic contaminant behaviour during rapid infiltration of secondary wastewater at the Phoenix, 23rd Avenue Project, *Wat. Res.* **18**(1984) pp. 463-472.
8. Carlson R. R., Linstedt K. D., Bennett E. R. and Hartman R. B., Rapid infiltration treatment of primary and secondary effluent, *JWPCF* **54** (1982) pp. 270-280.
9. George B. D., Leftwhich D. B., Klein N. A. and Claborn B. J., Redesign of a land treatment system to protect groundwater, *JWPCF* **59**(1987) pp. 813-820.
10. Ho G. E., Gibbs R. A., Mathew K. and Parker W. F., Groundwater recharge of sewage effluent through amended sand, *Wat. Res.* **26**(1992) pp. 285-293.

11. Kamarek A., Arohi A. and Michail M., Municipal wastewater reuse via soil aquifer treatment for non-potable purposes, *Water Sci. Technol.* **27**(1993) pp. 53-61.
12. McCarty P. L., Reinhard M. and Rittman B. E., Trace organics in groundwater. *Envir. Sci. Technol* **15** (1981) pp. 40-51.
13. Muszakat L., Raucher D., Magaritz M., Ronen D. and Auniel A. J., Unsaturated zone and groundwater contamination by organic pollutants in a sewage effluent-irrigated site, *Groundwater* **31**(1993) pp. 556-565.
14. Nema P., Wastewater renovation through soil aquifer system (Ph.D. Thesis, University of Roorkee, India, 1998).
15. Ojha C. S. P., Ground water study in a part of IISc Campus (M.E. dissertation, Civil Engineering, IISc, Bangalore, India, 1984).
16. Powelson D. K. and Gerba C. P., Virus removal from sewage effluents during saturated and unsaturated flow through soil columns, *Wat. Res.* **28**(1994) pp. 2175-2181.
17. Powelson D. K., Gerba C. P. and Yahya M. J., Virus transport and removal in wastewater during aquifer recharge, *Water Sci. Technol.* **27**(1993) pp.583-590.
18. Rice R. C and Bouwer H., Soil aquifer treatment using primary effluent, *JWPCF* **56**(1984) pp. 84-88.
19. Suzuki T., Katsuno T. and Yamaura G., Land application of wastewater using three types of trenches set in lysimeters and its mass balance of nitrogen. *Wat. Res.* **26**(1992) pp. 1433-1444.
20. Thomas R. E. and Bendixen T. W., Degradation of wastewater organics in soil, *JWPCF* **41**(1969) 808-813
21. Zoller U., Non-ionic surfactants in reused water: Are activated sludge/soil aquifer treatments sufficient?, *Wat. Res.* **28** (1994) pp. 1625-1629.

UNCERTAINTY CONCEPTS IN STREAM WATER QUALITY MANAGEMENT MODELS

P.P. MUJUMDAR
Department of Civil Engineering, Indian Institute of Science, Bangalore - 560 012, INDIA
E-mail: pradeep@civil.iisc.ernet.in

Uncertainties in water quality management stem from a number of factors such as parameter and scenario uncertainty, stochastic input variables and a broad range of possible alternative formulations, which may include the role of risks related to the violation of water quality standards. A large number of mathematical models have been developed incorporating various forms of uncertainties to aid water quality decision making to arrive at optimal allocation of the assimilative capacity of the river system. Recent interest in addressing uncertainty not only due to randomness but also due to fuzziness has lead to use of fuzzy optimization. This paper presents a brief review of the uncertainty concepts used in mathematical models for water quality management decisions. Within the broad category of water quality management models, only modeling of uncertainty in a particular class of problems called the waste load allocation (WLA) problems is reviewed and perspectives for future research directions are provided.

1 Introduction

Water quality management problems are characterized by uncertainties of various types at different stages of the decision making. A large number of mathematical models have been developed incorporating various forms of uncertainties to aid water quality decision making to arrive at optimal allocation of the assimilative capacity of the river system. This paper presents a review of the uncertainty concepts used in mathematical models for water quality management decisions. Within the broad category of water quality management models, only modeling of uncertainty in a particular class of problems called the waste load allocation (WLA) problems is reviewed. Waste load allocation refers to the determination of the required pollutant removal (or treatment level) at a collection of point sources to ensure that water quality standards are maintained throughout the receiving water body. Optimal waste load allocation implies that the treatment vector selected not only maintains the water quality standards, but also results in the best value for the objective function defined for the management problem.

Attempts to address the uncertainty in waste load allocation problems [36] preceded the development of deterministic models. The type of uncertainty in WLA problems that has received much attention is that due to randomness associated with various components of a water quality system. Most stochastic models consider streamflow as a random variable and may add as random variables, the reaeration rate, biodegradation rate, travel time, waste strength, initial DO and BOD concentrations [8, 9, 10, 16, 18, 19, 20, 33, 34]. Another type of uncertainty prominent in the management of water quality systems, and that is recognized in recent years, is the uncertainty due to imprecision associated with the standards and management goals related to water quality. Desirable and permissible water quality criteria, and minimal pollutant treatment levels are set up depending on the environmental objectives. In a majority of the cases, establishing these limits is not precise. Thus, the multiple objectives in a water quality system are not only conflicting but are also imprecise to some extent. A relevant question would be how to cope with the imprecision rather than how to eliminate it. Recent developments in modeling for water quality management have addressed the problem of imprecision through the use of fuzzy systems theory.

Traditional approach to incorporating uncertainty into the waste load allocation problem has been to choose one set of design conditions that includes a particular low flow value, such as the seven day average low flow with a 10-year return period (7Q10), and an extreme (high) water temperature value. More recent work has recognized that the complexity of the uncertainty within the water quality modeling process can not be readily captured by considering flow as the sole source of randomness. This has lead to an explicit representation of a wider array of uncertain variables such as the in-stream biochemical oxygen demand (BOD) levels and dissolved oxygen (DO) deficits, the deoxygenation, reaeration and sedimentation rates, and the nonpoint pollutant sources. Such efforts have resulted in the formulation of more complex optimization models [8, 16] and solution approaches that combine simulation and optimization models.

The present review is concerned only with water quality *management* models, and not with the broader area of water quality modeling, which addresses modeling of physical, chemical and biological processes governing the pollutant transport in streams. An exhaustive and lucid review of uncertainty concepts in water quality modeling has been presented by Beck [1] earlier. An in-depth description of the water quality management models is provided by Loucks [38], Thomann [59], Loucks *et al.* [39] and James [24]. A brief review on water quality management optimization models is also provided by Somlyody [53]. Early systems applications to waste load allocation include ReVelle *et al.* [45], Loucks *et al.* [37] and ReVelle *et al.* [48].

The basic mathematical models, pollution transport simulation models and the optimization models within which the simulation models are imbedded, usually employ reaches along a river, defined in such a way that the physical environment within a reach is homogeneous. The confluence of a tributary, the presence of a point source of pollution or a marked change in the physical environment typically define a new reach. Figure 1 shows a schematic of a waste load allocation problem indicating river reaches and dischargers (point sources of pollutants). Most models reviewed in this paper may be related to the general system shown in Fig. 1. Among all water quality indicators, the dissolved oxygen (DO) related aspects of water quality management is more popular among researchers for two reasons: First, the extremely important role that DO plays in water quality determination. There exists no better general indicator of water quality level. Second reason is the ease with which the BOD-DO response mechanism can be cast in a form directly amenable for use in a linear optimization model.

The trend of development suggests that in the area of water quality management there is a growing interest in a variety of methodological alternatives for incorporating uncertainty into the analysis of decision making. This review is accordingly divided into the following class of models based on the method of incorporating uncertainty: Chance Constrained Programming, Stochastic Dynamic Programming, Monte Carlo Simulation, Risk Equivalent Models, and Fuzzy Decision Models.

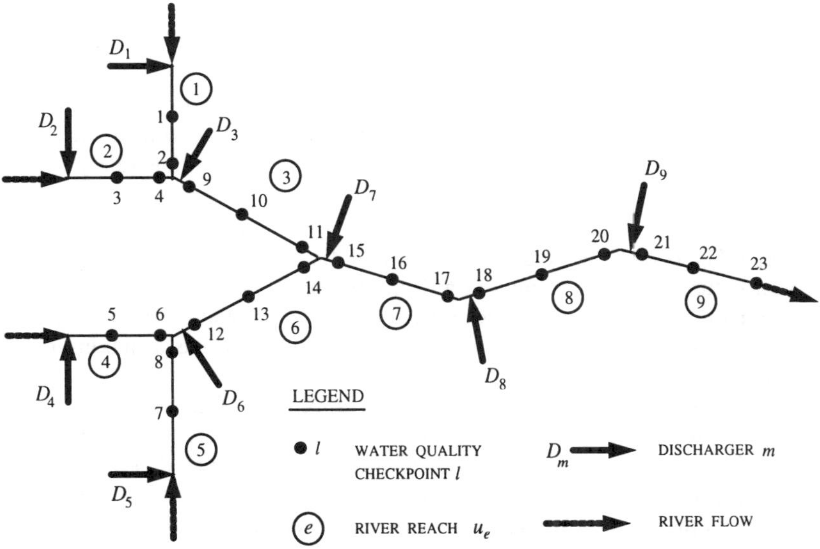

Figure 1. Schematic Diagram of the Stream Water Quality System

2 Chance Constrained Programming (CCP)

In the chance constrained optimization approach, constraints that ensure acceptable water quality are formulated as probabilistic constraints. To each of these probabilistic constraints is assigned an acceptable reliability level that must be achieved. Then, for a given design set of reliability levels, the model is transformed to an equivalent deterministic optimization model.

A general form of the chance constraint, $P[\mathbf{A}\ \mathbf{X} \leq \mathbf{b}\ ] \geq \alpha$, where **A** is a mxn coefficient matrix, **X** is a nx1 decision vector and **b** is a mx1 vector of right-hand side elements in the constraints, is expressed in the water quality management context [16] as,

$$P[(QE_{ij} + \sum_{k=1}^{i} q_k e_{ij}^{k}) / (Q + \sum_{k=1}^{i} q_k) \geq D_{max}] \leq \beta$$

where, Q is the streamflow, E_{ij} is the dissolved oxygen deficit in the stream due to the action of streamflow alone (i.e., not including deficit due to point source effluents), e_{ij}^{k} is the in-stream deficit in reach i, check point j due solely to effluent k, D_{max} is the stipulated maximum DO deficit and β is the pre-assigned probability. The term on the left hand side of the inequality within the square brackets, viz., $(Q\ E_{ij} + \Sigma\ q_k\ e_{ij}^{k}) / (Q + \Sigma\ q_k)$ indicates the DO deficit at the check point j in reach i. The exact nature of this term may vary from one problem to another, depending on the mathematical model used to describe the BOD-DO relationships. The limit on the summations, viz., k=1 to k=i, indicates that each reach has a maximum of one discharger.

Under chance constrained optimization approach, functions that represent pollution control cost, reliability of pollution control decisions or other policy issues, such as equity among waste dischargers, are optimized. For pre-assigned reliability levels, the chance constrained problem is transformed to an equivalent deterministic optimization models. For this approach, characterization of the joint probability distribution of the coefficients and the right hand side constants of the constraint sets is required. Because of limited stream and water quality information for many river systems, however, it is often difficult to describe the joint probability distribution of the coefficients. This approach, therefore, usually requires simplifying assumptions about the input information and distributions of the coefficients of the constraints.

Lohani and Thanh [33] introduced an innovative approach for waste load allocation in a stochastic environment. The objective of the proposed stochastic programming model was to minimize the cost of waste water treatment subjected to a chance constraint for the reliability of meeting water quality standards. The solution of the deterministic equivalent of chance constrained optimization problem gave the optimum fraction removal levels so that the entire assimilative capacity of a river system might be used. The developed model was applied to Hsintien river in Taiwan. Burn and McBean [8] developed a stochastic optimization model, to maximize the total sum of Dissolved Oxygen (DO) concentration at given check points subjected to a budget limit for treatment cost. An advantages of this formulation is that its dual problem is a chance constraint problem that can be directly converted to an equivalent deterministic problem. The model considers deoxygenation rates, reaeration rates, flow travel times, and Biochemical Oxygen Demand (BOD) loading as random variables. For conversion to a deterministic equivalent, the authors employ first order uncertainty analysis to predict the means and variances of functions of random variables. The model can cope with a wide range of uncertainties.

Fujiwara *et al.* [18, 19, 20] developed chance constrained optimization models for the problem of finding optimal waste load allocation for a river system while maintaining the risks of violation of water quality standards in different reaches below the prescribed bounds. Fujiwara *et al.* [18] developed a recursive procedure to express the dissolved oxygen deficits at a water quality monitoring points in a river system in terms of the fraction removal levels of the pollutants discharged into the river system. This model does not require prior knowledge of initial DO deficits at the beginning of a reach, which are necessary in the Lohani and Thanh [33] model. Streamflow is treated as a random variable with known probability distribution. The developed procedure was applied to the Hsintien river Taiwan. This procedure was extended by Fujiwara *et al.* [19] to a chance constrained model, in which the flows in main stream and tributaries, and flow due to storm water are assumed as random variables to determine the most economical level of wastewater treatment at each discharger along a river. The model considered the risk of violation of the stream water quality explicitly. Fujiwara *et al.* [20] further developed a simulation based optimization model by modifying the Burn and McBean [8] formulation. This model uses the concept of Monte Carlo simulation technique instead of first order uncertainty analysis used by Burn and McBean [8] to handle numerous sources of randomness in the input parameters of the model. To maintain violation of water quality standards within maximum allowable probability levels, an iterative procedure of optimization and simulation is used. The chance constrained problem with equal weights on DO concentration at each check point is solved with a constraint on treatment budget. A simulation analysis is used to calculate the probability of violating DO deficit standards at each check point. If this probability is beyond an acceptable level at certain points, then weights on DO concentration at these points are increased and the chance constrained programming is solved with the new weights. This procedure is repeated until the solution is acceptable.

Ellis [16] developed a chance constrained stochastic programming model for water quality management of a river system. The model determines the least cost allocation of BOD fraction removal levels, subject to probabilistic restriction on maximum allowable instream DO deficit. Randomness associated with parameters and variables like river flow and effluent flow rates, reaeration rate, de-oxygenation rate, sedimentation rate, photosynthetic input-benthic depletion rate, and nonpoint source BOD input rate are considered in the problem formulation. A new chance constrained programming approach with variant and imbedded chance constraints is presented with an example application. This method imbeds a chance constraint within a chance constraint programming. A joint chance constrained formulation is also presented which illustrates the possibility for prescription of an overall system reliability level.

Burn and Lence [7] presented a novel approach for including uncertainty in water quality management. The uncertainty inherent in the water quality management process is accounted for with a set of "scenarios" for the transport and impact of a given pollutant. Each scenario represents a combination of hydrologic, meteorologic and pollutant loading design conditions. The scenarios are simultaneously included in a waste load allocation model for pollutant discharges that determines the required removal levels at the pollutant sources. This approach represents a departure from the probabilistic approach to including uncertainty in water quality management in that no statement as to the probability of exceedence for the collection of conditions that constitutes a scenario is required. Rather, the emphasis is on the selection of a solution to the waste load allocation problem that is in some way robust to the range of conditions that might occur in the water body. The uncertainty is considered to exist in the inputs and parameters of the model as opposed to the model structure.

Traditional chance-constrained programming (CCP) and simulation-optimization methods of incorporating input information uncertainty in pollution management models are unsuitable for complex river systems with several critical water quality segments. Characterization of the joint probability distribution of the management model is often difficult because stream information is limited and the model formulation is generally difficult to understand. To overcome such weaknesses, a multiple realization chance constrained water quality management model has been recently developed by Takyi and Lence [58]. The multiple realization model includes several scenarios of design conditions simultaneously in an optimization model to overcome weaknesses of chance constrained programming and simulation-optimization models, by not requiring the joint probability distributions of the stochastic model coefficients and by producing non-inferior solutions. Heuristic and neural network techniques are developed to reduce the computational time required to solve the multiple realization model, through identification and utilization of only potentially important stream and water quality information that influence the optimal solution. Results showed that the heuristic technique is computationally efficient when less than 1000 realizations are included in the model, while the neural network method was found to be suitable when several realizations are needed to adequately represent the stochastic water quality system. The work presents a new approach based on extensions of the multiple realization model [42, 63] and neural network applications [43] for evaluating the trade-off between economic cost and reliability of pollution control decisions under input information uncertainty. The simulation-optimization approach utilizes Monte-Carlo simulation or a long record of historical information to generate several possible scenarios of hydrologic-, hydraulic- and pollution loading conditions of the water quality system. An optimization model of a waste load allocation problem is solved for each of the scenarios of realizations of the water quality conditions. The solutions generated are analyzed to provide policy information to decision makers. Such analysis often involves generating cumulative probability distributions of the solutions on the basis of the magnitudes and ranks of the objective functions and decision variables. For some systems, this approach may produce inferior decisions at certain reliability levels [42, 55].

In a multiple realizations modeling approach a number of possible scenarios of the uncertain input information are generated on the basis of a Monte Carlo simulation or the history of water quality record, and incorporated into a single optimization model. For real system applications, several large optimization models that exact large computational resources need to be solved. The realizations included in the optimization model must adequately represent the stochastic nature of the anticipated stream and pollution discharge conditions. The solution of the optimization model based on all possible realizations corresponds to a nominal reliability of one, with respect to the realizations used to represent the stochastic water quality management system. An optimal solution at the next lower reliability level is obtained by solving the optimization model again with all but one particular realization. The next solution of the model includes all but two particular realizations. This process continues successively until the entire objective function and nominal reliability trade off is obtained. The CPU time required on most commonly available computers is quite high if a large number of realizations are required to adequately represent a stochastic environmental management system.

The fundamental difference between a simulation model and the Multiple Realizations model is that, at a given reliability level, the former solves an optimization model on the basis of a single realization while the latter includes a set of realizations in the optimization model. Although the CCP and multiple realization approaches can explicitly solve for the trade-off curve between the two objectives of total treatment cost and reliability, neither approach is efficient for solving nonlinear problems and CCP has the additional disadvantage of requiring simplifying assumptions about the input information and its distributions.

3 Stochastic Dynamic Programming (SDP) Models

Conventional waste load allocation schemes determine static treatment levels at individual point sources. Variable strategies allow for different operations, dependent on season, streamflow, temperature and current water quality levels [3, 21, 23, 44]. Variable treatment strategies are readily accommodated in a Dynamic Programming (DP) framework, but an important caution must be exercised: the operational strategies imply time variable discharges but the Markov transition matrix used in a SDP model assumes steady state conditions. This apparent temporal inconsistency is, however, a minor issue. Specifically if the travel time in a reach is considerably smaller than the time required for treatment process modification, then a steady state assumption is reasonable.

In the SDP models, the transition function becomes a one-step Markov transition matrix. That is, the probability of being in state s at stage n is independent of the state of the system at stages other than stage n+1. While the state at n may depend on external factors, such as the control applied, it only "remembers" one stage into the past. In the deterministic case, the Markov transition matrix collapses to a strict mapping of a state and control option at one stage to a single state at the next stage with a probability of one.

The recursive relationship between stage n and stage n-1 of the stochastic dynamic programming is written, for the water quality problem, as,

$$V^{n}(S) = \underset{x \, \varepsilon \, \mathbf{X}}{\text{Min}} \; [\, r(x) + \Sigma \, P[S^*/S,x] \, V^{n-1}(S^*) \,]$$

where $P[S^*/S,x]$ is the probability that the system will transition to state S^* at stage n-1, if the control x is applied to state S in stage n, $\mathbf{X}$ is a vector of possible controls, r(x) is the cost associated with control (e.g., treatment level) x and $V^n(S)$ is the optimized cost stage n for a given state S. The water quality state S may be defined by the concentration level of a water quality indicator (e.g., DO deficit). The state transformation to S^* in stage n to S in stage n-1 (in a backward recursion) is obtained by an appropriate pollutant transport model.

In the DP approach, generally, the river reaches correspond to the stages; water quality parameters are the state variables and control actions in the dynamic programming represent the treatment levels. Water quality simulation models connect the state-stage pairs, and minimum total system cost serves as the objective function. Lohani and Hee [35] used streamflow as a random variable in a DP model. Their study solved the waste load allocation problem using a chance constrained approach. The use of DP does not demand linearity, so that highly advanced water quality simulation models, such as QUALE2E (developed by the US Environmental Protection Agency, EPA) can be used to estimate the DO levels in streams. However, constraints and objective functions that necessitate looking across time periods are difficult to accommodate in the DP models. While these constraints can sometimes be incorporated as additional state variables, the corresponding increase in problem size soon becomes computationally intractable.

A stochastic optimization/simulation model was developed by Ellis *et al.*[18] for arriving at least-cost treatment sequences for a centralized liquid industrial waste treatment facility. The stochastic dynamic programming is used as a screening device identifying unit treatment processes, which were then analyzed with stochastic simulation. The aim of the model is to produce an acceptable effluent stream quality, given a probabilistically generated influent waste regime.

A particularly useful concept in analyzing uncertainty in mathematical models is the use of Type I and Type II uncertainty. Type I uncertainty is the result of the choice of an incorrect model which has correct (deterministic) parameters; Type II uncertainty results from the choice of the correct model with incorrect (or uncertain) parameters. In real situations, such a partition is seldom feasible. However, in the analysis of uncertainty when a given model is being analyzed for sources of uncertainty such a distinction is very useful [5]. Cardwell and Ellis [10] used stochastic dynamic programming for the solution of waste load allocation problem formulated in a stochastic framework. The optimization model employed water quality simulation models to determine the dissolved oxygen distribution. Parameter and model uncertainties were considered in the problem formulation. The uncertainty in the selected input parameters (Type II uncertainty) to the simulation models gave rise to the stochastic dynamic programming approach for the solution procedure. The effects of the model uncertainty (Type I uncertainty) on the control decisions were employed by incorporating several DO simulation models into the optimization procedure.

Takyi and Lence [56] developed a non stationary Markov chain model for derving a seasonal risk equivalent water quality management strategy. The Markov chain approach determines the risk of water violation based on the transition probability matrices for streamflows at adjacent gauging stations, a selected water quality management model, and the Baye's theorem of conditional probability. Unlike the Rossman [47] and Lence *et al.* [31] models (discussed subsequently under the section 'Seasonal Risk Equivalent Models'), which calculate the risk of water quality violations based only on a known and limited set of flow data for a given river basin, the Markov chain approach estimates the likelihood of other flow combinations at all gauging stations. A major advantage of a Markov chain approach is its ability to explicitly incorporate probabilistic relationships between any pair of spatial or temporal data, like the probabilistic relationship between reaeration rate and streamflow, streamflow and stream velocity, etc. An application of Markov chain approach is demonstrated for controlling the BOD wastes in an example river basin on the Willamette River in Oregon, USA.

Integration of water quantity and quality management is recently considered by deAzevedo *et al.* [14]. Emphasis is given to simulation based assessment of strategic planning alternatives through the combined use of water allocation and water quality routing. Uncertainty from temporal and spatial variability and inadequate data associated with model parameters is addressed. Water quality performance measures are stream standard compliance reliability, water quality index, spatial and temporal uniformity of water quality. Water allocation performance measures include total reliability, resiliency and vulnerability. Several management alternatives combining various reservoir release policies with differing levels of waste water treatment are considered.

4 Monte Carlo Simulation

Unlike in the explicit stochastic optimization techniques (such as the CCP and the SDP) where the probability distributions of the random variables are incorporated in the optimization problem, in the implicit stochastic approach the optimization model itself remains deterministic and the uncertainty is addressed externally through Monte Carlo simulation [65]. In this section, literature related to implicit stochastic models for waste load allocation that use a Monte Carlo simulation in conjunction with a deterministic optimization model is reviewed.

Poor output information from the coupled simulation-optimization models arise because of uncertainty from the following sources: (a) stochasticity of pollution discharge and background stream data, (2) insufficient stream and water quality data to evaluate model parameters, and (3) lack of perfect knowledge of pollution transport processes and simplification of the water quality management system [62]. Although some recent work [10] addresses uncertainty from source 3, generally there is a greater tendency to gain more knowledge about pollution transport mechanisms and to develop sophisticated water quality simulation models than to incorporate uncertainty of the transport processes into a decision-making framework.

Burges and Lettenmaier [5] used first-order analysis to estimate the uncertainty in a deterministic water quality model due to uncertainty in parameters like travel time of the pollutants, reaeration rate constant, and temperature. The method is applied to simplified Streeter-Phelps equations for DO and BOD modeling. Monte Carlo simulation technique is used to check the accuracy of the results obtained by the first-order analysis. The results shows that, uncertainty in the BOD decay rates and travel time dominates near critical time. Brutsaert [4] demonstrated the use of Monte Carlo simulation for water quality modeling. The model developed accounts for stochasticity of the input parameters. Triangular probability density functions are shown to be useful, in case insufficient information is available to define meaningful frequency distributions of input parameters. The model output is presented as probability distributions of stream quality parameters. Warwick [64] developed a first order uncertainty technique to quantify the relationship between field data collection and a modeling exercise. He compared results from a traditional Monte Carlo simulation approach with those of the first order analysis, using a simple Streeter Phelps DO model, and concluded that the observed bias (over estimation) of the first order approach was mostly caused by the filtering (or removal) of unrealistic outcomes within the Monte Carlo framework.

Shih [50] developed a simulation-optimization scheme for the determination of policies in regional water quality management subject to specific water quality standards. A series of simulation models describing the statistics of water quality control phenomena was developed. Stochastic quadratic programming has been used for optimization. A simulation model for the description of probabilistic nature of the stream quality is also developed. The developed models are applied to the San Antonia river basin in Texas, USA. Spear and Hornberger [54] extended a regionalized sensitivity analysis to study the control of systems for which there is a good deal of uncertainty in the mathematical model used to describe the system. The method is based on a binary classification of Monte Carlo simulation results as being either satisfactory or unsatisfactory. The method is applied to the problem of regulating the discharge from a lagoon with the object of preventing DO from falling below a predetermined standard. The results suggest that even modest levels of uncertainty in the process parameters can have a considerable effect on the performance of the control decisions. Joshi and Modak [25] used the seven state variable QUAL 1 water quality simulation model in a DP approach and added a Monte Carlo uncertainty component. Burn [6] used a combined simulation-optimization approach for waste load allocation in a river network with multiple dischargers. A Monte Carlo simulation model was used for generating a series of water quality responses that led to the formulation of a constraint set for the optimization model. Repeated solution of the optimization model yielded probability distributions for the waste load treatment cost and the fraction removal levels. This information is used for policy implementation.

Burn and Lence [7] proposed a multiple-scenario based model to include the uncertainty in the water quality management process. Each scenario represented a possible combination of hydrologic, meteorologic and pollutant loading design conditions. The four optimization formulations considered are minimization of maximum violation (MMV), minimization of maximum regret (MMR), minimization of total violation (MTV), and minimization of total regret (MTR). All these formulations minimize the respective quantities over all the water quality checkpoints and all possible scenarios. MMV model minimizes the maximum of violations of dissolved oxygen (DO) stan1dard. MMR model minimizes the maximum of violations of DO level from the DO standard relaxed by an amount equal to the violation of DO standard obtained when each scenario is solved individually using MMV model. MTV and MTR models minimize, respectively, the sum of water quality violations and the total regret at each of the checkpoints for all scenarios. All these models constrained the amount of money spent on treatment to be no more than the budgeted amount for waste treatment.

The First Order Reliability Method (FORM) has been shown [61] to provide a close approximation of the cumulative distribution functions for the probability of failure with respect to dissolved oxygen (DO) standards and generally requires less computational time as compared to Monte Carlo simulation [41]. Another advantage of the FORM approach is that information about the output sensitivity to uncertainty in the random inputs can be obtained. FORM is based on the principle that any engineering system can be expressed in terms of its demand (load) and capacity (resistance). In the water quality context, the pollution load and a particular water quality standard correspond to the system's load and resistance respectively [15, 28].

In the optimization model, the limit on reliability level is written as, $h(\mathbf{Y},\mathbf{Q_s},\mathbf{X})\geq\alpha$, where $h(\mathbf{X},\mathbf{R_s},\mathbf{Y})$ is the estimated reliability for a given vector of waste treatment levels, $\mathbf{Y}$, water quality standard at a critical location s, $\mathbf{R_s}$, and vector of random variables in the water quality response model, $\mathbf{X}$, and α is the specified reliability of the system. The reliability estimate is obtained from the general reliability software package RELAN.

Vasquez *et al.* [61] present an efficient approach for obtaining waste load allocation solutions that provide the optimal trade off between treatment cost and reliability. This approach links a Genetic Algorithm (GA) with the first order reliability method (FORM) for estimating the probability of system failure under a given waste load allocation. The GA-FORM optimization approach is demonstrated for the case study of managing water quality in the Willameter River in Orgeon. The random variables used to generate the reliability estimates include streamflow, temperature and reaeration coefficient values. The results indicate that the GA-FORM approach is nearly as accurate as the approach that links the GA with Monte Carlo simulation and is far more efficient. The trade off between total treatment cost and reliability is shown to be most sensitive to the uncertainty in the reaeration coefficient. This sensitivity also increases at increased reliability levels. In a recent work, Maier *et al.* [40] have shown the use of FORM in estimating the reliability, resilience and vulnerability of a water quality management system, and have demonstrated it with the case study of the Willamette River, Oregon. Traditionally these performance measures are estimated using simulation. When complex system-response models are used, simulation can be computationally intensive, especially when persistence among the data needs to be taken into account. The authors develop an efficient method based on the FORM to estimate these performance measures. Reliability, resilience and vulnerability are determined for different DO standards. The DO is simulated using the QUAL2EU water quality response model. The approach developed does not need Monte Carlo simulation for estimating the performance measures, and, the authors conclude that when the number of random variables is not too large, the FORM-based approach is likely to be more attractive than the use of Monte Carlo simulation when the system response model is computationally intensive.

5 Risk Equivalent Seasonal Models

Seasonal waste water discharge programs employ different effluent standards during different times of the year to take advantage of the variation in receiving water susceptibility to adverse impacts. These programs try to achieve the maximum economic benefits possible without increasing the risk of water quality impairment. A general procedure followed in the seasonal models consists of treating seasonal flow as a random variable, assuming Markov chain like behavior for the random variable between seasons and using a non-linear programming model to find seasonal discharge limits that minimize the waste water treatment while limiting the annual risk of water quality violations. The information needs of the risk equivalent seasonal waste discharge programs are generally greater than those of nonseasonal programs. While nonseasonal waste discharge programs are typically designed on the basis of critical stream conditions, seasonal programs require long historical records of stream data for all times of the year.

Rossman [47] developed a seasonal waste load allocation model to achieve the maximum economic benefits without violating water quality standards. This model describes an approach for designing risk equivalent seasonal discharge limits for single-discharger stream segments, where risk is defined as the probability of incurring one or more water quality violations in any given year. The work presents a practical technique, compatible with existing waste load allocation practice, that can be implemented using readily available streamflow and water quality data. The developed procedure is applied for a case study for controlling ammonia toxicity in the Quinnipiac River, Connecticut, USA.

Lence *et al.* [31] developed risk equivalent seasonal discharge programs for multidischarger system. This approach is similar to that of Rossman [47], but it also accommodates river segments with several dischargers. Two management objectives are proposed as a surrogate for minimizing the seasonally varying waste treatment effort: the minimum average uniform treatment and the maximum total discharge objective. The designed seasonal waste load allocation maintains risk equivalence with a non seasonal waste discharge program and optimizes one of these two objectives. Application of the model is demonstrated with the case of Willamette River basin in USA.

Wotton and Lence [66] presented a modified seasonal waste discharge program for managing BOD and DO in river systems that have ice covers during certain periods of the year. The program designs a set of seasonal uniform treatment removal levels such that the average percent removal over the year is minimized and the risk of water quality violation is equal to that which would occur under a nonseasonal waste discharge program. The uniform treatment levels during the ice covered period are evaluated by simulating water quality based on re-aeration coefficients that are nearly zero. A sensitivity analysis showed that the water quality responses resulting during the ice-covered period are not the most critical conditions, and that there is no real advantage in separating the ice covered period from other times of the year, for the case study of St. John River. Lence and Takyi [32] applied a modified regionalized sensitivity analysis applied for assessing the effect of unreliable stream records on the design of seasonal discharge programs. Uncertain input parameters include flow and temperature data at different times in the year and at different locations in the stream. It is argued that the degree to which uncertain stream conditions affect the management model outcome depends on the water quality goals and the length of the seasons examined. To provide robustness to the water quality management models, Takyi and Lence [57] used the Chebyshev criteria to develop direct regulation water quality management models that maximize excess water quality above the water quality goal at all checkpoints along the river. Such models may be robust (in terms of achieving water quality) to uncertainties in input information. The reserve capacity at a location in a river is defined as the excess water quality over the water quality goal at that location. The objective of the model is to maximize the minimum reserve capacity. Three formulations led to three different models. These models are called Maximize Minimum Reserve Capacity (MMRC) models. The MMRC1 model maximizes the minimum reserve capacity subject to water quality constraints and limits on waste discharge. The other two formulations include additional constraints on available budget and level of equity respectively. The MMRC formulations assign optimal fraction removal levels of waste so that the water quality is improved at the worst checkpoints. The developed models are demonstrated for controlling BOD wastes in an example river basin on the Willamette River in Oregon, USA.

Uncertainties in water quality management stem from a number of factors such as parameter and scenario uncertainty, stochastic input variables and a broad range of possible alternative formulations, which may include the role of risks related to the violation of water quality standards. Recent interest in addressing uncertainty not only due to randomness but also due to fuzziness has lead to use of fuzzy optimization. The following section briefly reviews the water quality decision making models that use the concept of fuzzy decision making.

6 Fuzzy Decision Models

The concept of fuzzy decision was first introduced by Bellman and Zadeh [2]. Imprecisely defined goals and constraints are represented as fuzzy sets in the space of alternatives. The confluence of fuzzy goals and fuzzy constraints is defined as the fuzzy decision. The fuzzy decision Z, is thus defined as the fuzzy set resulting from the intersection of a fuzzy goal F and a fuzzy constraint, C.

That is,

$$Z = F \cap C$$

Goals and constraints are treated identically in fuzzy optimization. Representing the fuzzy goals and constraints by fuzzy sets, F_i, i = 1,2,...n_F, the resulting fuzzy decision is defined as,

$$Z = \bigcap_{i=1}^{n_F} F_i$$

In terms of the corresponding membership functions, the resulting decision is,

$$\mu_Z(\mathbf{X}) = \text{Min} [\mu_{Fi} (\mathbf{X})]$$

The optimal solution **X*** is given by,

$$\mu_Z(\mathbf{X}^*) = \lambda^* = \max [\mu_Z(\mathbf{X})]$$

This is illustrated in Fig. 2 for a single fuzzy goal F and a single fuzzy constraint, C.

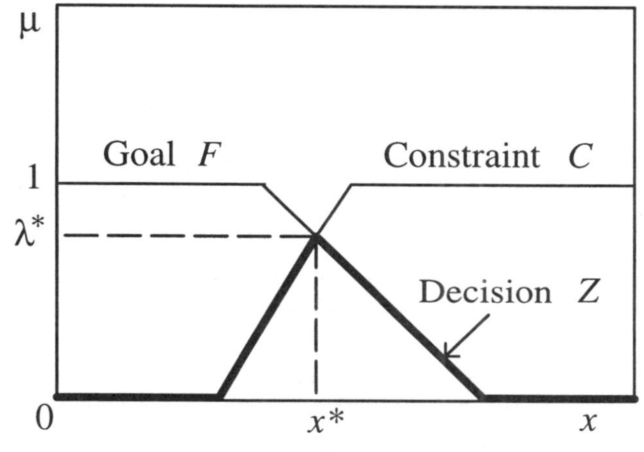

Figure 2. Fuzzy Decision

The space of alternatives, **X** (that is, the decision space) is restricted by precisely defined crisp constraints (e.g., minimum waste treatment level, maximum acceptable DO deficit etc.). Incorporating the crisp constraints, $g_j(\mathbf{X}) \leq 0$, $j = 1,2,\ldots n_g$, the fuzzy optimization model is written as follows:

$$\text{Max } \lambda$$

s.t.

$$\mu_{F_i}(\mathbf{X}) \geq \lambda \qquad \forall i$$

$$g_j(\mathbf{X}) \leq 0 \qquad \forall j$$

$$0 \leq \lambda \leq 1$$

Jowitt and Lumbers [26] used fuzzy sets to describe the water quality standards through linguistic description of water quality. For example, the pollution concentration has been represented by the linguistic terms like *high*, *medium* and *low* concentration, and the resulting water quality by *poor* and *good*. The linguistic terms *high*, *medium* and *low* are treated as fuzzy sets with appropriate membership functions and related to the linguistic description of water quality. For a particular event of pollution concentration, which is also described linguistically, the associated fuzzy description of water quality is obtained by the operation of composition between the fuzzy sets of concentration levels and the fuzzy relation of the concentration and the water quality. The combination of several linguistic statements relating the concentration levels and the associated quality is achieved through the union operation of the separate relations between the concentration levels and quality descriptions. An extension of this method is also presented that considers several pollutant indicators to define the water quality. Another early study on fuzzy sets in water quality management was reported by Koo and Shin [27].

Hathhorn and Tung [22] addressed the waste load allocation problem in a fuzzy optimization framework. Fuzzy optimization method provides a new methodology to solve the multiobjective optimization problems. For the waste load allocation problem, two objectives, namely, maximization of the waste discharge, and minimization of the largest difference in an equity measure between the various dischargers, are considered. The first objective function is derived by adding the sum of the decision variables representing effluent waste discharge and DO deficit at each discharge location for all the locations. For the second objective function, two forms of equity measures are considered in the study. First form of the equity measure is defined as the fraction removal level of the dischargers, and the second form is defined as the effluent concentration. The largest difference between the equity measures is defined as a decision variable which is greater than or equal to the difference of the equity measures of all pairs of dischargers in the system. Minimization of this largest difference forms the second objective of the waste load allocation problem. These two objectives are expressed as fuzzy sets with either linear or logistic membership functions. The multiobjective problem is then transformed to a fuzzy linear programming problem with constraints due to water quality requirements, definition of largest difference in equity, and the limitations on fraction removal levels.

Chang and Wang [12] addressed the solid waste management problem as a multiobjective problem formulated in a fuzzy framework. The three objectives related to the net management benefits, recycling targets, and anticipated combustion temperature during incineration are considered as fuzzy goals. The important constraints in the optimization problem are the financial constraints on capacity limitations of incinerators and shipping equipment, and the upper bounds on recycling of the waste. The imprecise objectives are quantified through the use of specific membership functions. The optimization problem has been solved using nonlinear fuzzy goal programming. The model is applied to a case study involving the solid waste management system for the Tainan city in Taiwan. The optimal solution obtained through the fuzzy optimization procedure could evaluate the optimal extent to which the various objectives are compatible and suggest the operational guidelines for solid waste management in the city.

Chang *et al.* [11] present a model for water pollution control in river basin that uses interactive fuzzy interval multiobjective programming. The objective of the work is to show how different uncertainties can be quantified and combined through the use of interval numbers and membership functions for water pollution control in a river basin. Uncertainty due to the environmental and economic parameters is modeled using interval numbers, and that due to the imprecise objectives identified by decision makers is modeled using fuzzy membership functions. The resulting interactive fuzzy interval multiobjective mixed integer programming approach is used to evaluate optimal waste water treatment strategies for pollution control in a river basin. Application of the model is illustrated with a case study for water quality control in the Tseng-Wen river basin in South Taiwan.

Sasikumar and Mujumdar [48] developed a fuzzy waste load allocation model for water quality management of a river system using fuzzy multiple-objective optimization in a deterministic framework. An important feature of this model is its capability to incorporate the aspirations of dischargers and pollution control agency. The imprecision associated with specifying the water quality criteria and fraction removal levels was modeled in a fuzzy frame work. The goals related to the pollution control agency and dischargers are expressed as fuzzy sets. The membership functions of these fuzzy sets are considered to represent the variation of satisfaction levels of the pollution control agency and dischargers in attaining their respective goals. The management problem is formulated as a fuzzy multiobjective optimization problem that can be solved using linear programming technique. Two formulations, namely, the *Max-Min* and the *Max-Bias* formulations are proposed for the waste load allocation.

Chen and Chang [13] proposed a new formulation of multiobjective mixed integer programming model with an integrated nonlinear graded membership functions as an alternative form of the objective function. Genetic Algorithm (GA) is used to solve the problem. Three objectives were identified in this work: maximization of assimilative capacity in the river, minimization of treatment cost for water pollution control, and maximization of economic value of river flow corresponding to recreation aspects. To determine the tolerance intervals of the membership functions, several model runs based on the planning scenario with single objective optimization are independently performed. The model application was demonstrated with the practical example of the Tseng-Wen river basin in south Taiwan.

By using fuzzy sets theory, failure events can be defined in a more flexible form than the usual crisp form. The concept of fuzzy probability [52] aids in deriving a new reliability analysis which includes subjective assessment of engineers without introducing additional corrections [51]. The concept of fuzzy random variables has been discussed by Kwakernaak [29, 30]. Sasikumar and Mujumdar [49] addressed the water quality management of a river system in a fuzzy and probabilistic framework. The two uncertainties, randomness and fuzziness, are treated simultaneously in the management problem. The event of low water quality at a check point in the river system is considered as a fuzzy event. The risk of low water quality is then defined as the probability of the fuzzy event of low water quality. A fuzzy set of low risks that considers a range of risk levels is introduced, defining appropriate membership functions. The concept of probability of a fuzzy event is used and the probability distribution functions of the DO deficit at various check points is computed from the solution of the fuzzy optimization model. The CDF, $F_{Cl}(c_l)$, of the DO deficit at the check point l is expressed as,

$$F_{Cl}(c_l) = P(C_l \le c_l)$$

Use of a pollution transport model, such as the Streeter-Phelps model to relate the DO deficit at check point l to the BOD loading at upstream locations, renders the DO deficit C_l to be expressed in terms of river flow, effluent flow and the treatment levels, as,

$$C_l = (A_l Q + B_l)/(Q+V_l)$$

where A_l is a constant depending on the BOD and DO deficit at the beginning of the first reach, B_l is a function of the vector **X** of fraction removal (treatment) levels, and V_l is a constant depending on the deterministic component of the total flow (river flow plus the effluent flow), at check point l.

The CDF of the DO deficit is then written as,

$$F_{Cl}(c_l) = 1 - P[Q \leq (B_l - V_l c_l)/(c_l - A_l)] = 1 - F_Q[(B_l - V_l c_l)/(c_l - A_l)]$$

where $F_Q(.)$ is the known cumulative distribution function of the streamflows.

Risk of low water quality is related to the distribution of the DO deficit and the memebership functions of the fuzzy sets of low water quality at different check points. To account for the uncertainty in the standards for determining a failure, occurrence of failure itself has been treated as a fuzzy event, in the paper. The fuzzy definition of low water quality ensures that there is no single 'threshold' value which defines a failed state. Indeed, all discrete water quality concentrations are treated as 'failures' of different degrees. The fuzzy set of low water quality maps all water quality levels to 'low water quality' and its membership function denotes the degree to which the water quality is low. The question, "How low is low water quality?" is being addressed through such a fuzzy set. A very high water quality, for example, will have a membership value of zero in the fuzzy set of low water quality. Similarly, through the fuzzy set of low risk, we address the question, "How low is low risk?".

The membership functions themselves are subjective statements of the perceptions of the decision makers. For example, the membership function for the low water quality indicates the perception of the decision maker of the degree of 'low quality', to a given level of water quality. A related question would then be, "How does one fix the lower and upper bounds for the membership functions?". These bounds are again subjective, and may depend on the particular problem being solved. To address such uncertainty in fixing the lower and upper bounds of the membership functions (which, it must be noted, have nothing to do with the *definition of failure*, but deal only with the *degree of failure*), the fuzzy membership functions themselves may be treated as fuzzy in the model and may be modeled using gray numbers [11].

7 Perspectives for Future Research

Almost four decades of developments in uncertainty modeling for water quality management have only opened up new avenues for incorporating uncertainties in models. In all management models, setting up standards, limits on constraints and even defining management goals and objectives are subjective exercises and introduce uncertainty due to imprecision. Up to early 90s probability has been the only uncertainty considered in water quality decision making. The implication of probability, as symbolized by the concept of randomness, is based on the 'chance' that exists in a real world event. Fuzziness, however, takes in another aspect of uncertainty, which represents ambiguity that can be found in linguistic description of a concept in decision making. In mathematics, the probability density function and membership function are used to illustrate the notion of 'probability' and 'fuzziness' respectively.

A vast repository of literature exists in modeling uncertainty due to randomness through the use of probability theory. In most models for water quality management, however, the only stochastic variable considered is stream flow. It has, for long, been realized that even for the simplest of the management problems of maintaining the DO deficits, many other parameters, most notably the BOD loading, reaeration and deoxygenation rate coefficients and temperature also vary randomly. Including randomness of these variables also in the models is often done through an implicit optimization model using Monte Carlo simulation (a notable exception being the imbedded chance constraint model developed by Ellis [16]). Explicit consideration of randomness in most variables influencing the pollutant transport would require understanding the degree of correlation between these variables and estimating their joint probability distributions. Both these by themselves are research topics worth pursuing, but inclusion of such joint probabilities in optimization models for decision making is a greater challenge that requires a significant computational effort, and development of specific algorithms for solutions will become necessary. While most decision models still focus on the BOD-DO models, with the progress in research in transport of other pollutants, such as the toxic metals, reactive chemicals etc., incorporating such models into the systems models for decision making will soon become a desirable necessity.

Applications of the fuzzy sets theory for modeling uncertainty due to imprecision in water quality management is relatively recent. For realistic applications, simultaneous consideration of both types of uncertainty, viz., the uncertainty due to randomness and that due to imprecision, in a single integrated model is critical. A challenging task would be to imbed the fuzziness in stochastic models. Development of the fuzzy systems theory has also opened up the question of precision, or indeed the lack of it, in our ability to assign probabilities to critical events. The representation of knowledge in conventional Bayesian decision analysis is in the form of precisely specified distributions and is the same no matter how weak the information source for this knowledge. A Bayesian analysis therefore may inadvertently impart too much precision to the input information and to the results. The concept of imprecise probability addresses this problem of excessive precision and a number of methods incorporating this concept have emerged A challenging research direction would be to use the concept of imprecise (or fuzzy) probabilities in the water quality management problems.

References

1. Beck M. B., Water quality modeling: a review of the analysis of uncertainty, *Water Resour. Res.* **23** (1987) pp.1393-1442.

2. Bellman R. E., and Zadeh L. A., Decision-making in a fuzzy environment, *Manage. Sci.* **17** (1970) pp. B-141 - B-164.
3. Boner M. C., and Furland L. P., Seasonal treatment and variable effluent quality based on assimilative capacity, *Res. J. Water Pollut. Control Fed.* **54** (1982) pp. 1408-1416.
4. Brutsaert W. F., Water Quality Modeling by Monte Carlo Simulation, *Water Resour. Bull.* **11** (1975) pp. 229-236.
5. Burges S. J., and Lettenmaier D. P., Probabilistic methods in stream quality management, *Water Resour. Bull.* **11** (1975) pp. 115-130.
6. Burn D. H., Water quality management through combined simulation optimization approach, *J. Environ. Eng.* **115** (1989) pp. 1011-1024.
7. Burn D. H. and Lence B. J., Comparison of optimization formulations for waste-load allocations, *J. Environ. Eng.* **118** (1992) pp. 597-612.
8. Burn D. H. and McBean E. A., Optimization modeling of water quality in an uncertain environment, *Water Resour. Res.* **21** (1985) pp. 934-940.
9. Burn D. H. and McBean E. A., Linear stochastic optimization applied to biochemical oxygen demand - dissolved oxygen modelling, *Can. J. Civ. Eng.* **13** (1986) pp. 249-254.
10. Cardwell H. and Ellis H., Stochastic dynamic programming models for water quality management, *Water Resour. Res.* **29** (1993) pp. 803-813.
11. Chang N. B., Chen H. W., Shaw D. G. and Yang C. H., Water pollution control in the river basin by interactive fuzzy interval multiobjective programming, *J. Environ. Div.* **123** (1997) pp.1208-1216.
12. Chang N. and Wang S. W., Managerial fuzzy optimal planning for solid waste management systems, *J. of Environ. Eng.* **122** (1996) pp. 649-658.
13. Chen H. W. and Ni-Bin Chang, Water pollution control in the river basin by fuzzy genitic algorithm-based multiobjective programming modeling, *Wat. Sci. Tech*, **37** (1998) pp. 55-63.
14. De Azevedo G. T., Gates T. E., Fontane D. G., Labadie J. W. and Porto R. L., Integration of water quantity and quality in strategic river basin planning, *J. Wat. Resour. Plng. Mgmt.* **126** (2000) pp. 85-97.
15. Duckstein L. and Bernier J., System framework for engineering risk analysis, *Proc. Engg. Found. Conf., Risk-based decision making in water resources,* New York (1986), pp. 90-110.
16. Ellis J. H., Stochastic water quality optimization using imbedded chance constraints, *Water Resour. Res.*, **23** (1987) pp. 2227-2238.
17. Ellis J. H., McBean E. A. and Farquhar G. J., Stochastic optimization/simulation of centralized liquid industrial waste treatment, *J. Environ. Engg.* **111** (1985) pp. 804-821.
18. Fujiwara O., Gnanendran S. K. and Ohgaki S., River quality management under stochastic stream flow, *J.Environ. Eng.* **112** (1986) pp. 185-198.
19. Fujiwara O., Gnanendran S. K. and Ohgaki S., Chance constrained model for river water quality management, *J. Environ. Eng.* **113** (1987) pp. 1018- 1031.

20. Fujiwara O., Gnanendran S. K. and Ohgaki S., River basin water quality management in stochastic environment, *J. Environ. Eng.* **114** (1988) pp. 864-876.
21. Hart J. E., Defore R. and Chiesa S.C., Activated sludge control for seasonal nitrification, *Res. J. Water Pollut. Control Fed.* **58** (1986) pp. 358-363.
22. Hathhorn W. E. and Tung Y. K., Bi-objective analysis of waste load allocation using fuzzy linear programming, *Water Resour. Mgmt.* **3** (1989) pp. 243-257.
23. Herbay, J. -P., Smeers Y. and Tyteca D., Water quality management with time varying riverflow and discharger control, *Water Resour. Res.* **19** (1989) pp. 1481-1487.
24. James A., An introduction to water quality modeling (Second edition, John Wiley and Sons, New York, 1993).
25. Joshi V. and Modak P., A reliability based discrete differntial dynamic programming model for river water quality management under financial constraint. In *Systems Analysis in Water Quality Management,* ed. by Beck, M. B., (Advances in Water Pollution Control, Pergamon Press, Oxford, 1987), pp. 15-22.
26. Jowitt P. W. and Lumbers J. P., Water quality objectives, discharge standards and fuzzy logic. In Optimal Allocation of Water Resources *Proceedings of the Exeter Symposium July 1982*, IAHS Publ. no: **135** (1982) pp. 241-250.
27. Koo J. K. and Shin H. S., Application of fuzzy sets to water quality management, *Water Supply* **4** (1986) pp. 293-304.
28. Kundzewicz Z. W., Renewal theory criteria of evaluation of water resource systems, Reliability and Resilience. *Nordic Hydrol.* Nordic Association of Hydrology, Denmark, **20** (1989) pp. 215-230.
29. Kwakernaak H., Fuzzy random variables-I. Definitions and theorems, *Information sciences* **15** (1978) pp. 1-29.
30. Kwakernaak H., Fuzzy random variables-II. Algorithms and examples for the discrete case, *Information sciences* **17** (1979) pp. 253-278.
31. Lence B. J., Eheart J. W. and Brill E. D. Jr., Risk equivalent seasonal discharge programs for multidischargers streams, *J. Water Resour. Plng. and Mgmt.* **116** (1990) pp. 170-186.
32. Lence B. J. and Takyi A. K., Data requirements for seasonal discharge programs: An application of a regionalized sensitivity analysis, *Water Resour. Res.* **28** (1992) pp.1781-1789.
33. Lohani B. N. and Thanh N. C., Stochastic programming model for water quality management in a river, *J. Water Pollution Control Fed.* **50** (1978) pp. 2175- 2182.
34. Lohani B. N. and Thanh N. C., Probabilistic water quality control policies, *J. Environ. Div.* **105** (1979) pp. 713-725.
35. Lohani B. N. and Hee K. B., A CCDP model for water quality management in the Hsintien River in Taiwan, *Int. J. Water Resour. Dev.* **1** (1983), pp. 91-114.

36. Loucks D. P., Surface water quality management models. In *Systems Approach to Water Management* ed. by Asit K. Biswas (McGraw Hill Kogakusha Ltd., New York, 1976).
37. Loucks D. P. and Lynn W. R., Probabilistic models for predicting stream quality, *Water Resour. Res.*, **2** (1966) pp. 593-605.
38. Loucks D. P., ReVelle C. S. and Lynn W. R., Linear programming model for water pollution control, *Manage. Sci.* **14** (1967) pp. B-166 - B-181.
39. Loucks D. P., Stedinger J. R. and Haith D. A., Water Resource Systems Planning and Analysis (Prentice Hall, Inc., Englewood Cliff, New Jersey, 1981) pp. 427-545.
40. Maier H. R., Lence B. J., Tolson B. A. and Foschi R. O., First-order Reliability Method for Estimating Reliability, Vulnerability and Resilience, *Water Resour. Res* **37** (2001) pp. 779-790.
41. Melching C. S. and Anmangandla S., Improved first-order uncertainty method for water quality modeling, *J. Environ. Engg.* **118** (1992) pp. 791-805.
42. Morgan D. R., Eheart J. W. and Valocchi A. J., Aquifer remediation design under uncertainty using a new chance constrained programming technique. *Water Resour. Res*. **29** (1993) pp. 551-561.
43. Ranjithan S., Eheart J. W. and Garrett J. H. Jr., Neural network based screening for groundwater reclamation under uncertainty, *Water Resour. Res.* **29** (1993) pp. 563-574.
44. Reheis H. F., Dozier J. C., Word D. M. and Holland J. R., Treatment costs saving through monthly variable effluent limits, *Res. J. Water Pollut. Control Fed.* **54** (1982) pp. 1224-1230.
45. ReVelle C. S., Loucks D. P. and Lynn W. R., A management model for water quality control, *J. Water Pollut. Control Fed.* **39** (1967) pp. 1164-1183.
46. ReVelle C. S., Loucks D. P. and Lynn W. R., Linear programming applied to water quality management, *Water Resour. Res.* **4** (1968) pp. 1-9.
47. Rossman L. A., Risk equivalent seasonal waste load allocation, *Water Resour. Res.,* **25** (1989) pp. 2083-2090.
48. Sasikumar K. and Mujumdar P. P., Fuzzy optimization model for water quality management of a river system, *J. Water Resour. Plng. and Mgmt.* **124** (1998) pp. 79-87.
49. Sasikumar K. and Mujumdar P. P., Application of fuzzy probability in water quality management of a river system, *International Journal of Systems Science* **31** (2000) pp. 575-591.
50. Shih C. S., Stochastic water quality control by simulation, *Water Resour. Bull.* **11** (1975) pp. 256-267.
51. Shih and Wangsawidjaja, Mixed fuzzy-probability programming approach, *Computers Ops. Res* **59** (1995) pp. 283-290.
52. Shiraishi N. and Furuta H., Reliability analysis based on fuzzy probability, *J. Engg. Mech.*, **109** (1982) pp. 1445-1459.

53. Somlyody L., Use of optimization models in river basin water quality planning, *Wat. Sc. Tech.* **36** (1997) pp. 209-218.
54. Spear R. C. and Hornberger G. M., Control of DO level in river under uncertainty, *Water Resour. Res.* **19** (1983) pp. 1266-1270.
55. Takyi A. K. and Lence B. J., Incorporating input information uncertainty in a water quality management model using combined simulation and optimization. (Paper presented at International UNESCO Symposium on Water Resources Planning in a Changing World, Karlsruhe, Germany, 1994).
56. Takyi A. K. and Lence B. J., Markov chain model for seasonal-water quality management, *J. Water Resour. Plng. and Mgmt.* **121** (1995) pp. 144-157.
57. Takyi A. K. and Lence B. J., Chebyshev model for water-quality management, *J. Water Resour. Plng. and Mgmt.* **122** (1996) pp. 40-48.
58. Takyi A. K. and Lence B. J., Surface water quality management using a multiple-realisation chance constraint method. *Water Resour. Res.* **35** (1999) pp. 1657-1670.
59. Thomann R. V., Systems Analysis for Water Quality Management (Mc Graw Hill, Inc., New York, 1972).
60. Upton C., A model of water quality management under uncertainty, *Water Resour. Res.* **6** (1970), pp. 690-699.
61. Vasquez J. A., Maier H. R., Lence B. J., Tolson B. A. and Foschi R. O., Achieving water quality stream reliability using genetic algorithms, *J. Environ. Engg.* **126** (2000) pp. 954-962.
62. Vicens G. J., Rodriguez-Iturbe I. and Schaake J. C. Jr., A Bayesian framework for the use of regional information in hydrology, *Water Resour. Res.* **11** (1975) pp. 404-415.
63. Wagner B. J. and Gorelick S. M., Reliable aquifer remediation in the presence of spatially variable hydraulic conductivity: From data to design, *Water Resour. Res* **25** (1989) pp. 2211-2226.
64. Warwick J. J., Use of first-order uncertainty analysis to optimize successful stream water quality simulation, *J. Am. Water Res. Assoc.* **33** (1997) pp. 1173-1185.
65. Whitehead P. G. and Young P. C., Water quality in river systems: Monte Carlo analysis, *Water Resour. Res.* **15** (1979) pp. 1305-1312.
66. Wotton C. L. and Lence B. J., Risk-equivalent seasonal discharge programs for ice-covered rivers, *J. Wat. Resour. Plng. Mgmt.* **121** (1995) pp. 275-282.

WATER RESOURCES AND THEIR MANAGEMENT FOR SUSTAINABLE AGRICULTURAL PRODUCTION IN INDIA

P.B. S. SARMA

Civil Engineering Department, Indian Institute of Technology, Delhi
INDIA
E -mail: pbssarma@hotmail.com

Water is an essential life supporting natural resource and plays a dominant role in agricultural production. The imperative need for producing the required food grains to meet the needs of increasing population calls for intensive agriculture using high yielding varieties, fertilisers, modern technologies and most importantly, irrigation. Recognising the importance of the role of water in increasing and stabilising agricultural production, irrigation facilities were developed at a rapid rate in the post independence era of Indian history. The gross irrigation potential was increased from about 22.6 m.ha in 1950-51 to 94.73 m.ha by1999-2000. As a result, the food grain production had a fourfold increase from 51 m. tons in 1950-51 to 206 m. tons in 1999-2000. This led to transformation of India from a country of ship to mouth existence to food grains exporting country, in addition to creation of food stocks of over 34 m. tons in 1999-2000. However, there is no scope for complacence. There is a need for producing as much as 325-350 m. tons of food grains by 2025 AD to meet the food requirement of the growing population. In the wake of the development of large-scale irrigation projects, several serious problems of degradation of soils, water, and environment cropped up threatening the sustainability of agricultural production. As we enter the next millennium, it is appropriate to take stock of the situation and analyse the strengths and weaknesses of our policies and practices and assess the opportunities and the threats and identify R&D efforts needed in irrigation water management to ensure sustained and increased agricultural production. The future scenario indicates that while availability of the water resources for agricultural sector will be considerably reduced, the quality of available water will also be reduced. This scenario dictates that the sustainability perspective underscores the water resource management in agriculture.

In this article, the irrigation development, and the associated agricultural production over the past fifty years was briefly reviewed to assess the role of irrigation in increasing and stabilising the food grain production in India. The future scenario of the agriculture as well as the water resource availability for agriculture is examined. The current R& D efforts were reviewed critically and the scope for maximizing water resource utilisation through adoption of scientific water management technologies such as water harvesting, conjunctive use, use of marginal quality of water, computer aided decision support systems for efficient water management, use of micro irrigation systems etc., were discussed through case studies. Future efforts required in R&D in improving productivity of agricultural lands are also identified.

1 Introduction

1.1 Irrigation and Agriculture

Water is an essential life supporting natural resource. Human population always settled and thrived in regions close to water bodies. It is the single vital input in agriculture and has a major contribution in ensuring stability, self-sufficiency, and sustainability of our food grain production. The importance of irrigation in increasing and sustaining higher levels of agricultural production was well recognised in India from time immemorium as evidenced from the following quotations:

> *"No grain is ever produced without water, but too much water tends to spoil the grain. An inundation is as injurious to growth as dearth of water"*
>
> --- Narada *Smrithi" XI.19*

Irrigation water not only aids germination of the seeds and emergence of seedlings but also all the other phases of plant growth and maintains thermal equilibrium of the plant. Water provides a means of transport of plant nutrients from the root zone to the plant body. In rain fed crops, one single life saving irrigation at the crucial stage of the crop growth makes all the difference of the crop's good production or total failure. A comparison of yields from irrigated and unirrigated crops clearly brings out the fact that irrigation increases the average productivity of crops (Table 1). Further, irrigation also helps stabilising the crop yields and mitigating the ill effects of drought.

Table 1. Impact Of Irrigation on Productivity of Crops--Average (State) Yields Of Irrigated And Un -Irrigated Crops 1993-94 Q/Ha. [41]

Crop	State	Irrigated	Unirrigated
a. ***Paddy*** .	Punjab	32.40	20.25
	Tamil Nadu	37.18	12.16
b. ***Wheat.***	Punjab	37.80	21.30
c. *Jowar*	Tamil Nadu	19.68	9.25
	Maharashtra (rabi)	10.46	5.22
d. ***Maize.***	Punjab	22.40	13.34
	Karnataka	28.29	23.95
e. ***Groundnut.***	Gujarat	17.68 (summer)	4.30 (kharif)

Besides this importance of irrigation the significance of flow measurement, control and distribution of water were also well known in ancient times :

> *" The whole country was very prosperous because of two crops grown in the year with irrigation facilities.... The district Officers measure the land*

and inspect the sluices by which water is distributed into the branch canals so that every one may enjoy his fair share of the benefit"
---- Megasthenes, the Greek Ambassador in the Court of Emperor Chandragupta.300B.C.

In view of the limited availability and competing demands for water, it is imperative to utilise it with utmost efficiency. It is also necessary to recognise the warning bells sounded by the Vice President of the World Bank that "if wars were fought on oil in the 20th century, wars will be fought on water in the 21st century" [187].

1.2 The Issue of Sustainability

Many civilisations such as those around Mesopotamian plains, and the valleys of the Yellow River, Indus and Nile rivers have flourished and ultimately perished due to failure of judicious use of water on a sustainable basis. It is increasingly being stressed at various international for a that the combination of demand management, reduced waste and proper management of claims against the water supplies would make a major contribution to the sustainability of water resources [61,219,231]. In particular, sustainability of irrigated agriculture is now a subject of serious global concern. The International Commission on Irrigation and Drainage (ICID), in its Hague declaration [60], emphasized the urgent need of efficient water management in agriculture. The National Academy of Agricultural Sciences (NAAS), has in 1996, brought out a policy document on the Agricultural Scientists' Perceptions of National Water Policy [105,106]. Sustainability of agricultural production depends, besides irrigation, on several other factors such as climate, hydrology, Soils, and land use, demography, ecology, environment, and other related aspects. In this article the focus is on the water management for sustainable agricultural production.

As we enter the next millennium, it is appropriate to take stock of the situation and analyse the strengths, weaknesses and the threats of our policies and practices and look for the opportunities in order to ensure sustained and increased agricultural production to meet the food requirements of the future generations.

In this article, a brief description of the water resources of India, irrigation development and the associated agricultural production, and the future scenario of the agriculture as well as the water resource availability for agriculture is examined. The strengths, weakness, opportunities including the uncommon opportunities, the scope for maximizing water resource utilisation through water management technologies like water harvesting, conjunctive use, use of marginal quality of water and adoption of scientific approaches such as computer aided decision support systems for efficient water management., use of micro irrigation systems, efficient on farm water management, etc., were discussed through case studies. The R&D efforts needed were also identified.

2 Water Resources and Agricultural Production in India

2.1 Water Resources

Agriculture is the major user of water resources in India. Currently (2001 A.D.), the share of agriculture is around 83%. Availability of water resources relative to crop's water needs during different periods of its growth affects directly its productively, and indirectly influences the use of other inputs. Hence it is necessary to asses the water resources availability and their progressive development and utilisation. This needs to be done both spatially and temporally using analytical procedures and models.

2.1.1 · Rainfall

Rainfall is the major source of water over the country barring a limited quantity of snow that occurs in the Himalayan region in the north. The rainfall over India is characterised by wide spatial and temporal variations. The annual rainfall varies from a mere 311 mm in western Rajasthan to over 11400 mm in Meghalaya with an average value of 1195 mm for the entire country. More than 73 per cent of this rainfall occurs during 4 to 5 months of monsoon (June-October). About 70 percent of the geographical area of the country experiences annual rainfall of 750 mm or more (Tables 2).

2.1.2 River Flows

The annual precipitation occurring over the geographical area of 329 m.ha of the country amounts to 4000 Billion Cubic Meters (BCM). This generates runoff from 12 major river basins (catchment area greater than 20,000 ha) and 48 medium river basins (catchment area between 2,000 and 20,000 ha). The total annual surface flow is estimated to be 1953 Billion Cubic Meters (BCM). But over 90% of the annual runoff in the peninsular region and over 80% in the Himalayan rivers occur during the monsoon months. Due to this and other constraints, the Central water Commission that out of the 1953 BCM, only 690 BCM çan be beneficially harnessed through the presently available technology (Table. 3) assesses it.

Table 2. Distribution of Geographic and Cultivated Area of India Under Different Rainfall Regions [70]

Rainfall Range Mm	Geographic Area m.ha	Available Rain water BCM*	Net sown Area m.ha	Volume of Rain water Received in Net sown area BCM*
100- 500	52.07	156.2	29	87.0
500-750	40.26	251.6	22	137.5
750-1000	65.86	576.3	34	297.5
1000-2500	137.24	2058.6	44	660.0
> 2500	32.57	957.3	14	411.5
Total	328.00	4000.0	143	1593.5

* 1 BCM (Billion Cubic Meters)=10^9 m^3.

2.1.3 Storage Reservoirs

The total live storage created or under creation in India during the last 50 years is about 249 BCM / year which is about 1/8 of the average annual flow of 1953 BCM. The estimated ultimate storage potential is 500 BCM / year including inter basin transfers. USA, which has nearly the same water resource potential (1700 BCM) as in India, created storage of about 450 BCM / year, during the past 40 years [57]. The role of this storage capacity in reducing the vulnerability of the agricultural activity to the vagaries of the weather in USA is obvious. Thus there is an immense need to build up the water storage capacity in India to the maximum possible extent to ward off severe droughts.

2.1.4 Tanks

Tanks are age-old indigenous, water harvesting systems of ancient India. These are devised to capture, store, and distribute water for irrigation besides meeting the domestic needs of the population. As per minor irrigation census carried out in 1987, there are about 1.5 million tanks in various States accounting for storage of 35-45 BCM. The States of Andhra Pradesh, Karnataka, Madhya Pradesh, Maharashtra, Tamil Nadu, and West Bengal have large number of wells and agriculture depends on this source of water. During the past 200 years, these systems have faced neglect and most of them have become inefficient [94].

Table 3. Surface Water Resources Potential and Status of Live Storage in the River Basins of India

(Unit b.cu.m.)

Sl No	Name of the River Basin	Average annual potential in the river	Estimated utilisable flow excluding ground water	Live storage		
				Completed projects	Ongoing projects	Proposed projects
1	Indus (upto border)	73.31	46.00	13.85	2.45	0.27
2	a. Ganga	525.02	250.00	36.84	17.12	29.56
	b. Brahmaputra	629.05	24.00	1.10		
	b. Barak & Others	48.36				
3	Subemarekha	12.27	6.81	0.65	1.65	1.59
4	Brahmani and Baitami	28.48	18.30	4.76	0.24	8.72
5	Mahanadi	66.88	49.99	8.49	5.39	10.96
6	Godavari	110.54	76.30	12.51	10.65	8.28
7	Krishna	69.81	58.00	34.48	7.78	0.13
8	Pennar	6.32	6.86	0.38	2.13	
9	Cauvery	21.36	19.00	7.43	0.39	0.34
10	Tapi	14.88	14.50	8.53	1.00	1.99
11	Narmada	45.64	34.50	6.60	16.72	0.46
12	Mahi	11.02	3.10	4.75	0.36	0.02
13	Sabarmati	3.81	1.93	1.35	0.12	0.09
14	West flowing rivers of Kutch, Saurashtra including Luni	15.10	14.98	4.31	0.58	3.14
15	West flowing Rivers South of Tapi	200.94	36.21	17.35	4.97	2.54
16	East Flowing & Rivers from Mahanadi to Godavari	17.08		1.63	1.45	0.86
17	East flowing rivers between Godavari and Krishna	1.81	13.11			
18	East flowing Rivers between Krishna and Pennar	3.63				
19	East flowing Rivers between Pennar and Cauvery	9.98	16.73	1.42	0.02	0
20	East flowing Rivers South of Cauvery	6.48				
21	Area of North Ladakh not Draining into Indus	NA	NA	NA	NA	NA
22	Rivers Draining into Bangladesh	8.57	NA	NA	NA	NA
23	Rivers Draining into Myanammar	22.43	NA	0.31	NA	NA
24	Drainage areas of Andman, Nicobar and Lakshadwee island	NA	NA	NA	NA	NA
	Total Say	1952.87 1953.00	690.32 690.00	173.71 174.00	75.43 76.00	132.32 132.00

2.1.5 Inter Basin Transfers

There are several river basins in India where the water resources are excessively surplus, while there are many river basins where the available water resource is too small. Technically, it is feasible to link these areas through inter basin transfer of water The National Water Development Agency (NWDA) estimates that about 200-250 BCM of water can be transferred and this could facilitate to bring an additional 25 million ha. under irrigation [82].

2.1.6 Groundwater

Groundwater is a naturally occurring renewable resource and is the major source of irrigation as well as drinking water in India. The total annual recharge to groundwater is estimated to be 431.8 BCM [31]. Of this, 361 BCM are available for irrigation after setting aside the balance for domestic and industrial uses. The ultimate irrigation potential from groundwater, estimated based on various factors such as crop water requirements; availability of cultivable land where ground water is accessible, climatic factors, etc. is 64.05 m.ha. (Table 4). 1992 developed about 36.25 m.ha of irrigation potential developed from groundwater resources. Thus the total utilisable water resources were estimated by the Central Water Commission is 1122 BCM (690.s.w + 432 g.w). As of 1997-98, the requirements for water resources for use in different sectors such as domestic, industrial, agricultural and other uses are around 520 BCM. Projections for the years 2010, 2025, and 2050 are given in Table. 5.

Table 4. Groundwater Resources, Their Development and Irrigation Potential Created in India

Sl No	State	Total replinishable Ground water resources	Provision for domestic, Industrial and other uses	Available Ground water resource for Irrigation	Utilisable Irrigation Potential for development	Irrigation Potential Created from Ground Water	Weighted Average Delta for Irrigation	Balance Ground Water resource for future use
		m.ha-m/yr	m.ha-m/yr	m.ha-m/yr	m.ha	m.ha	M	m.ha-m/yr
1	Andhra Pradesh	3.529	0.529	3.000	3.960	1.929	0.05-1.47	2.291
2	Arunachal Pradesh	0.144	0.022	0.123	0.018	0.002	-	0.123
3	Assam	22.472	0.371	2.101	0.900	0.180	1.283	2.007
4	Bihar	3.352	0.503	2.849	4.947	1.428	0.4-0.65	2.303
5	Goa	0.022	0.003	0.019	0.029	0.002	0.57	0.017
6	Gujarat	2.037	0.306	1.732	2.756	1.841	0.45-0.71	1.015
7	Haryana	0.853	0.128	0.725	1.462	1.588	0.39-0.60	0.117
8	Himachal Pradesh	0.037	0.007	0.029	0.069	0.015	0.385	0.024
9	Jammu & Kashmir	0.443	0.067	0.376	0.708	0.012	0.39-0.6	0.372
10	Karnataka	1.619	0.243	1.376	2.573	0.728	0.18-0.74	0.946
11	Kerala	0.790	0.131	0.659	0.879	0.157	0.53-0.83	0.558
12	Madhya Pradesh	5.089	0.763	4.326	9.732	1.874	0.40	3.612
13	Maharashtra	3.787	1.240	2.547	3.652	1.290	0.43-1.28	1.773
14	Manipur	0.315	0.047	0.268	0.369	0.001	0.65	0.268
15	Meghalaya	0.054	0.008	0.046	0.064	0.010	0.65	0.044
16	Mizoram	-	*Not	Assessed	-	-	-	-
17	Nagaland	0.072	0.011	0.062	*Not	Assessed	-	0.062
18	Orissa	2.000	0.300	1.700	4.202	0.393	0.34-0.44	1.557
19	Punjab	1.866	0.187	1.679	2.917	5.12	0.52	0.103
20	Rajasthan	1.271	0.199	1.071	1.778	1.505	0.46-0.60	0.529
21	Sikkim	-	*Not	Assessed	-	-	-	-
22	Tamil Nadu	2.639	0.396	2.243	2.832	1.963	0.37-0.93	0.887
23	Tripura	0.066	0.01	0.056	0.081	0.020	0.63	0.037
24	Uttar Pradesh	8.382	1.257	7.125	16.799	14.000	0.20-0.50	4.441
25	West Bengal	2.309	0.346	1.963	3.312	1.320	0.33-0.75	1.488
26	Union Territories	0.041	0.019	0.007	0.005	0.001	-	0.006
27	**Total All India**	**43.189**	**7.093**	**36.08**	**64.05**	**35.38**	**0.56**	**24.58**

Table 5. Estimated Sectorial Water Requirements (Medium) For The Years 2010, 2025 And 2050 A.D. [82]

** Range of "Low" and " High" estimates.   Units: (km^3 or BCM / year)*

	2010 A.D.	2025 A.D.	2050 A.D.
Irrigation	536	688	1008
Domestic	41.6	52	67
Industry	37	67	81
Power	4.4	12	40
Others	74	119	219
Total	693 (644 - 733)*	942 (861 -1027)*	1422 (1221-1681)*

2.2 Irrigation Development

Irrigation development is conditioned by the climatic and physiographic factors of a region. For instance, in the arid and semi- arid plains of northern India, the flood flows of the perennial rivers like the Indus and Ganga are diverted through inundation channels for use in agriculture. In the low rainfall regions of the peninsular India, the storm water is captured in numerous tanks and used for irrigation. In the alluvial plains, and the weathered regions of the hard rock areas, Where groundwater is close to the surface, water was harnessed through wells for irrigation. Large scale irrigation projects such as the Mettur dam on Kaveri, the Godavari *anicut*, and the Ganga canal head works at Haridwar are some of the projects that are existing for over 150 years. Most of these projects were mainly designed as "protective " works against the prolonged dry spells and the vagaries of the monsoon, for stabilising agricultural production.

The ultimate irrigation potential of India is estimated to be 140 m.ha. [82]. This includes 58.5 m.ha. through major and medium irrigation projects, and 17.4 m.ha. through minor (CCA<2000 ha.) irrigation projects from surface water and 64.05 m.ha. through groundwater. With about 25 m.ha. of irrigation potential from inter basin transfers, the ultimate potential will be 165 m.ha. As of 1997, there are 352 major, 1037 medium irrigation and a large number of minor irrigation projects operating in the country [14].

Recognising the importance of irrigation in increasing and stabilising the agricultural production at higher levels, and poverty alleviation, the successive Governments of Independent India gave priority to development of irrigation potential through its various five year and annual plans. Not only in India, but all over the world there was a spurt in irrigation development during the past 50 years. In particular, the Asian region is predominant in this respect (Table. 6).

Irrigation from groundwater is substantial and accounts for about 49 % of the total irrigated area in the country. During the past four decades, there has been a phenomenal increase in the growth of groundwater abstraction structures due to implementation of technically viable schemes for development of the resource, backed by liberal funding from institutional finance, improvement in availability of electric power diesel, and Govt. subsidies, etc.(Table. 7.)

Of the 94.73 m.ha of irrigation potential developed by 1999-2000, major and medium irrigation projects contribute to 35.35 m.ha., while minor irrigation from groundwater and surface water schemes account for 59.38 m.ha.[20]. However, a large gap of 10 m.ha, exists between the irrigation potential created and utilized (Table 8).

Table 6. A Comparison of Development of Net Irrigated Area in India, China, Asia and The World [43]

(million hectares)

Year	India	China	Asia	World
1961	24.7	30.4	90.17	138.81
1970	30.44	38.11	109.45	167.33
1980	38.48	45.50	132.20	209.23
1990	45.14	47.97	153.60	242.19
1996	57.0	49.8	183.33	263.28
1998	59.00	52.58	191.10	271.40

2.3 *Agricultural Production*

The Indian agriculture has undergone a sea change during the second half of the 20th century .The green revolution of the mid sixties and the rapid pace of irrigation development are the two most important factors that contributed to this. With the advent of high yielding varieties (HYV) of crops during the third five year plan, emphasis on use of chemical fertilizers and irrigation has increased for exploiting

Table 7. Groundwater Structures, Pump Sets and Area Irrigated By Groundwater in 1950-51 and 1993-94 [31]

Item	1950-51	1993-94
Dug wells	3.86 million	10.12 million
Shallow tube wells	3,000	5.38 million
Deep tube wells	Negligible	68, 000
Electric pump sets	, ,	9.34 million
Diesel pump sets	66,000	4.59 million
Area irrigated	6.5 m.ha	36.4 m.ha

the full potential of the HYVs. The area under HYV increased steadily from 1.8 m.ha in 1966-67 to 76.5 m.ha in 1996-97. Likewise, consumption of fertilizers also increased from 1.1 m. tons in 1966-67 to about 23.6 m. tons by 1999-2000. There was a marked shift from a low input, subsistence agriculture to one of high input, high productivity, intensive agriculture. Thus with the rapid, near fourfold, expansion of irrigation facilities, from the beginning of the first five year plan (1950-51), in combination with the use of high yielding varieties, good rain fall for the twelfth consecutive year, higher levels of fertiliser application, and plant protection, the food grain production in India had a fourfold increase from 51 m. tons in 1950-51 to 206 m. tons in 1999-2000 (Fig.1). This lead to transformation of India from a country of "ship to mouth existence" to a "grain exporting" country, besides having an unprecedented food grain stock of about 44 m. tons It may, however be noted that only 40% of the area under food grains is irrigated (Table 9).

India has now become the world's largest producer of milk, and cashew nuts, second largest producers of rice, wheat, sugarcane, groundnut and vegetables, third largest producer of wheat, cotton, sorghum, potatoes, Coffee, rapeseed- mustard and the sixth largest producer of maize (Table10). Irrigation is the major factor that made this possible.

This progress is impressive considering that only thirty years ago, Indian agriculture was one of subsistence and hardly had any position in the world. However, this should not give place for complacence. It is estimated that about 325-350 m. tons of food grains are required to be produced by 2025 AD [74, 84] to meet the food requirement of the growing population. It is to be noted that of late, therates of growth of irrigation facilities and food grain production are decreasing. This is indicative of limitations on availability of suitable sites for further construction of irrigation projects as well as financial resources besides resistance from social forces. Further, due to increasing pressure for the land from the ever-increasing population as well as industrial growth, further land availability for agriculture is extremely limited.

2.4 Low Productivity of Irrigated Areas

Productivity of irrigated areas in India leaves much to be desired. While as much as 10 tons/ha/year could be produced in research and demonstration farms, the average productivity is as low as 4 tons/ha/year. The order of the ranking in producvity of various crops in India Vis-a Vis the other countries in the world is also given in Table10. This Information serves as an eye opener that in spite of the impressive rankings as indicated in column 2 of Table 10, the productivity in terms of yield per hector of almost all the crops are way behind the best productivity values achieved elsewhere in the world (Table.10). A comparison of area, production and yield figures of food grains in India, China, Asia and the average values for the world are given in Table 11.

Table 8. Plan-wise Irrigation Potential Created and utilised in India (1951-2000)

(Cumulative Values)

Plan period	Potential Created, m.ha.					Potential Utilised, m.ha.				
		Minor					Minor			
	Major & Medium	Surface Water	Ground Water	Total	Total	Major & Medium	Surface Water	Ground Water	Total	Total
Pre Plan up to 1951	9.70	6.40	6.50	12.90	22.60	9.70	6.40	6.50	12.90	22.60
I Plan (1951-56)	12.20	6.43	7.63	14.06	26.26	10.98	6.43	7.63	14.06	25.04
II Plan (1956-61)	14.33	6.43	8.30	14.75	29.08	13.05	6.45	8.30	14.75	27.80
III Plan (1961-66)	16.57	6.48	10.52	17.00	33.57	15.17	6.48	10.52	17.00	32.17
Annual Plans (1966-69)	18.10	6.50	12.50	19.00	37.10	16.75	6.30	12.50	19.00	35.75
IV Plan (1969-74)	20.70	7.00	16.50	23.50	44.20	18.39	7.00	16.50	23.50	41.89
V Plan (1974-78)	24.72	7.50	19.80	27.30	52.02	21.16	7.50	19.80	27.30	48.46
Annual Plans (1978-80)	26.61	8.00	22.00	30.00	56.61	22.64	8.00	22.00	30.00	52.64
VI Plan (1980-85)	27.70	9.70	27.82	37.52	65.22	23.57	9.01	26.24	35.25	58.82
VII Plan (1985-90)	29.92	10.99	35.62	46.61	76.53	25.47	9.97	33.15	43.12	68.59
Annual Plans (1990-92)	30.74	11.46	38.89	50.35	81.09	26.31	10.29	36.25	46.54	72.85
VII Plan (1992-97)	32.96	NA	NA	56.60	89.56	28.44	NA	NA	52.32	80.96
1999-2000	35.35	NA	NA	59.38	94.73	30.47	NA	NA	54.23	84.70
Targets for IX Plan (1997-2002)	45.16	NA	NA	66.62	106.6	39.18	NA	NA	57.24	94.59

Tables 10 and 11 clearly indicate that India has a long way to go to achieve the best yields obtained elsewhere in the world. The reasons for this situation are not far to seek [10, 155, 156 and 178]. Of late, it is recognised that due to unscientific practices of water management, improper cropping patterns, inequity of water availability among crops, water logging and soil salinity afflict parts of the irrigated areas. It is reported that as much as 2.6 m.ha of area is effected by waterlogging [11] while about 8.38 m.ha is effected by soil salinity [107]. This led to severe reduction in productivity. In general, large irrigation project command areas are characterised by

i) wide spatial variations in soils, climate, crops grown and their water and nutrient requirements, people's preferences and customs etc.
ii) traditionally irrigation systems were designed mainly for protective irrigation

and not productive irrigation

iii) the water delivery system is supply based rather than demand based,

iv) severe inequity, lack of timeliness, adequacy, and reliability of water

v) supplies in the irrigation systems can not meet the requirement of the modern intensive agriculture based on high yielding crop varieties and multiple cropping with increased fertiliser use,

vi) soil - water -plant -environment interactions are not reflected either in planning or in design of irrigation systems,

vii) water balance in the command area is not considered nor the possible impact of irrigation on the groundwater conditions ever studied at planning stage,

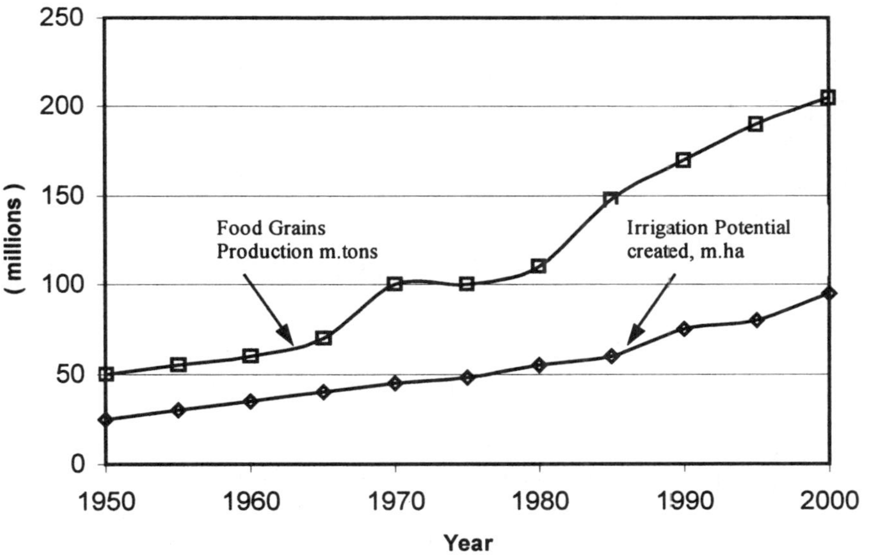

Figure 1. Progressive Development of Irrigation Potential and Food Grain Production in India 1950 - 2000

Table 9. State Wise Area, Production and Yields of Food Grains and Coverage under Irrigation in India (1998-99)

State	Area. m.ha	% of total area under food grains in India	Productio n (m.tons)	% of total production in India	Yield (kg/ha)	% of coverage (1995-96)
Andhra Pradesh	7.19	5.7	14.40	7.1	2003	56.7
Assam	2.67	2.1	3.43	1.7	1288	19.5
Bihar	8.96	7.1	12.91	6.4	1441	47.3
Gujarat	3.90	3.1	5.57	2.7	1426	29.3
Haryana	4.49	3.6	12.12	6.0	2700	77.3
Himachal Pradesh	0.84	0.7	1.49	0.7	1760	18.5
Jammu & Kashmir	0.88	0.7	1.52	0.7	1728	38.8
Karnataka	7.38	5.9	9.98	4.9	1352	21.4
Kerala	0.39	0.3	0.69	0.3	1768	51.4
Madhya Pradesh	17.79	14.2	19.80	0.8	1113	30.8
Maharashtra	13.09	10.4	12.75	6.3	974	12.7
Orissa	5.38	4.3	5.81	2.9	1080	28.1
Punjab	6.12	4.9	22.91	11.3	3741	96.8
Rajasthan	13.46	10.7	12.93	6.4	961	23.6
Tamil Nadu	4.45	3.5	10.14	5.0	2278	60.4
Uttar Pradesh	20.52	16.4	40.15	19.8	1957	65.0
West Bengal	6.54	5.2	14.37	7.1	2198	28.0
Others	1.31	1.0	2.07	1.0	1580	NA
All-India	125.36	100.0	203.04	100.0	1620	40.6

viii) unscientific water management practices, and low water use efficiency,
ix) unrealistically low water charges
x) poor monitoring, and inadequate funding to OMR activities,
xi) large gap in the irrigation potential created and utilised. As much 10.0 m.ha of the created potential remains unutilised by 1999-2000(Table 8), and
xii) marked deterioration of the environment.

These problems are mainly caused by factors that can be attributed to planning and design constraints, operational constraints, and management constraints.

Table 10. Rankings of Area, Production and yields of selected crops grown in India, 1999

Crop	Production Rank & Quantity m. tons	Area. Rank & size m.ha	Yield T/ha	Yield Rank	World Ave. Yield t/ha.	Max. Yield t/ha & Country	Max. Production m.tons & Country
Rice (Paddy)	(2) 131.2	(1) 44.8	2.90	46	3.85	10.0 Australia	220.5 China
Wheat	(2) 70.8	(2) 27.4	2.50	39	2.70	7.98 Belgium	114.0 China
Maize	(6) 10.5	(4) 6.3	1.67	88	4.313	23.0 Kuwait	240.0 USA
Sorghum	(3) 9.5	(1) 10.4	1.44	69	1.44	12.5 Jordan	15.1 Nigeria
Sugarcane	(2) 283.2	(2) 4.2	68.0	34	65.68	122.2 Ethiopia	333 Brazil
Groundnut	(2) 7.3	(1) 8.00	0.91	93	1.3	6.0 China	12.0 China
Potatoes	(3) 22.5	(3) 1.3	17.3	50	16.36	44.8 Netherlands	55.3 China
Coffee, Green	(3) 0.27	(7) 0.78	0.95	11	0.58	2.50 Martinique	1.6 Brazil
Cashew nuts	(1) 0.44	(1) 0.72	0.61		0.605	0.57 Kenya	0.44 India
Mustard	(3) 5.78	(2) 6.6	0.88	40	1.54	66 Algeria	9.7 China

Yield : tons/ha

Table 11. Comparison of Area, Production and Yield of food grain crops in India, China, Asia and the World 1998-1999

Country/ Region	Total cropped area. m.ha	Wheat			Rice (grain –not rice)	Paddy husked		Total Food grains (with paddy grains)		
		Area	Prod.	Yield	Area	Prod	Yield	Area	Prod	Yield
India	180.60	27.4	70.8	2.58	44.8	131.2	2.93	101.6	230	2.26
China	535.54	28.8	114.4	3.96	31.7	200.5	6.32	93.6	457	4.88
Asia	1611.34	98.5	261	2.66	138.5	541	3.90	325.5	1021	3.13
World	4938.30	215	584	2.71	155	596	3.85	680	2064	3.04

A : Area, m.ha. ; P : Production, m.tons, Y : Yield tons/ha.

2.4.1 Planning and Design Constraints

These constitute i) improper assessment of command area size, ii) improper estimation of crop water demands, iii) inefficient water scheduling policies and operating rules, iv) inadequate distribution system to meet the crop water demands, v) mismatch of the water deliveries and requirements, vi) inadequate control of wastage, losses and unauthorised uses, vii) absence of control structures and measuring devices, and viii) improper fixing of outlet sizes ,

2.4.2 Operational Constraints

These include: i) deliveries are based on "duty" and not on crop water requirements, (ii) timing of irrigation supplies usually is arbitrarily fixed and not linked to plant water needs, (iii) water supply in *Warabandi* system is made in terms of time proportional to area instead of in terms of proportionate quantities, (iv) lack of timely and adequate maintenance of water distribution and drainage network, (v) absence of lining of the critical segments of the conveyance system , (vi) non-adoption of appropriate methods of irrigation, and vii) high cost of system management.

2.4.3 Management Constraints

It is a common observation that irrigation systems are rarely "managed"; rather, they are, "administered". In general, there is no co-ordination between the Irrigation Departments and the Agricultural Departments. There is an imperative need to professionalize management of irrigation project areas to ensure reliability, adequacy and timeliness of irrigation water supplies and induce operational efficiency.

2.5 Threats for Irrigated Agriculture

Degradation and shrinking of resource base, falling groundwater levels, need for high inputs, high energy use, water use illiteracy and associated misuse, inadequate legal framework etc., cause threats to irrigated agriculture. Some of these are discussed briefly.

2.5.1 Future Shortages of Water

Agriculture, as the single major user of water (currently the share of agriculture is about 83%), stands to face the stiff competition, in future, from other sectors such as industry, environment, municipal use, recreation etc, where water has higher economically productive value. Availability of fresh water will ultimately become the key constraint for food production. It is estimated that by 2025 A.D. The share of water available for agriculture would be reduced to 73 percent. Hence it is essential to reorient irrigated agriculture to produce more with less water as well as to learn to use water of marginal quality.

2.5.2 Misuse of Water

Apart from poor water management practices guided by non-scientific factors, resource illiteracy, social and political factors also contribute to low productivity of irrigated areas. For instance, many farmers believe that the more the irrigation water is applied, higher will be the yields. Particularly they are guided by the unreliability of water supplies and high subsidy associated with irrigation water supply. These farmers are generally ignorant of the fact that excessive irrigation not only leads to wastage of water through seepage, and runoff, but also causes poor aeration of the root zone besides washing out most of the nutrients from the fields. There is need to organise a massive water use literacy programmes to educate about the ill effects of misuse of irrigation water and adoption of proper water management practices.

Social and political forces tend to corner the benefits to few individuals or group of farmers wile distorting the planned distribution of water. For instance, it is reported that in Maharashtra, only 3% of total cropped area occupied by sugarcane is claiming 75% of the irrigation water, while other crops are denied even a life saving irrigation. Influential farmers also tend to grow high water requiring crops like paddy and sugarcane even on light soils where percolation losses cause excessive wastage of water. This is in spite of the guidelines provided by the concerned project authorities.

2.5.3 Pollution of Soil and Water Resources

Anthropogenic activities involving rapid pace of urbanisation, industrialisation and development of irrigated agriculture have made environmental pollution a great concern. The faulty on-farm water management of irrigation water particularly in the absence of appropriate drainage facilities led to water logging and build up of soil salinity in many parts of Haryana, Uttar Pradesh, Rajasthan, Gujarat, etc. It is estimated that about 8.6 m.ha. of land is affected by waterlogging and soil salinity in India [67]. This is besides seepage from septic tanks, leakage from sewage system, leachate from compost pits, solid waste dumps and effluent disposal pits, disposal of untreated sewage into rivers, and nutrient enriched return flows of irrigation water cause severe pollution to both surface and groundwater resources.

2.5.4 Inadequate Legal Frame Work

There is hardly any legal framework to ensure equity and efficiency of irrigation water use. It is necessary to establish an effective water rights regime that incorporates economic, technical, social and organisational requirements to ensure equity and efficiency in water distribution and use. Constitution of Water Users Associations (WUAs) with adequate legal provisions, and entrusting them with the responsibility planning, implementation and monitoring of efficient use and management of irrigation water is a right step in this direction. This approach is yet to be adopted widely. Except Andhra Pradesh, Maharashtra and Gujarat, many states have yet to make a beginning in this direction.

2.5.5 Social Misconceptions

Due to the differences in perceptions, and lack of adequate scientific exercise, and analysis, some of the well meaning environmentalists, NGOs seem to be strongly convinced that large scale irrigation development is the main culprit for the socio-economic ills and environmental deterioration. Such an approach not only causes inordinate delays in the irrigation development, but also the stream of benefits will be delayed besides the cost and time over runs. In the long-term interest of the nation, it is necessary to appreciate that there can be no two opinions about irrigation being the main catalyst for the self-sufficiency in food production, and that there is no viable alternative to large-scale irrigation projects to serve this purpose.

2.5.6 Conflicts and Food Security

The extent to which agricultural research and improved technologies, particularly water management, can alleviate poverty, improve food security, and encourage sustainable use of natural resources and the extent to which conflicts are avoided are mutually interdependent. Thus the interaction between the conflict and the impact of agricultural water management research deserve more attention from the research community.

3 Status of Research in Irrigation Water Management at Project Level

Since irrigation has to play a key role in enhancing the agricultural production, sustainability of irrigated agriculture through efficient eco-friendly irrigation management practices assumes great significance. The Indian Council of Agricultural Research (ICAR), with its associated National Agricultural Research System (NARS) has focussed its attention to these issues through establishment of three Water Technology Centres, a Directorate of Water Management Research, three coordinated research projects, and supporting several research programmes at various State Agricultural Universities (SAUs) and traditional universities. The Ministry of Water Resources, Govt.of India also supports research in irrigation water management through the Indian National Committee of Irrigation and Drainage (INCID).

Water resources development and management for sustainable agriculture requires a formal framework to i) quantify, analyse and assess available water resources from different sources relative to their requirements, ii) plan and evaluate alternative water resource management strategies before arriving at specific plans and operational policies, and iii) evaluate the impact of current and proposed strategies on water use, crop productivity, and sustainability of soil and water resources. Thus irrigation management needs to be made more knowledge intensive and the capabilities for decision making at all levels need to be improved.

The answer to how to plan, manage and operate an irrigation system cannot be separated from real-world conditions. So, even though some specific and simple problems can be addressed directly and in isolation, a real field case study based approach is a primary requirement for the development of the above framework. This has been the general approach adopted by many a researcher. This section presents the progressive evolution of the work on the development of the above framework starting from simple and empirical models to more complex physically based process models, their integration into irrigation systems models at various levels, and to the present research on irrigation decision support systems involving the use of Geographic Information Systems (GIS) and complex regional models and databases.

3.1 Water Resources Assessment

3.1.1 Surface Water Resources

Rainfall, stored water from surface runoff, river flows, soil moisture and ground-water are the various sources of water for agriculture. The availability of these resources relative to a crop's water needs during different periods of its growth affects its productivity directly and also indirectly by influencing the use of other inputs. The studies on water resource assessment have to focus on the assessment of the above resources across a spectrum of spatial and temporal scales through use of remote sensing, analytical procedures and models.

3.1.1.1 Analysis of rainfall

About 70% of average annual rainfall over India occurs during the 3-4 months of the monsoon season (June-September). Even during this period, the rainfall is not uniformly distributed. Typically every region experiences a minimum of 2 to 3 dry spells during the monsoon season, some of them extending beyond 3 to 4 weeks. Understanding the rainfall characteristics can lead to increasing crop productivity through proper crop planning [223] and water management. The main rainfall characteristics of interest are: i) Onset of effective monsoon to initiate crop sowing operations [224,225], ii) frequency and duration of wet spells and dry spells (for crop and irrigation planning), [179] iii) amounts of weekly rainfall at specified probabilities of occurrence (for water conservation and irrigation management) [118], and (iv) extreme events (like probable maximum precipitation for hydraulic systems design [141], v) daily rainfall simulation for drainage design [124], and vi) drought management [147,148].

3.1.1.1.1 Onset of effective monsoon (OEM)

Besides the vast rain fed area of about 100 m.ha, there are also large chunks of areas within the irrigation project areas that can not receive water from the system and remain under rain fed agriculture. Agricultural operations start with the onset of south -West monsoon in rain fed areas in India. While the India Meteorological Department prepared charts and maps showing normal dates of onset of monsoon over India, these charts are not adequate for planning agricultural activities such as seedbed preparation and sowing. Over the years, many definitions and criteria for identifying the dates of onset of "effective" monsoon (OEM) were proposed. However, most of these criteria are based on only on the amount of rainfall received over a period at the beginning of monsoon. But in practice, it is not only the total rainfall but also the resulting soil moisture conditions govern the initiation of sowing operations. The following criteria for identifying the date of OEM were developed to overcome this limitation [225]:

i) The first day's rain in a seven days spell should not be less than the average daily pan evaporation (e mm) for the season,

ii) The total rain during the seven days spell should be at least equal to that required to recharge the soil moisture to a level adequate for initiating sowings and ensuring crop germination. This condition is met if the total rain is more than an "Effective Onset Rainfall" (EOR) given by,

$$EOR = 0.75\ (FC\text{-}WP).\ BD'.\ d + 5\ e + RO \qquad (1)$$

In which,
FC = Moisture content at field capacity (w/w),
WP = Moisture content at wilting point (w/w),
BD' = Relative bulk density of soil with respect to water
d = Effective seeding zone depth or ploughing depth whichever is more, mm, and,
RO = Run-off, mm (assumed to be 10-15% depending on the condition of land Preparation before the rains)

iii) At least three out of 7 days are rainy days (having rainfall > 2.5 mm /day).

However, if above three criteria are satisfied in a week, but followed by a prolonged dry spell then it should be considered as a pre-monsoon spell and not OEM. These criteria were tested and adopted for determining the date of OEM for several crops and locations across the country.

3.1.1.1.2 Wet spells, dry spells and critical dry spells

After the onset of effective monsoon, the next rainy spell is called a wet spell if it meets the following criteria [179,225]:

i) A rainy day with at least 5 e rainfall or
ii) A spell of two consecutive rainy days with rainfall totalling at least 5 e, or
iii) A week having at least three rainy days with rainfall totalling 5 e or more

The intervening period between any two wet spells is a dry spell. If the duration of any dry spell is long enough it can become a "Critical Dry Spell (CDS)". Several concepts were proposed to characterise the CDS starting from identifying a spell longer than a specified period [156,164, and 165] or a threshold value of duration related to the soil moisture status in the root zone [224], in combination with the fact that these periods coincide with the critical growth stages of the crops [179].

The occurrence of dry spells and wet spells are also studied as an alternating renewal process [126]. The model consists of two components, one for the rainfall occurrence (wet/dry process) and another for the rainfall amounts in the wet spells. The truncated negative binomial distribution was fit to the lengths of wet and dry spells and the two-parameter gamma distribution was fit to the rainfall amounts. The model was applied to the Mahi irrigation project area, where it simulated the monthly rainfall and the wet and dry spells adequately. Predictions of the lengths of wet spells, which is critical for decisions related to withholding canal releases during the monsoon season, were particularly accurate.

3.1.1.1.3 Probable weekly rainfall for planning agronomic operations and supplementary irrigations

Probabilistic analysis of weekly rainfall is normally used for planning crop management and irrigation scheduling and rainwater harvesting. Information about the lowest assured weekly rainfall at different probability levels, the mean weekly rainfall and weekly potential evapotranspiration for the main crops during *kharif* and *rabi* seasons respectively needs to be analysed. Further, it is necessary to conserve excess rainwater during the periods when the weekly rainfall amounts are sufficiently high. Gamma distribution is considered representative of the short duration rainfall values like weekly values and is routinely used in most studies. Weekly rainfall forecasting can also be made using stochastic models. Weekly rainfall is treated as a first order Markovian process and the probability transition model derived for the area, can be used to forecast the next week's rainfall from data of current week's rainfall [133].

For taking a decision of the life saving supplemental irrigation in rain fed areas, the probability of such an irrigation being effective is an essential factor, as a rain that occurs immediately after the irrigation would mitigate all its beneficial effects. Accordingly, the probability of applying a supplemental irrigation and its benefits are to be analysed. This concept was proposed and tested in the case study of Kandy area, Punjab [224].

3.1.1.1.4 Probable maximum precipitation (PMP)

Studies of probable maximum precipitation (PMP) are useful for planning and design of hydraulic structures commonly used in irrigation and drainage systems. A procedure for estimation of the PMP value using Gumbel's theory was developed [141]. Based on this procedure one-day PMP values for coastal Andhra Pradesh using daily rainfall data from 93 rain gauge stations were estimated and iso -PMP maps were prepared. The Water And Power Consultancy Services (WAPCOS) of the ministry of Water Resources, Govt. of India developed the PMP values for several river basins based on analysis of depth -area- duration- precipitable moisture, etc.

3.1.1.2 Estimation of surface runoff

Estimation of runoff generated in a catchment in response to storm rainfall is the central problem addressed by hydrologists for surface water resources assessment. This may be done, by adopting i) deterministic approach using physical or empirical relationships between rainfall and runoff, or ii) stochastic approach by extracting information from observed variability of the runoff series. Both deterministic and stochastic models for the rainfall-runoff relationship have been widely used to assess the runoff yields of watersheds to design runoff storage reservoirs and farm ponds, and forecast inflows into irrigation reservoirs.

3.1.1.2.1 Deterministic rainfall-runoff models

The amount and distribution of, antecedent moisture condition land slope, infiltration characteristics and land use affect runoff. Numerous mathematical models have been developed for predicting the time distribution of runoff resulting from single rainfall events. Both mathematical and conceptual models were employed for estimation of runoff from a watershed. A majority of the deterministic models for runoff from single rainstorms are based on the assumption that the watershed is a simple linear, deterministic, lumped system, which transforms the effective rainfall into direct surface runoff. Among the mathematical models, the instantaneous unit hydrograph derived by using Fourier transforms resulted in near perfect prediction of observed runoff [26,139,144,161, 215]. For purposes of planning agricultural activities and runoff harvesting, daily runoff values are estimated by using the Soil Conservation Service [217] method [226,227]. The method requires selection of an appropriate curve number (CN) on the basis of land use, soil type, farming treatment and hydrologic condition, which is adjusted for antecedent moisture conditions in terms of the total rainfall of the preceding five days.

For Indian watersheds, the equation for daily runoff (Q), from daily precipitation (P) is given by [5]

$$Q = [R-0.3S]^{**2} \;/\; [R+0.7S] \;;\; \text{if } P > 0.3S$$
$$= 0 \qquad \text{Otherwise,}$$
for non-black soil regions and,

$$Q = [R-0.1S]^{**2} \;;\; \text{if } P > 0.1S \tag{2}$$
$$= 0 \qquad \text{Otherwise,}$$
for black soil regions.

The retention parameter, S, varies among watersheds (because of variations in soils, land use and management practices) and with time (because of changes in soil moisture content) and is related to curve number (CN) by the relationship:

$$S = 254\,(\,100/CN - 1\,) \tag{3}$$

The value of CN, the curve number, varies with antecedent moisture condition (AMC). The curve number values, CNI, CNII, and CNIII, corresponding to three moisture conditions are given in the Handbook of Hydrology [5]. A major limitation of this method is that the soil moisture varies continuously rather than in just three discrete levels as assumed in the SCS method. This was modified to adjust for AMC based on actual soil moisture conditions derived by employing a moisture accounting method proposed by Sharpley and Williams [190]. Using CN1 and CN3, the retention parameter can be directly related to the soil moisture content according to the equation:

$$S = S1\,[1 - FFC/\{FFC + \exp\,(w_1 - w_2 * FFC\,)\}\,] \tag{4}$$

where S1 is the value of S associated with CN1; FFC is the fraction of field capacity; and w1 and w2 are called shape parameters. FFC is given by the equation:

$$FFC = (MC1 - PWP) \,/\, (FC - PWP) \tag{5}$$

Where MC1 is the soil water content in the root zone 1 and FC PWP are the field capacity and permanent wilting point respectively. Values of w_1 and w_2 are obtained from the simultaneous solution of the two equations obtained from Eqn. (4)

according to the assumptions that S = S2 (when CN=CN2) and FFC = 0.5; and S = S3 (when CN=CN3) and FFC = 1.0.

$$w_1 = ln\ [(1.0/(1\text{-}S3/S1)\ \text{-}1.0\ ]\ + w_2 \quad (6)$$

$$and\ \ w_2 = 2\ [\ ln\ \{0.5/(1\text{-}S2/S1)\text{-}\ 0.5\}\ \text{-}\ ln\{1.0/(1\text{-}S3/S1)\ \text{-}1.0\ \}] \quad (7)$$

This approach was tested for a number of cases and was found to improve the result substantially [37,130] .The runoff model and the procedure for selecting CN and other model parameters can be integrated in an expert system environment to facilitate runoff assessment from a wide range of watersheds in India [177].

3.1.1.2.2 Stochastic models for estimation of inflows into irrigation reservoir

Box-Jenkins time series models [29] are popular for this approach, and were used to forecast weekly reservoir inflows for the case study of Jayakwadi Irrigation Project in Maharashtra. Data of weekly inflows into the reservoir for 18 years were de-trended and normalised, and the periodicity was also removed before identifying the stochastic component of the time series. Among the auto-regressive (AR), moving average (MA) and auto-regressive moving average (ARMA) models tested, the ARMA model provided the best fit for the stochastic component of the observed inflow series for Jayakwadi reservoir. The use of the model in a current season for making one and two step (week) ahead forecasts of inflows into the reservoir using observed data of previous week's inflows was demonstrated in irrigation planning and scheduling [51,52 and 53]

3.1.1.2.3 Development of water harvesting technologies

Efficient rainwater management remains the top priority in India where two thirds of agriculture is rain fed. Adopting harvesting of rainwater can substantially augment irrigation potential in the semi-arid regions of India. It was estimated that the rainwater harvesting potential in India is as much as 24m.ha.m [70]. The major components of water harvesting technology, besides in-situ conservation in crop lands, are inducement of runoff from land surface, conservation of harvested water in storage tanks and using the conserved water most efficiently to provide a life saving irrigation to crops [58, 212, 213, 214, 215, 226 and 227].

<u>Runoff Inducement</u>: treating the topsoil so as to inhibit infiltration can increase Runoff. The surface treatments include shaping and smoothening of top surface, compaction of the top surface, use of chemicals such as sodium and calcium salts, bentonite or a mixture of bentonite and sodium chloride, asphalt and bitumen as well as using polyethylene sheets to cover the surface. Experimental studies to

evaluate the relative performance of these various methods indicated that the water harvesting efficiency of the surface treatment with polyethylene was highest compared to all other treatments. However, from the economic point of view, treatment with sodium carbonate is considered more efficient. The water harvesting efficiency for such a case ranged between 40 to 81% depending on rainfall [212, 213 and 215].

Design of water harvesting tanks: Water harvesting tanks serve to store runoff during periods of excess rainfall bursts. Design of water harvesting tanks and planning for supplementary irrigation based on the crop and climatic features is highly location specific. Verma and Sarma [226,227] developed a generic procedure for the hydrologic as well as hydraulic design of water harvesting tanks for rain fed farming and computation of the associated benefit cost ratio, which can be adopted for other rain fed areas. Tanks designed on the basis of seasonal rainfall and used for pre-sowing irrigation of wheat were found to be the most beneficial for catchment areas varying from 1-100 ha in Kandy region of Punjab. The probability level of the lowest assured runoff corresponding to the lowest annual cost per unit of available water increased with increasing tank capacity and varied from 40-80%.

Design of appropriate irrigation system. Having created valuable water storages, it is imperative to use the same with utmost prudence. Thus, planning of the timing of use of this water and the appropriate irrigation system become very important. In a case study of the Kandy area in Punjab, this was illustrated by designing an appropriate irrigation system that is also economically viable. The irrigation system comprised a diesel operated pumping set, PVC piping system for conveyance and furrow method of irrigation water distribution [224].

3.1.2 Groundwater Resources

Groundwater constitutes a major source of irrigation. Recharge resulting from large-scale irrigation development through major and medium irrigation projects contributed substantially to the increase of the groundwater resources. Assessment of dynamic recharge is the central problem of assessment and use of groundwater resources available irrigation. Analytical tools developed for assessment of recharge to groundwater from surface irrigation systems involve broadly three approaches: i) Use of empirical norms and regional water balance to assess regional groundwater recharge in irrigation project command areas, ii) Development of simple analytical models to predict the local response of water table to surface recharge, and iii) Development of numerical groundwater flow models in the vertical plane to predict water table response for the real surface and subsurface conditions of the command area.

3.1.2.1 Empirical methods for regional water balance

The Central Groundwater Board developed guidelines for estimation of recharge to groundwater bodies and modified them from time to time [9]. The norms provide empirical coefficients for recharge from various sources, rainfall, seepage from main canals, branch canals, distributaries, and return flows from canal and groundwater irrigated fields. They provide estimates of gross recharge from the various sources. However, the entire gross recharge is not available as storage in the aquifer. Some of it leaves the basin as groundwater outflow. Estimating the outflow requires understanding the regional water balance. The use of empirical norms together with the regional water balance to estimate net regional groundwater has been standardised and applied to a number of irrigation projects namely, Mahi Right Bank Canal (MRBC) [116,119,200, 201, 202, 203], Sri Ram Sagar Project [130], Sharda Sahayak [168,169], and Bhakra Canal irrigation project [46]. The regional water balance is an expression of conservation of mass over a period of time for a physical system. The groundwater basin, with its properly identified boundaries, is considered as the system. The annual regional water balance of a groundwater basin is given by

$$P + I - (ET + E + Q_p + Q_s + Q_g + ET_{gw}) = \Delta S_g + \Delta S_m + \Delta S_s \qquad (8)$$

Where all items (annual values), are expressed as depth in mm. over the area of the basin, are as follows:

P = total precipitation over the area,
I = Irrigation water applied over the area from all sources,
ET = evapo transpiration of the crops and other vegetation,
E = evaporation from uncultivated areas,
Q_p = Quantity of groundwater pumped,
Q_s = surface runoff from the area,
Q_g = groundwater outflow across the basin boundaries,
ET_{gw} (a) = additional evaporation from groundwater by capillary contribution,
ΔS_g = change in groundwater storage,
ΔS_s = change in surface water storage ,
ΔS_m = change in soil moisture storage

The annual duration generally considered is from 1st June to 31st May, as May and June are the hottest and driest months, the soil moisture storage and the surface storages are at their minimum and also experience negligible changes over the annual period considered. Changes of groundwater storage are assessed from the observations of the water levels for the pre monsoon (June) and post monsoon (October) months. Information on rainfall, canal flows are taken from official records. Surface runoff can be estimated using the Curve Number method. Evapotranspiration of various crops can be estimated using the standard FAO procedures. Evaporation from uncultivated areas can be determined using the Jalota and Prihar [62] procedure. The net recharge to the groundwater basin is then determined as the gross recharge to groundwater less the subsurface outflow and the groundwater draft [202].

3.1.2.2 Analytical models for assessment of recharge to water table aquifers

Seepage from the vast networks of canal systems in Irrigation projects and seepage from the water stored in farm ponds and lakes contribute to considerable recharge to groundwater. As a sequel, the rising water table in irrigated areas is a common experience. It is necessary to assess this recharge not only to control the rise of groundwater table but also for proper planning of conjunctive use of surface and groundwater. Analytical approach to solve the partial differential equation of groundwater flow helps estimation of the time history of the growth of groundwater mound in unconfined aquifers beneath a recharging area. Typically, canals and farm ponds are idealised by representing them with linear (semi- infinite) strip basins and rectangular recharging areas respectively. These equations were traditionally solved assuming homogeneous, isotropic aquifers of infinite aerial extent [56]. However, in reality, aquifers have finite dimensions and are bounded by different types of boundaries, such as an outcrop / dyke (barrier boundary) or a water body such as drain / canal, (recharge boundary). A general methodology for assessing recharge to finite water table aquifers with different types of boundaries was developed [110,111,112,113,114, 115]. Finite aquifers bounded by impervious, recharge and mixed boundaries for strip and rectangular basins were considered. For example, the problem of the case of recharge to finite unconfined aquifers from a strip basin (aquifer bounded by recharge boundaries Fig.2 a), the flow equation can be written as

$$\partial / \partial x(H \partial H / \partial x) = (\mathrm{e/K})\, \partial H / \partial t - \mathrm{p/K} \tag{9}$$

$$\partial^2 \mathrm{s} / \partial x^2 = (1/\mathrm{a})\, \partial s / \partial t \ - 2\, \mathrm{p/K} \tag{10}$$

Equation 1 is a non-linear partial differential equation, which, on linearisation using Hantush' s [56] procedure, and writing in terms of drawdown, s, becomes Equation (10).

where

$a = K h_l / e$

$s = H^2 - h^2$

$$h_l = (H+h)/2$$
$$p = p \qquad x_1 < x < x_2; \; for\ t>0$$
$$\quad 0 \qquad Otherwise$$

subject to :

$$I.C. \quad s(x,0) = 0$$
$$B.C. \quad s(0,t) = s(A,t) = 0 = s(B,t) \tag{11}$$

The finite Fourier sine transform of Eq. (10) is

$$ds(m,t)/dt + (a / A^2)(\Pi^2 m^2 s(m,t) - (2\, p\, a\, A/ Km\, \Pi)[\cos(m\, \Pi x_1/A) - \cos(m\, \Pi x_2/A)] \tag{12}$$

Which is an ordinary differential equation. On solving Eq (12) and taking inverse Fourier transform the solution becomes

$$s(x,t) = (8\, pA^2 / k\, \Pi^2) \sum_{m=1}^{\infty} [\, m^{-3}\{1 - \exp(-a\, \Pi^2 m^2 t / A^2)\} * \sin\{m\, \Pi (x_2 - x_1)/2A\} \sin\{m\, \Pi (x_2 + x_1)/2A\} \sin(m\Pi\, x/A)\,] \tag{13}$$

Similar sets of solutions were developed for other cases also. All the existing solutions were shown to be special cases of the generalised theory proposed. The solutions were evaluated by comparison with existing numerical and analytical results as well as field data. Application of these solutions is more realistic than using solutions available for infinite aquifers. The efficacy of the solutions and the accuracy of the results were critically examined by other researchers, who concluded that "*the analytical solution proposed by Rao and Sarma for predicting the growth of groundwater mounds resulting from seepage to bounded aquifer provide a quick method of predicting maximum mound height and does not require advanced mathematical technique. The solution offers relatively easy means of assessing to mound height*" [48]. This approach is not only simple and but also yields results closer to observed field conditions. The solutions are frequently used by various agencies, such as, Office of Water Resources Research in the State of California in USA.

In related studies [110,114], it was proposed that even in infinite aquifers, the horizontal extent of the groundwater mound is limited for finite duration of recharge from strip basins (like canals). A method of identifying the extent of the groundwater mound using solutions of the equation of groundwater flow for finite aquifers was also suggested.

3.1.2.3 Numerical models of water table response to recharge

The limitations of analytical models, namely, fixed geometries of recharging systems and homogeneous aquifer conditions, can be overcome by solving the non-linear equation of groundwater flow numerically. The effects of surface (near by canals, drains) and subsurface conditions (lithology, depth to water table) on recharge of leaky aquifers by seepage from extensive irrigation systems could also be more realistically included [35,122]. The equation is solved numerically in the vertical (x-z) plane to include the effects subsurface conditions and by fixing the water table as the lower boundary.

3.1.2.4 Design of special groundwater wells

3.1.2.4.1 Skimming wells

In many coastal and inland areas fresh groundwater occurs underlain by saline or brackish water. When a well driven into the upper fresh water zone is pumped for irrigation or other use, the drawdown near the well causes an up coning or mounding of the interface. In such aquifers skimming wells are designed to pump the overlying fresh water alone while preventing mixing of the saline water with the fresh water. The phenomenon of up coning of fresh water-saline water interface during pumping from skimming wells is normally studied using the principles of hydraulics of groundwater wells [211]. A new approach to the theoretical analysis was developed for studying the phenomenon of up coning below the skimming wells [170]. This approach views the problem as defining the shape of the groundwater mound that is formed in the saline aquifer alone, assuming that the saline aquifer is subject to an effective recharge. The approach was validated by comparison with existing theoretical solutions and with published experimental results. The geometry of a multiple skimming well system was determined by examining a number of geometrical configurations with a common pumping point and delivery point. Such systems are also useful for vertical drainage [16, 188, and 211].

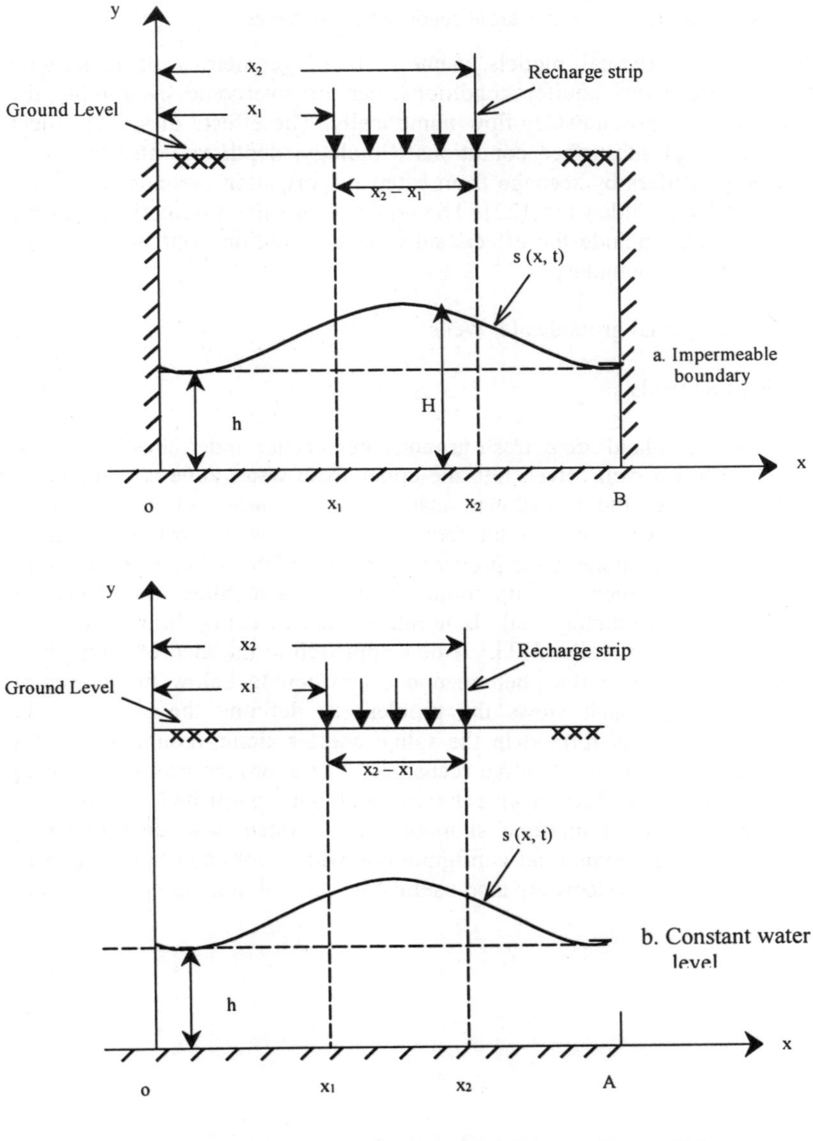

Figure 2. Groundwater Recahrge from a Strip Basin to Finite Unconfined Aquifers

3.1.2.4.2 Cavity wells

In situations where a water-bearing stratum is under artesian conditions underlain by a hard clay layer, it is possible to puncture the clayey aquiclude and draw water without having to penetrate the aquifer. Such wells are common in the plains of Uttar Pradesh, Bihar and West Bengal and are called cavity wells or non-penetrating wells. Cavity wells frequently fail because of collapse of the roof of the cavity. In order to avoid such failures, these wells must be designed both hydraulically and structurally. A stress analysis approach (Fig. 3) based model was developed to analyse the failure phenomenon and determining the limiting discharge at which the roof of the cavity fails [109, 152, and 153]. Design charts were developed based on simulations with the above model.

3.1.2.4.3 Draw down around partially penetrating wells in a confined aquifer

Analytical solution could be derived for the first time for the case of steady flow around partially penetrating wells in a non-leaky, confined aquifers using the finite Fourier transforms [173]. The solution obtained is consistent with the form of the solutions for other cases of flow towards partially penetrating wells. The solution was reduced to a compact form by introducing a correction factor. Values of the correction factor for a wide range of common field conditions were also presented in the form of tables to facilitate easy computations for design purposes.

3.1.2.5 Survey and renovation of sick tube wells

Farmers and public agencies often experience tube well failures. These may be due to inadequate design of the wells, improper construction procedures, selection of improper material for well casing and well screen, lodging of foreign material in the well corrosion of the pipe, incrustation of the screens and other unforeseen causes. The physical conditions inside sick tube wells can be photographed using a 3-dimensional bore hole camera system and measures can be suggested for renovating the wells. Several studies were carried out to investigate sick wells in U.P., Punjab, Haryana, Tamil Nadu and Gujarat states [145,156,158,159, 160, and 171]. These renovated wells obviate additional expenses for new wells and improve irrigation facility.

3.1.3 Water Supply and Demand Assessment of Irrigation Projects

Assessment of an irrigation project's ability to meet present and future irrigation demands is carried out by combining the regional water balance studies with estimation of crop water requirements. Such studies can help in obtaining a first order assessment of the current operating procedures and adequacy of the available water resources in the command areas of major irrigation projects. This also facilitates a logical matching of weekly / monthly / seasonal or annual irrigation demands with the canal and groundwater supplies and management strategies can be evolved. For example, a strategy based on conservation and the use of groundwater in the monsoon season for full exploitation of the reservoir storage potential in the subsequent seasons, was developed and tested [116,119,140, 143,157 and 199]. This strategy can be adapted to other irrigation projects as well. Based on such study policies for alternate cropping patterns, conjunctive use of surface and groundwater were suggested for MRBC Irrigation project, Gujarat.

3.2 Planning and Management of Irrigation Operations

The overriding goal of research on planning and management of irrigation operations is to provide quantitative decision tools for irrigation planning, management and operations. The general direction of research is to integrate process knowledge (simulation models) of agricultural, surface and groundwater hydrology, soil-plant-atmosphere interactions, hydraulics of canal and groundwater flows, and irrigation agronomy with operations research and other techniques of resources management (optimisation models). While doing so advantage of developments in related computer and information technologies such as expert systems and GIS can also be taken. The general approach is to develop a formal theoretical framework to formulate and analyse each water management problem addressed so that it can be applied over a wide range of environments. This framework is then incorporated into formal computer aided decision support systems (DSS). Specifically questions related to planning and management of irrigation operations at the following levels need to be addressed:

i) Irrigation scheduling of crops (field level)
ii) Irrigation canal and reservoir operation (system level)
iii) Conjunctive use of canal water and groundwater in irrigation project areas, and
iv) Non point source (NPS) pollution caused by use of agricultural chemicals in irrigation project areas.

At the central core of development of the DSS to solve the water management problem at all these levels is a simple model of soil water balance in the crop root zone. This model facilitates assessment and monitoring of the moisture regime in the root zone, which guides the irrigation decisions. The other simulation and optimisation models, expert systems and GIS are built around this model and derive their inputs from it to evolve the DSS.

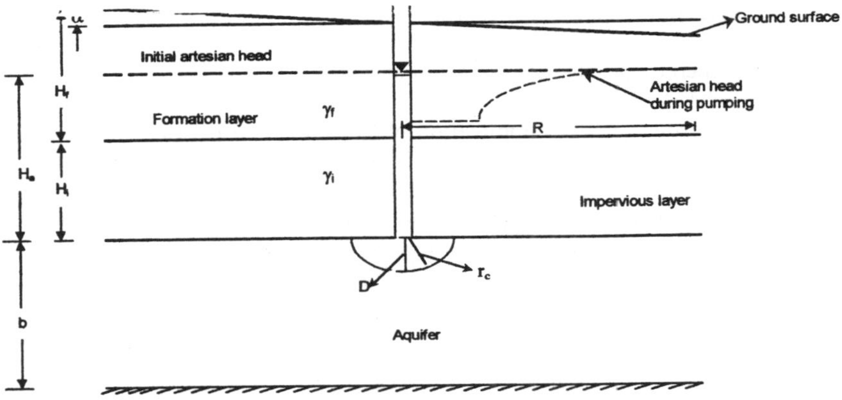

Definition sketch of a Cavity Well

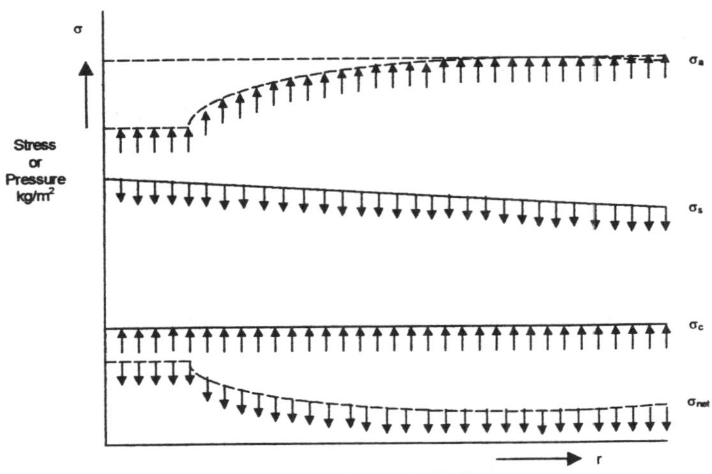

σ_s Overburden soil stress
σ_c Safe compressive stress of impervious soil
σ_{net} Net stress
σ_a Artesian pressure

Figure 3. Definition Sketch of Analysis of Stresses of Around a Cavity Well

3.2.1 Irrigation Scheduling of Crops

3.2.1.1 Scheduling irrigation for a single crop

Irrigation to crops involves application of 'right' amount of water to the crop at the 'right' time so as to obtain high yield of good quality produce, with high water use efficiency and with least damage to the environment, all at low cost of operation. The primary questions of irrigation scheduling are when to irrigate and how much water to apply at each irrigation for a given crop. Irrigation scheduling procedures are different for the cases of i) adequate water supply and ii) limited water supply.

3.2.1.1.1 Irrigation scheduling under adequate water supply.

When adequate water supply is available, the objective of irrigation scheduling is to completely avoid periods of water deficit, so that the crop yield will be at its potential value and the water does not become the factor limiting the crop growth. Irrigations are applied to replenish the soil reservoir whenever the soil water content of the root zone falls to a level at which it begins to have an adverse impact on the crop yield. The agronomists or crop scientists seek to maximise the output of the individual fields for a given water supply using the empirically derived scientific knowledge of crop response to available soil water. The main factors, which govern the irrigation scheduling in this case, are climate, soil, and type of crop and its stage of growth. Numerous agronomic studies are reported in literature for the irrigation schedules for a large varieties of crops of cereals, pulses, oil seeds, fiber and horticultural crops and cropping systems [6,7,38,42, 93,132,133,134, 174] The studies generally focussed on determining the depth and intervals of irrigation, crop water requirements during different stages of the crop growth, soil moisture extraction pattern from the root zone, optimal soil moisture regime to be maintained in the root zone at various growth stages, etc.

These procedures are generally guided by one of the following criteria:

i) the critical growth stages of the crop [92] ,
ii) the IW/CPE ratio (ratio of irrigation water applied to the cumulative value of pan evaporation) [8 ,92] ,
iii) soil-water depletion [53,120, and 121] ,
iv) assessment of crop ET using the climatic factors, and the crop factors derived through experiments conducted locally for each crop using Penman-Montieth method [38, 42],
v) the soil moisture tension values observed in the field [2] and
vi) visual features of the crop (plant drooping, change in leaf colour, rolling of leaves, etc., and plant indices such as relative leaf water potential, leaf water Content, leaf water diffusion resistance etc.

The Indian Council of Agricultural Research (ICAR) supports a number of All India Co-ordinated Research Projects (AICRP) that deal with this subject. These procedures are also used to develop Irrigation schedules for growing crops in locations with high water table conditions where the capillary rise of water would partially meet the crop water needs [88,198], irrigation and nutrient management [47, 89,138,193,195,196], irrigation scheduling for crops grown on saline and sodic soils [2 133] irrigation and pest incidence [66,77,197]. Scheduling of irrigations is aimed at obtaining maximum yield and high water use efficiency.

Water management for paddy crop. Among the cereal crops, paddy requires heavy quantities of water of the order of 1200 to 2200 mm. depending upon the soil and climatic conditions [32,59]. Most of this is expended as evapotranspiration (ET) and percolation. The approximate proportion of these two forms of losses varies from 25% ET and 75% percolation to 40% ET and 60% percolation [7,92]. Paddy is grown on 42 m. ha. of which 22m. ha are under irrigation, 12 m.ha are under low land paddy while the rest is under upland paddy. Hence any savings of water affected for this crop would be extremely valuable. Various research efforts were directed towards this issue. The depth of submergence regimes for transplanted paddy was a major issue of research in water management for rice under adequate water supply. Continuous submergence with 5-10cm of water would lead to heavy percolation and evaporation losses. Experimental studies in different locations in the country indicate that for most locations that rotation of three days of submergence and three days of no irrigation after disappearance of the ponded water until hair cracks appear on the soil leads to nearly the same yields as the case of continuous submergence while affecting as much as fifty percent saving of irrigation water [7,45].

Another approach is intermittent submergence (standing water). In this approach, the standing water is maintained during the critical growth stages of the crop such as active tillering and upto flowering and afterwards maintaining saturation only. This approach leads to saving of about 35% of water compared to continuous submergence without loss of yields. Another water saving approach is adopting direct broadcast (wet seeding) of pre-germinated paddy seeds in puddled clay fields. This approach is reported to give yields at par with transplanted rice and is associated with 50% savings of water [7]. The exact approach is location specific and is related to the local soil and climate [208, 209].

Percolation from rice fields can also be reduced to near half by intensive puddling, shortening of the land preparation period from 20-30 days to 10- 12 days [30]. The potential of shallow water table to contribute to the crop ET) is also substantial and thus affects savings in irrigation water [6, 89].

Development of crop factors: Development crop factors for a different crop at various growth stages under various conditions of soil types, groundwater, and climate were a subject of several major investigations. The crop factors for several crops grown in north India are now available [8,132 and 134]. These crop factors are used in assessing crop water requirements and hence the irrigation requirements.

The results of these empirical studies are extremely useful for developing irrigation schedules and guiding the local farmers. Almost all the results are dependent on the local soil and climatic conditions and hence are location specific and not easily amenable for adoption elsewhere. Further, these are incompatible with the water delivery constraints of the irrigation system . In order to integrate the scientific practices of irrigation agronomy with the irrigation system management practices, there is a need to generalise the empirical results of the field experiments through mathematical models and use them with optimization procedures to account for the constraints of the irrigation system operation.

3.2.1.1.2 Irrigation scheduling under limited water supply.

When water supplies are limited, crop water deficits in some periods of the growing season are unavoidable. But, crop response to deficits at different periods of the growing season is not uniform and deficits in some critical periods of growth have higher impact on yield than in others. Thus under limited water supply conditions, the irrigation scheduling problem becomes one of distributing the deficits optimally over the crop growing season [12] . The problem is complex and any attempts for its solution require the integration, in a unified framework, of information on soil and crop responses to timed inputs of water with resource allocation and decision making procedures of systems analysis and operations research. Such a framework has been developed to solve the problem of irrigation scheduling of crops. It consists essentially of three components:

i) a soil water balance model to partition the water inputs (rainfall and irrigation) into different components, namely runoff, deep percolation, evaporation and transpiration, to quantify the available water in the crop root zone at any time during its growing season [30, 50, 53, 117, 149, 154, and 175].

ii) a dated water production function model of crops to relate crop yield with water used at different periods of crop growth [39, 55, 120, 207, 220,

iii) an optimal irrigation programming model that integrates (i) and (ii) above with operational constraints of water supply systems to evaluate alternative irrigation schedules and choose the optimal schedule for given soil, weather, crop conditions [121, 126, 128 and 216].

<u>*The Soil Water Balance Model*</u>. The water balance is simply a statement of the law of conservation of mass. In its simplest form, the soil water balance equation states that, change in volumetric water content (ΔS) in the soil body over a specified period of time is equal to the difference between the amount of water added and the

amount of water withdrawn during the same period. ΔS will be positive if gains exceed losses and negative other wise. The terms of the water balance are normally expressed in units of depth, e.g., mm or mm/mm. For a given soil volume of root zone, and considering only vertical transfer, the gains and losses of water can be itemized. For this purpose, the soil reservoir is divided into two layers (Fig. 4): i) an active root zone of depth RD, and ii) a passive zone below the active zone, extending upto the maximum root depth, RDM. The active root zone is characterized by the presence of roots. While extraction of moisture in the form of both evapotranspiration and drainage occur from this zone, only drainage would occur from the passive zone. In the initial phases, the two, layers are distinct and their depths are governed by the rate of root growth. Once the roots grow to the maximum depth, the entire root zone becomes a single active zone of depth RDM. The factors influencing the soil water balance in the two zones are different. The water balance of the active root zone is governed by the values of rainfall (R), runoff (Q), irrigation (IRA), actual evapotranspiration (AET), and percolation to the lower zone (P). The water balance of the lower zone is governed by the net influx (Percolation into the zone minus deep percolation, DP out of the zone) plus the capillary upward flux if any, into the zone. In simple terms, the water balance is stated as:

Inflows - outflows = Change in storage.

Thus, for the active root zone, the average soil moisture content MC1, at the end of the Ith day can be calculated from the daily soil water balance equation for this zone as (See Fig. 4)

$$MC1_I = (MC1_{I-1} * RD_{I-1} + R_I - Q_I + IRR_I + DRD * MC2_{I-1} - P_I - AET_I)/RD_I; \quad I = 1,2,\ldots,N, \; N = \text{number of days in the crop season.} \qquad (14)$$

And for the passive zone, the average soil moisture content MC2 at the end of the Ith day is given by

$$\begin{aligned} MC2_I &= MC2_{-I}; & \text{if } P_I = 0 \;\&\; DP_I = 0 \\ &= MC2_{I-1} + (P_I - DP_I)/(RDM - RD_I) & O.W. \end{aligned} \qquad (15)$$

in which,

$$DRD = RD_I - RD_{I-1}$$

$$\begin{aligned} P_I &= R_I - Q_{UI} + IRA - AET_I - (FC - MC1_{I-1}) * RD_{I-1} - (FC - MC2_{I-1}) * DRD; \quad \text{if } P_I > 0 \\ &= 0 \qquad \text{Otherwise} \end{aligned} \qquad (16)$$

$$DP_I = P_I - (FC - MC2_{I-1})*(RDM - RD_I)\;;\quad if \quad DP_I > 0$$
$$= 0 \qquad Otherwise \tag{17}$$

For irrigation management applications, daily soil water balance is considered adequate. As root growth varies with time, the value of root depth at any time (identified as the number of days after planting (DAP)), is estimated by using Borg and Grimes equation [27] given by

$$RD = RDM\,[0.5 + 0.5 * \sin(3.03 * DAP/DTM - 1.47)] \tag{18}$$

Where DTM is the maximum root depth attainable (a crop & soil specific value), mm, and

RD = Root depth attained at DAP, mm.

Dated water production functions. In general, the functional relationship between the water input and the crop yield is called a water production function. For purpose of irrigation scheduling, the timing of irrigation besides its amount is important for healthy crop growth. The dated water production functions incorporate effects of both timing and quantity of irrigation water application on crop yield. The time periods are arbitrarily chosen to coincide with the physiological growth stages of crops, or some convenient time period such as a month or a week. The water use related variable can be represented by any of the following: the actual depth of irrigation water applied (IW), or the total field water supply (FWS) given by the sum of Effective precipitation +Irrigation applied + Stored soil moisture, or Evapotranspiration (ET) or the deficit in evapotranspiration (1-AET/PET). Some of the dated water production functions suggested by various researchers are listed in Table 12.

Optimal Irrigation Schedules: Irrigation programming refers to the process of arriving at an optimal schedule of irrigation applications for a crop during its growth period under specified conditions of water availability and climate. This calls for exploration and evaluation of the effect of all possible irrigation regimes on the crop yield and arriving at an optimal regime using an appropriate mathematical optimisation programme. The dated water production functions that relate the yield response to timed irrigation applications over the crop season provide a means for this. As the dated water productions are in terms of ET (or deficit in relative ET) and not irrigation depths, the soil water balance model can be used to translate the irrigation depths in different intervals to corresponding ET values or soil moisture content.

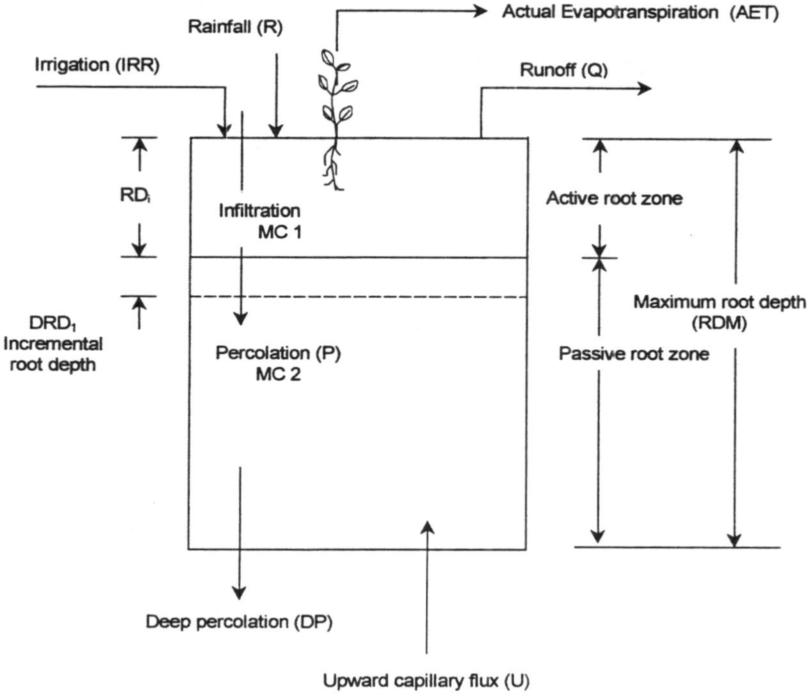

Figure 4. Components of Soil Water Balance in the Root Zone

The main difficulty addressed by models at (ii) and (iii) is that irrigation decisions in different intervals of the growing season are not independent. Each irrigation decision is a function of the available soil moisture, status of crop and available water supplies for the remaining period of the growing season. Information about all of these factors up to the time of the decision must be utilised before the irrigation decision is made [121]. Irrigation decisions are thus multistage, sequential and state dependent. This can be taken care of by carrying out the optimisation using dynamic programming with the state variables being defined by the soil moisture content and the water supply available for the remaining period of the growing season. In doing so, the model at (i) above is used to determine the soil

Table 12. Some Well Known Dated Water Production functions of Crops

Author(s)	Production function/ Relative crop yield	Independent variables
Hall and Burcher (1968)	$\prod_{i=1}^{a} a_1(d_1)$	Soil moisture deficit index (d_1)
Jensen (1968)	$\prod_{i=1}^{a} (AET/PET)^2$	Relative ET (AET/PET)
Hiler and Clark (1971)	$1 - \frac{A}{Y_M} \sum_{i=1}^{a} CS_1(1 - AET/PET)_1$	Relative ET deficit (1-AET/PET)
Hanks (1974)	$\prod_{i=1}^{a} (T/T_p)^2$	Relative transpiration (T/T_p)
Minhas *et al.* (1974)	$\prod_{i=1}^{a} (1 - 1 - AET/PET)_1^2 \ \vert^{tn}$	Relative ET deficit
Blank (1975)	$\sum_{i=1}^{a} A_1(AET/PET)_1$	Relative ET
Stewart *et al.* (1976)	$1 - \sum_{i=1}^{a} YRR_1(1 - AET/PET)_1$	Relative ET deficit
Sudar *et al.* (1979)	$\sum_{i=1}^{a} (1 - AET/PET)_1 YS_1$	Relative ET deficit
Rao *et al.* (1988a)	$\prod_{i=1}^{a} (1 - K_1(1 - AET/PET)_1)$	Relative ET deficit

Note : Y_M, potential crop yield; a_b λ_b CS_b, b_b A_b YRR_b YS_b K_1 are parameters (yield sensitivity coefficients to water deficits) for i^{th} stage, n is no. of stages.

moisture state variable and that at (ii) is used to define the objective function for optimization. The simple soil water balance model based on generally available soil parameters, dated water production functions derived from commonly available growth stage yield response factors of crops, and a realistic approach to assessing yield responses in weekly intervals were useful, practical advancements. The dynamic programming problem for a multistage irrigation scheduling process can be stated as follows:

Using the dated water production function proposed by Rao et.al [121], as

$$Y/Y_m = \prod_{I=1}^{I=N} [\,1 - K_I\,(1\text{-}AET/PET)_I\,], \quad With \quad AET_I = \sum_{I=1}^{M} e_{aij} \;\; ; \; and \; PET_I = \sum_{I=1}^{M} e_{pij}$$

where I = number of the growth stage,
j = the week number

the recursive equation of the dynamic programming problem is,

$$f_I\,(Q,W) = Max\,[1 - K_I\,(1\text{-}AET/PET)_I\,] * f_{I+1}\,(Q - X_{I+1} * W_{I-1}\,)$$
$$0 \le X_I \le Q \;\;,\; 0 \le Q \le Q_0 \;;\; I = N\text{-}1, N\text{-}2, N\text{-}3, \ldots\ldots 1 \qquad (19)$$

$$f_N\;(Q,W) = Max\,[1 - K_N\,(1\text{-}AET/PET)_N\,]\;;$$
$$0 \le X_N \le Q\;;\;\; 0 \le Q \le Q_0$$

This problem is solved subject to the specified constraints [121], to obtain the optimal water allocations, X_I (I=1, N) to the growth stages for specified water availability, Q0 and Moisture of the root zone, W_0, at the beginning of the season . The allocation to the weekly intervals is made by root zone water balance model.

3.2.1.1.3 Irrigation scheduling with crop growth simulation models

Several attempts to develop physiologically based continuous growth simulation models for several crops were reported in literature [4, 63, 64, and 91]. Basically these models simulate the various dynamic processes that lead to crop production and yield and their changes in response to the changes in environmental conditions of which water stress is one. Variations in light use efficiency and photosynthesis per unit leaf area are the major factors for biomass accumulation. However, the experimental data indicated that these two are minor factors while a change in light interception is the major factor for biomass accumulation [65]. The applicability of a rice crop growth simulation model for conserving water by making better use of rainfall and rotational irrigation was demonstrated [125]. It was shown that near potential yields could be obtained by proper choice of irrigation schedules even under relatively high water supply deficits.

3.2.1.2 Irrigation scheduling for multiple crops

Many farming situations are concerned with multiple crops grown in the same season. In such situations, a limited water supply implies that water is not adequate to produce potential yields of all the crops. This leads to competition for water between crops both at the seasonal and intraseasonal level. The problem of multicrop seasonal and intraseasonal allocation of water was addressed by Rao et. al. [127]. The single crop irrigation scheduling models [121] were extended to solve the field level multiple crop seasonal and intraseasonal irrigation water allocation problems. This was done by decomposing the problem into two levels, seasonal and intraseasonal (Fig.5). The single crop model provides the input to both levels. The effects of economic coefficients, areas cropped, and the relative sensitivities of crop growth stages to water stress are integrated to the decision taking mechanism. The sequential nature of irrigation scheduling is maintained by the optimisation at both seasonal and intraseasonal levels by dynamic programming. The final output is set of implementable weekly irrigation programmes for individual crops. The decomposition of the multi-crop water allocation problem reduced its dimensionality at the cost of solving the single crop model several times; thus the computational effort is reduced significantly.

3.2.1.3 Real-time irrigation scheduling

The models described above are useful for planning purpose and are based on estimated/probabilistic/forecast of weather and water supplies for the entire growing season. Irrigation schedules determined based on these models could be considered to be the planned or design irrigation schedules. But in a current season both weather and water supplies will be different from those assumed in deriving the planned irrigation schedules of crops. Thus in real-time, the planned schedule for a crop needs to be balanced with the uncertain information on weather and water supplies [128]. Since irrigation decisions are sequential, and the effect of each decision can be evaluated only at the end of the season as crop yield, they are determined in real-time by a two-stage process. First, irrigations are planned for the entire season at weekly intervals based on historical data of weather and water supplies using the single crop model developed earlier [121]. Then, the decisions are revised each week for the remaining season using real time data up to that week and solving the irrigation optimization model for the remaining weeks. Thus in each week an irrigation decision is made the entire planning horizon is kept in view (Fig.6).

3.2.1.4 Use of medium range weather forecasts in irrigation scheduling

Medium range weather forecasts provide information about the weather 3 to 10 days in advance. They are planned to be used widely in agricultural management including irrigation scheduling. A logical basis for the evaluation and use of medium range weather forecasts can be developed. In general, forecasts can be integrated into real –time operational management of dynamic systems, such as irrigation systems. The requirements are: i) the variables affecting the state of the

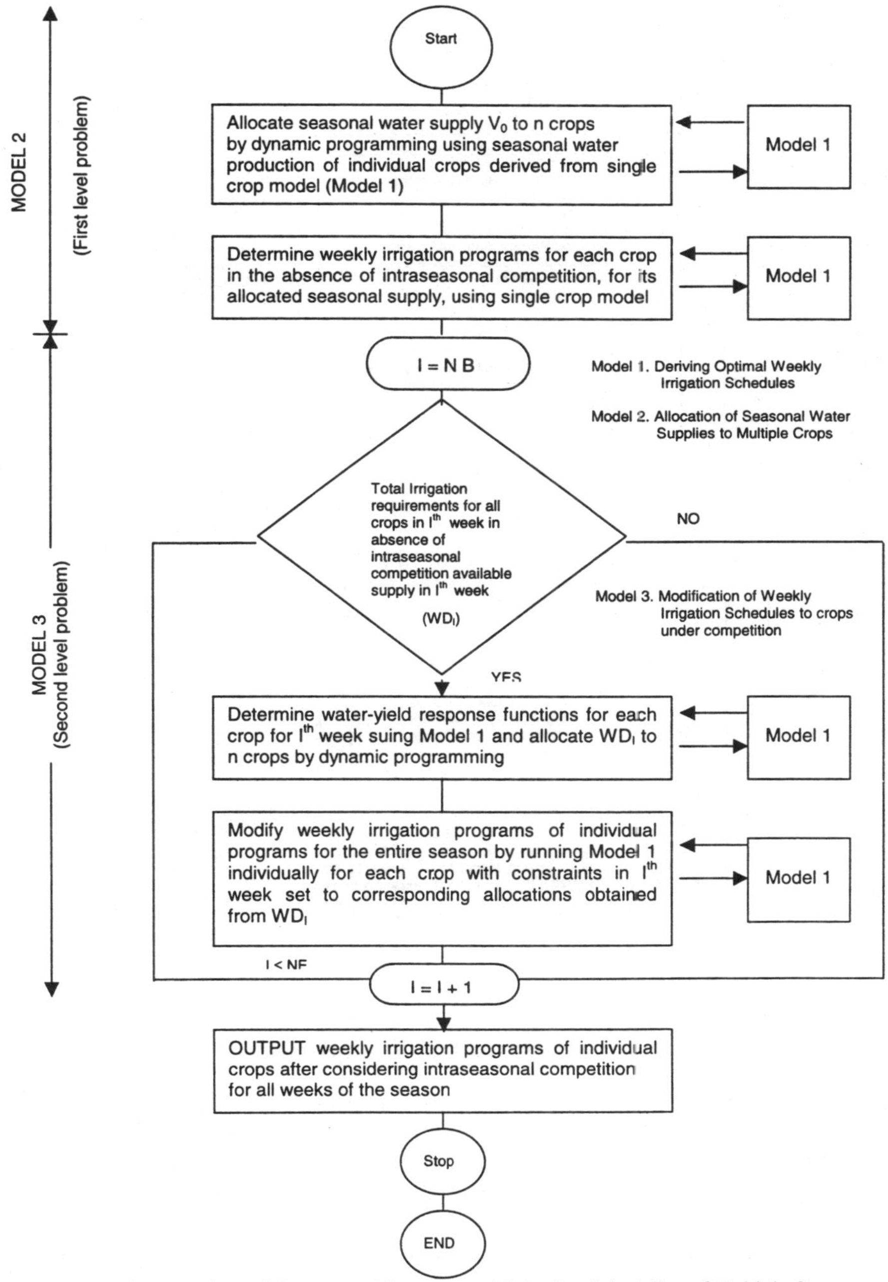

Figure 5. Computations of Seasons and Intraseasonal Irrigation Scheduling of Multiple Crops

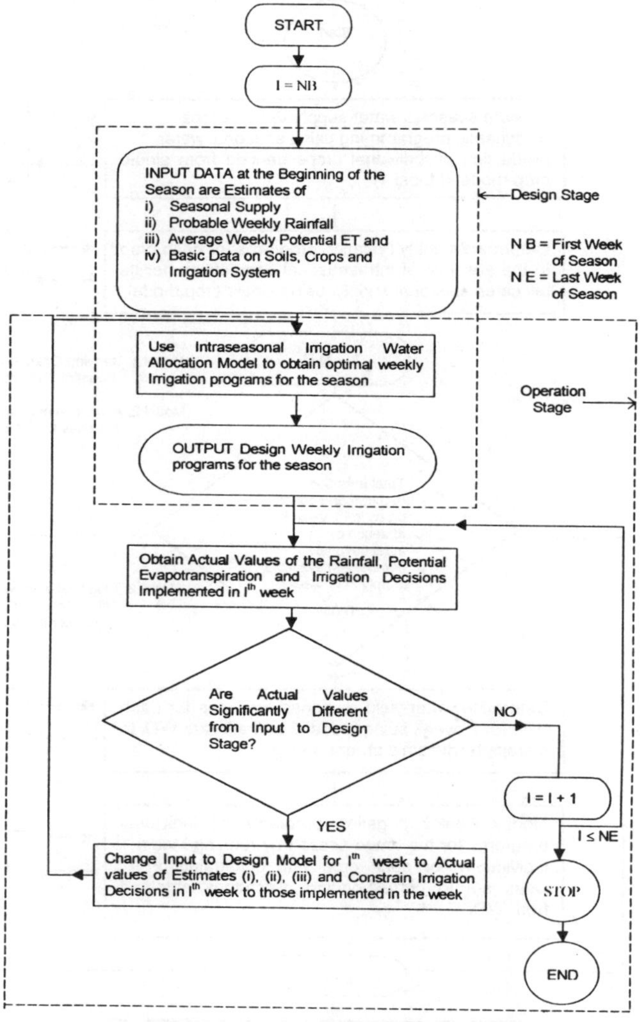

Figure 6. Scheme of Real Time Irrigation Scheduling

system and decisions are identified, ii) the forecasted variable affects the system, iii) this influence can be determined, and the state of the system updated iv) criteria for taking decision is linked to the state of the system and v) the constraints on the decision variables are specified. Historical rainfall data were used to examine the influence the 3 to 5 days advance information of rainfall will have on irrigation scheduling of crops (Table 13). The methodology is based on a two-layer daily soil water balance model to define the dynamic soil water state and irrigation scheduling criteria related to soil moisture depletion.

It was shown that if irrigation decisions are based on monitoring the soil moisture, then the forecasts do not lead to significant water savings for deep-rooted crops in soils with relatively high available water capacity. On the other hand, they may be useful in shallow soils and situations where small irrigation depths are applied more frequently [50].

Table 13. Number of Years (Out of 14) when there is a Shift in Irrigation Schedule by a week or more after use of 3 or 5-Day Forecasts in the Command Area of Jayakwadi Irrigation Project, Maharashtra [50].

Crop	No.of years with Shift of 1 week		No. of years with shift of two weeks	
	3- day forecast	5- day forecast	3- day forecast	5- day forecast
Sorghum	2	2	1	2
Cotton	4	3	3	5
Sugarcane	3	2	1	-
Banana	2	1	1	2

3.2.1.5 Expert decision support system for scheduling irrigations

Despite the advances in irrigation scheduling models, the uncertainties in model parameters on one hand and the farm-irrigation system interactions on the other, ensure that human judgement and expertise will continue to be a major source of decision support in irrigation management. Models will enhance this expertise to solve field problems by enabling quantification of information and evaluating alternatives. The most appropriate irrigation schedule can be arrived at by using quantitative model predictions together with local knowledge and experience of farmers and irrigation operators.

Such a balance between the rationality of concepts included in models and circumstantiality of practice is made possible by integrating models and heuristic knowledge in expert systems (ES) framework. Basically an expert system is an intelligent computer program that uses knowledge and inference procedures to solve problems that are difficult enough to require significant human expertise for their solution. The knowledge necessary to perform at such level plus the inference procedure used can be thought of as a Model of the expertise of the best practitioners in the field [44, 177]. Sarma *et al.* [177] have presented the development of such an integrated expert decision support system (IEDSS) for irrigation scheduling (Fig. 7). The IEDSS uses a simple soil water balance model. Using independent packages of heuristic rules (expert system, ES) facilitates the estimation of each component of the soil water balance (runoff, evapotranspiration, etc.,) for each model parameters. For example, the ES for runoff is based on a set of rules for assessing the values of Curve Number and soil hydrologic group. The ES for evapotranspiration includes rules for estimating crop coefficients, growth stages, and reference evapotranspiration. There are additional ES for estimating initial soil moisture content, field capacity and wilting point, all of which are important parameters of the soil water balance model. To increase confidence in IEDSS predictions of irrigation decisions, they were compared with decisions obtained by heuristics (qualitative) rules and farmer practice over 3 years. For this purpose two additional ES were developed, an ES for irrigation scheduling based on crop growth stages (ESGS) and the other based on qualitative assessment of soil moisture. The IEDSS predictions of irrigation timings led to the saving of one irrigation in all the years and were midway between the dates practised by the farmers (Table 14).

3.2.2 Planning and Operation of Irrigation Reservoirs

Irrigation reservoir operations are aimed at deriving maximum benefits from the water that can be stored in it and allocated to the crops. Water releases from the reservoirs are to be conveyed through a hierarchical distribution system of main canals, branch canals distributaries and field turn outs/outlets before they reach the cropped fields. The operations are complex but substantial increase in benefits can be obtained even from relatively small increases in operating efficiency. [75,76]. Many studies have dealt with the general problem of optimal reservoir operation and a detailed review of reservoir operations is presented by Yeh [234]. The problem of real time reservoir operation for irrigation of multiple crops was also studied [36, 221 and 222]. But these studies have focussed mainly on determining the reservoir releases. They did not simultaneously consider conveyance and distribution of the releases to the field turn outs and subsequent water allocation to crops on weekly / fortnightly basis. All the three stages of the irrigation operation problem, namely, determining the reservoir releases, transferring them to the field level, and allocating the field supplies to crops are important components of large irrigation systems.

As the irrigation release decisions are periodic and sequential, the consequences of the decision can be evaluated only at the end of the crop season, when the yield is known. This requires the entire planning horizon be kept in view while making the irrigation decisions in a current interval [128]. Hence the reservoir releases at the head works need to be planned carefully for the entire season after assessing the impacts of the decisions on crop yields. This can be done based on information of storage in the reservoir, expected inflows, target release requirements (based on crop water needs) and crop yield impacts of the intraseasonal irrigation decisions.

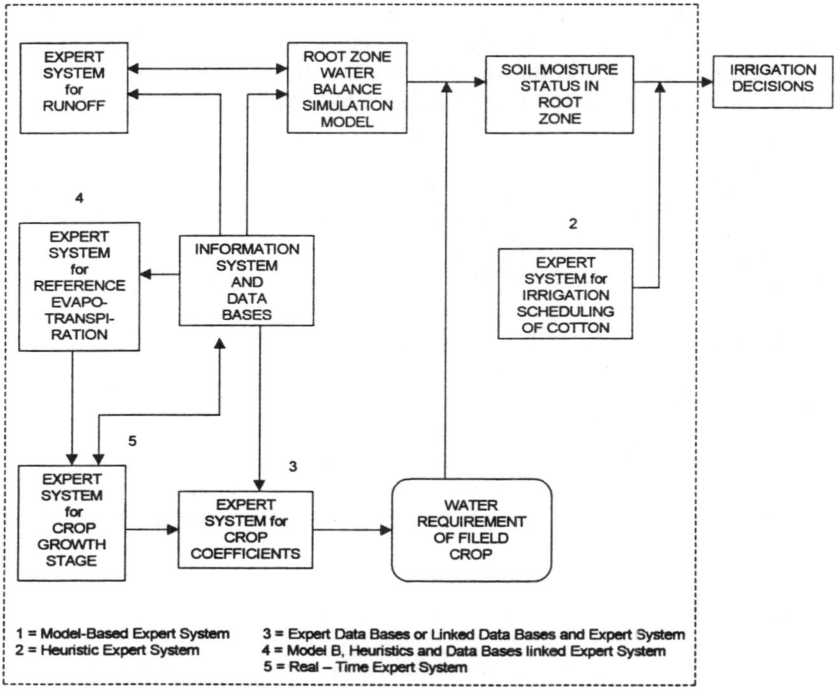

Figure 7. Schematic diagram of integrated expert decision support system for irrigation scheduling.

Table 14. Comparisons Of Irrigation To Cotton Crop By Using An IEDSS And The Farmers' Practice In The Command Area Of Paladugu Distributary, Nagarjuna Sagar Right Canal (NSRC) Command Area, Andhra Pradesh [37].

Year	Timing of irrigation (days after sowing)	
	IEDSS	Farmers' practice
1983-84	120	108 to 125
1984-85	115	100 to 131
1985-86	105	97 to 124

Further, in real time operation, the plan needs to be modified periodically based on current inflow data. This too shall be done keeping the entire planning horizon in view. A unified framework for the formulation and solution of the irrigation reservoir operation problem has been developed by posing the problem in two-stage [52].

i) planning seasonal and intraseasonal irrigation releases from the reservoir, and
ii) real time adaptive operation of the irrigation reservoir

The problem was decomposed into the following sub problems:

- Forecasting inflows into the reservoir using a stochastic hydrologic models [29],
- Preseason planning of periodic reservoir releases [52],
- Modifying the release plan in real time for the remaining season, at successive intervals using current data and revised forecasts [52],
- Transferring the planned releases in current interval to the field level, by accounting for the conveyance losses using a suitable canal hydraulic model [230]
- Allocating field level supplies in the current interval to crops using information on crop yield response to irrigation [52].

This set of problems was solved sequentially in each successive interval of the current season, from first to last interval using dynamic programming. The basis for these studies was provided by a number of earlier studies at irrigation project level dealing with quantifying available water supplies [119, 152, 155, 156, 163, 164, 165, 166 and 180], conveyance losses [230,], and field irrigation requirements [52, 167, and 181], and the field scale irrigation scheduling studies [120, 121, 127 and 128] discussed earlier.

3.2.2.1 Planning intraseasonal reservoir releases

The irrigation requirements at the reservoir head works were estimated in two stages [53]. At the first stage, the intraseasonal weekly/biweekly irrigation requirements for each crop were estimated using a two layer soil water balance model and the water requirements at the outlet level were calculated from information on the areas of crops. At the second level, the field level water requirements were transferred to the upstream nodes of the water distribution network after accounting for conveyance losses by a simple hydraulic model for flow in canals. The main questions examined were (i) whether there are significant year to year fluctuations in intraseasonal field level irrigation requirements of crops, and (ii) whether a reliable sequence of target weekly/biweekly reservoir releases can be derived for irrigation planning at the project level. This was done for the Jayakwadi irrigation project in Maharashtra, India. It was shown that for this project, a reliable deterministic weekly/biweekly sequence of irrigation release requirements at the reservoir head works could be derived.

3.2.2.2 Real-time reservoir operation

Real time adaptive operation of reservoir based canal irrigation systems is carried out in two phases (Fig.8), a plan phase [53] and operation phase [51,54]. In the plan phase, biweekly reservoir releases are optimised for a set of target releases [53] and mean inflows. This is done by minimising a water supply deficit function using dynamic programming. In the operation phase, real time releases are made according to plan for the current interval. The releases are transferred to the field level after accounting for conveyance losses. At this level, the allocations of available field supplies to competing crops are also optimised. The plan for the remaining intervals of the season is then modified by replacing the inflow for the previous interval with the observed inflow during the interval. The current interval inflow is replaced by the corresponding forecast inflow. The forecasts are obtained from a time series model of weekly inflows. These four steps, namely, (i) releasing supplies from reservoir according to seasonal plan in the current interval, (ii) determining the fraction of the releases available to crops at field level, (iii) allocating the field level supply to competing crops, and (iv) modifying the reservoir release plan for the remaining intervals of the year based on inflows and forecasts, operate sequentially in real time for each interval. As the season advances, this cycle repeats successively from the first to last intervals of the year. The application of the procedure is demonstrated using data from the Jayakwadi irrigation project in India [53].

The software for planning and operation of reservoir based irrigation systems is documented and a manual for its use was published by the Central Board of Irrigation and Power [52].

3.2.3 Conjunctive Use Of Canal Water And Groundwater In Irrigation Projects

In many large canal irrigation projects in India, the potential for additional groundwater development has increased significantly because of recharge from the irrigation systems [116, 119, 157, 168, and 169]. There is a high emphasis in recent times on conjunctive use of canal water and groundwater to exploit the additional potential to augment canal supplies and also to control the water table. Conjunctive water use refers to the integrated management of surface water (canal water) and groundwater resources in a region to meet specified objectives. This requires:

i) quantification of annual recharge and its spatial distribution to assess the potential for conjunctive use [46, 129, 142, 146, 200, 202, 203],

ii) simulation of the groundwater basin underlying the command area to analyse the impacts of irrigation and present and additional groundwater development on the water levels in the aquifer [21, 46, 130,131,142,172, 199, 201, 203 and 204], and

iii) identification of a conjunctive use strategy that is most suitable for the given hydrologic and agro-economic conditions [129, 130, 150, 186, 200, 202]

3.2.3.1 Estimation of recharge and its spatial distribution

The sources of recharge in canal irrigation project areas , besides rainfall, are:
i) seepage from canals and water distribution system, and

ii) percolation from cultivated and uncultivated fields.

Both of these components can be estimated using empirical norms developed by the Central Groundwater Board for rainfall recharge and conveyance losses in distribution system [9] or by using more physically based procedures. The empirical procedures were used to provide lumped regional and annual estimates of recharge for several irrigation project command areas [129,156, 168,169, 200, 202].

3.2.3.1.1 Seepage models

Canal flows are highly variable and the wetted area changes with depth of flow. Seepage losses therefore vary with flow conditions. To account for these variations a simple model to simulate flow conditions in canals was developed [230]. This model was used to estimate seepage losses for actual flow conditions in Jayakwadi project, Nagarjunasagar, MRBC and Godavari Central Delta Canal (GCDC) projects. Rating curves can also be developed using the results of the model application to relate canal discharges directly with seepage losses [33, 52, 53].

The seepage coefficient used in the calculation of seepage losses also varies with the subsurface water table and lithological conditions. Rao [222] developed a physically based procedure for including these conditions in seepage coefficient calculations. The procedure is based on simulation of vertical flows in the aquifer underlying the canal system. It was applied to the physical and lithological conditions of the aquifer underlying the command area of the MRBC and GCDC irrigation projects. It was shown that seepage losses could vary by a factor of 7 depending on the location of the water table and the depth of the aquifer. But only a fraction of these were vertical flows which recharge the underlying semi-confined aquifer.

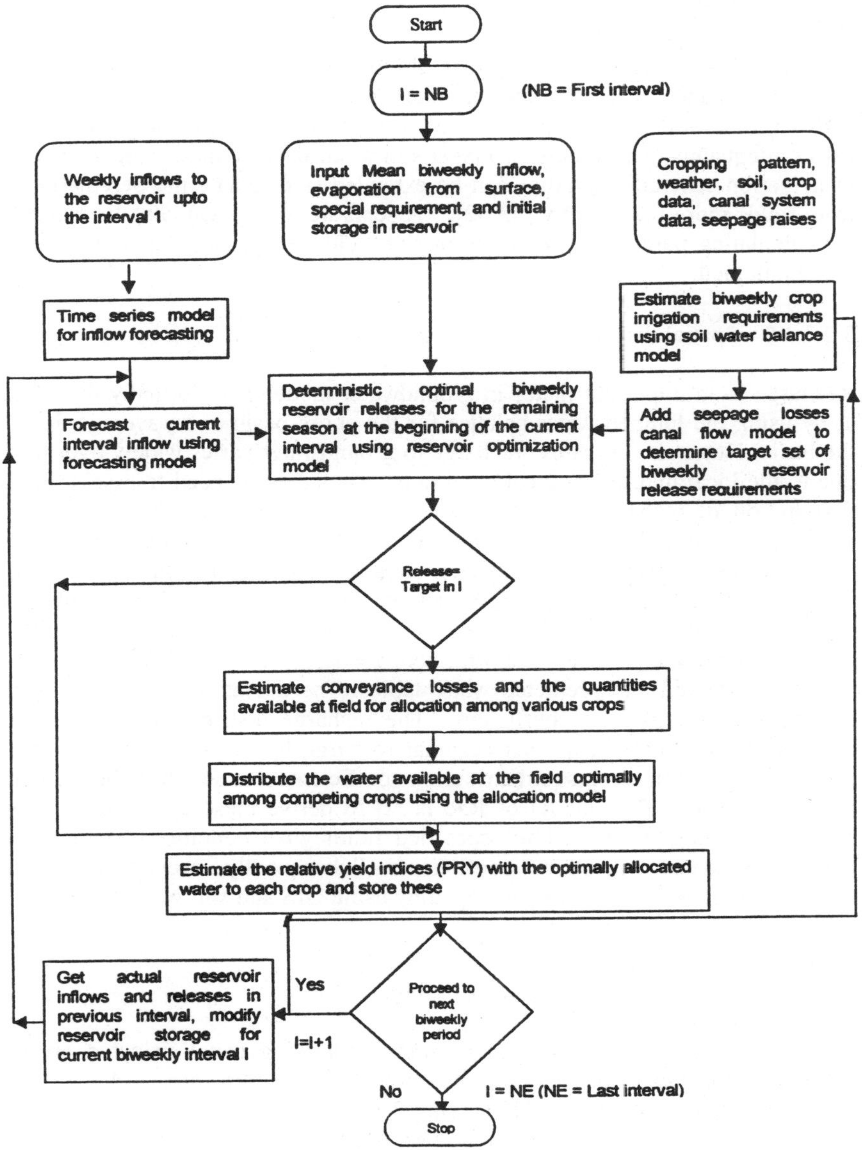

Figure 8. General Scheme for real time adaptive operation of irrigation operation

3.2.3.1.2 Percolation models

Deep percolation from irrigated areas ultimately joins the local groundwater body and thus enhances the groundwater potential. Percolation losses from fields vary with weather, soils and crop conditions. Empirical norms cannot account for these losses adequately. The simple two layer soil water balance models developed earlier in irrigation scheduling studies can be extended to calculate the percolation losses from fields more realistically [33,52]. A separate routine can also be incorporated for calculating percolation losses from rice fields where standing water conditions are maintained.

3.2.3.1.3 Spatial distribution of recharge

Recharge is an important factor in groundwater modelling. Normally this is arrived at by using the lumped water balance for the entire basin as an average depth over the basin. However, in reality, recharge to groundwater varies widely over the basin due to the local hydrogeological conditions. Thus there is need to assess the spatial distribution of recharge over the basin so as to incorporate this in groundwater models. Sondhi and Sarma [201] and Sondhi et.al, [202] developed the concept of 'recharge distribution coefficients' for an area. By this procedure, the empirical norms were first used to derive the annual lumped recharge. A set of recharge distribution coefficients for the area were derived by running a finite difference groundwater model for the area with zero recharge and calculating the deviations between observed and predicted water tables. The deviations were used to define the recharge distribution coefficients. The recharge distribution over the area is obtained by multiplying the net regional recharge by the above coefficients. The physical basis of the recharge distribution coefficients was sought to be increased by using the vertical flow model of Rao [122] to derive these coefficients. This was done by running the model for specified lithological conditions in the area and deriving seepage distribution coefficients. Subsequently, procedures for spatial distribution of recharge were developed by using GIS and soil water balance models [33].

3.2.3.2 Simulation of groundwater basins

Simulation models of groundwater basins underlying canal irrigated areas were developed for Indian Agricultural Research Institute farm area [143, 172], Mahi [201, 202, 204], Sriramsagar [131], Nagarjunasagar [108], Balasamund sub branch and Rana and Balasamund distributaries (BRBD) command area [46], Godavari Central Delta Canal [35] project areas. These are based on finite difference or finite element simulation of groundwater basins. The applicability of the models was sought to be increased by using expert systems technology for recharge and aquifer parameter estimation [129].

3.2.3.3 Estimation of aquifer parameters of a groundwater basin

One of the major difficulties in the development of a groundwater model is the identification/evaluation of the parameters of the physical system. Normally, very limited well-testing data are available for identifying the aquifer parameters and such data is not adequate for developing a dependable model for large regions. Sarma and Immaraj [172] presented a procedure the concepts of *inverse system analysis* to identify the values of parameters at a large number of locations in the basin. A constrained linear programming (LP) problem based optimization was adopted to obtain the suitable set of parameters that adequately represented the groundwater basin.

3.2.3.4 Development of conjunctive use strategies

Systematic procedures were developed to assess the potential for conjunctive use in irrigation projects [22, 116,117,150,210] and design conjunctive use plans.
Conjunctive use strategies were developed for the MRBC project in Gujarat using simulation models [199,200,202,203], Sriramasagar Irrigation Project Command area [131], and BRBD command area [46]. Linear programming and groundwater models were developed to derive optimal cropping patterns for conjunctive use and Bio drainage. The effects of each conjunctive use plan on the groundwater table were also evaluated.

3.2.3.5 Expert decision support system for conjunctive use

A balance between the rationality of concepts included in models and circumstantiality of practice can be made possible by integrating models and heuristic knowledge using expert systems (ES). Sarma et.al, [177] have presented the development of such an integrated expert decision support system (IEDSS) for conjunctive use (Fig. 9). The IEDSS uses the simple soil water balance model. The estimation of each component of the soil water balance and the groundwater model, (runoff, evapotranspiration, aquifer parameters, deep percolation.) is facilitated by using independently developed heuristic rules for each expert system, ES [37].

3.2.3.6 Biodrainage technology

It is estimated that nearly 24 per cent of the area irrigated by the major and medium irrigation projects in India has been degraded due to waterlogging and secondary salinization [105]. The problem of waterlogging becomes more complex in land locked regions with poor quality of groundwater as the conventional methods of surface/ subsurface/ or vertical drainage by tube wells is not feasible. The only alternative is "Biodrainage" namely, removing water in the upward direction into the open atmosphere as evapotranspiration through plants. This technology was successfully practiced as an effective measure of drainage, ,particularly in dry and arid regions like the Indira Gandhi Canal Project, Rajasthan [69]. A strategy is required for selecting a suitable socially acceptable and economically viable cropping and vegetation activity. A system analysis approach was developed for this purpose. Essentially, the strategy involves :

i) Analysis of the water balance of waterlogged region considering all inputs and outputs
ii) Assessing the groundwater recharge in the region and distributing the recharge spatially,
iii) Studying the groundwater regime of the region with an appropriate regional groundwater basin simulation model and identifying the areas with potential risk of waterlogging,
iv) Developing optimal cropping pattern using linear programming problem, so as to increase the evapotranspiration for the region to limit the water table to a favourable position , and to maximise net profits subject to local constraints of water availability, social acceptability, economic viability, etc., and evaluate alternative re-vegetation strategies to control the water table rise in such areas,
v) Formulating guidelines to determine the spatial distribution of different crops and vegetation in the region and study the long term impact of such a practice, on groundwater conditions..

The landlocked command area of BRBD in Hisar district, Haryana State, India covering an area of 86,370 ha. was selected as the case study area for developing the strategy and test the hypothesis [46, 185].

3.2.3.7 Drainage of agricultural lands

Drainage of agricultural lands is as important as irrigation for efficient water management. Poor or inadequate drainage leads to severe damage to standing crops, waterlogging and soil salinity, delays in cultivation of the subsequent kharif crop etc. Considerable research information is generated on various aspects of agricultural land drainage such as details of design [137 90], drainage materials, drainage coefficient [189], drainage structures [16] etc.

Singh *et al.*, [192] developed procedures for estimating 2-6 consecutive day rainfall values for different recurrence intervals of 2 to 20 years , to provide a basis for design of drainage systems. Another important factor in drainage design is the drainage coefficient . Sharma and Michael [189] developed a procedure for determining the drainage coefficient for the Tarapur block of the MRBC command area, Gujarat. The procedure involved a trial and error approach for determining the Curve Number (CN) value for estimation of runoff, determination of deep percolation through water balance analysis and assessing the changes in groundwater table levels.

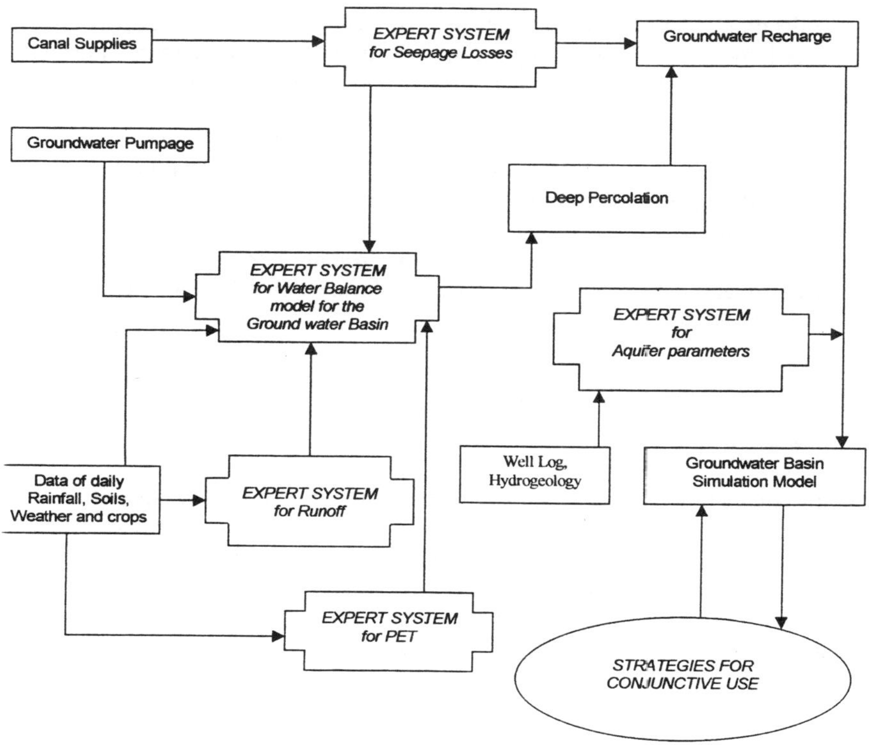

Figure 9. Integrated Expert Decision Support System for Conjunctive Systems

Wherever the topography is conducive , simple surface drains can be used to provide agricultural drainage . A drainage system can also be a combination of both surface as well as subsurface drainage . Such systems were designed for Delhi villages [23], and tea gardens in north east India[25]. In situations where pumping of groundwater is involved so as to lower the high water table, a battery of shallow wells can be used [211,188].

3.3 Non Point Source Pollution of Groundwater in Irrigation Projects

Quantifying the nitrate losses is important for devising measures to control them and ensure sustainability of soil and groundwater resources. This requires quantifying the different nitrogen transformation processes occurring in rice fields. These are urea hydrolysis, volatilisation, nitrification, mineralization, immobilisation, denitrification, uptake and leaching. All these processes are influenced by weather, soil properties, irrigation and other factors. Further, the concentration of nitrate in the leached water depends on the percolation rate. Thus, nitrate loadings to groundwater will also depend on the percolation losses from fields.

Chowdary *et al.* [34, 186] developed a coupled soil water balance-nitrogen balance model to estimate nitrate losses from rice fields. The model (Fig. 10) was tested with data of several locations. The model was also linked to a groundwater transport model to assess long term impact of groundwater pollution under different scenarios of agronomic practices, in the Godavari Central Delta Canal irrigation project (See also section 3.4.2).

3.4 Development of Decision Support Systems for Irrigation Management and Use of GIS

Linking the simulation models and expert system models to spatial databases through the use of GIS is useful for developing expert decision support systems for conjunctive use, monitoring non point source source source pollution and real time irrigation operation. The focus in these studies is on providing decision support to irrigation planners and managers by enabling them to use routinely collected spatial data and forecasts more effectively in decision-making.

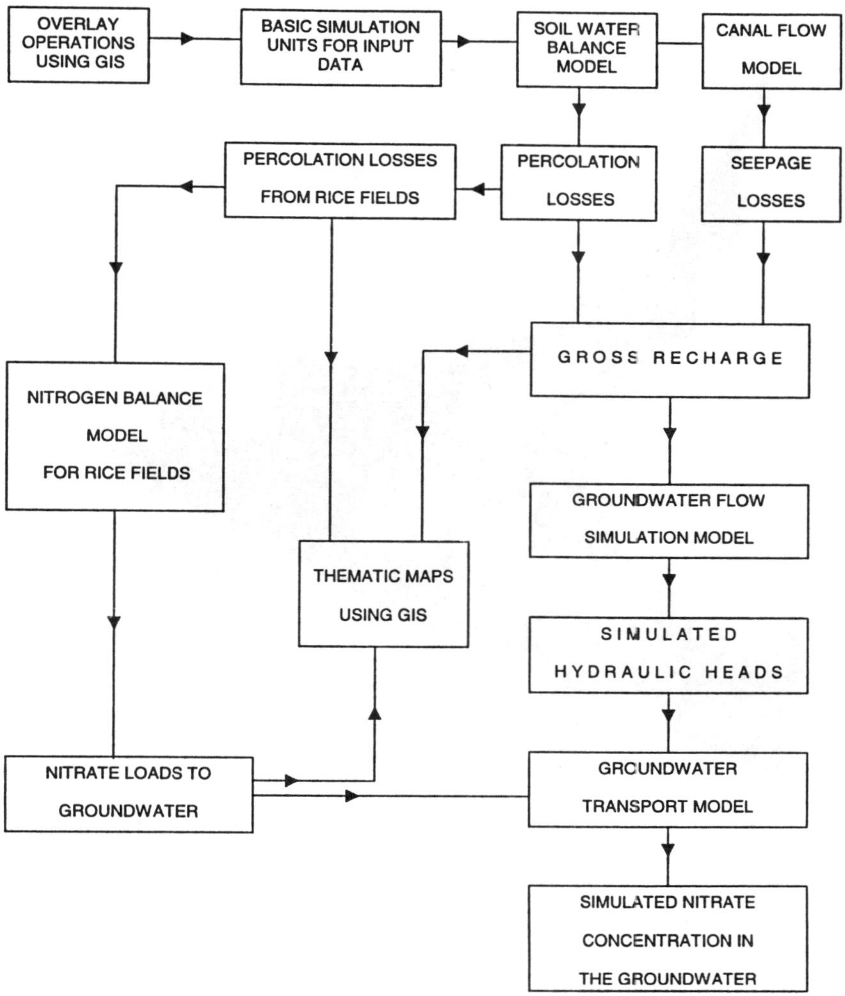

Figure 10. Scheme of Integrated Framework for Assessment of Non-Point Source Pollution of Groundwater in Irrigation Project Areas

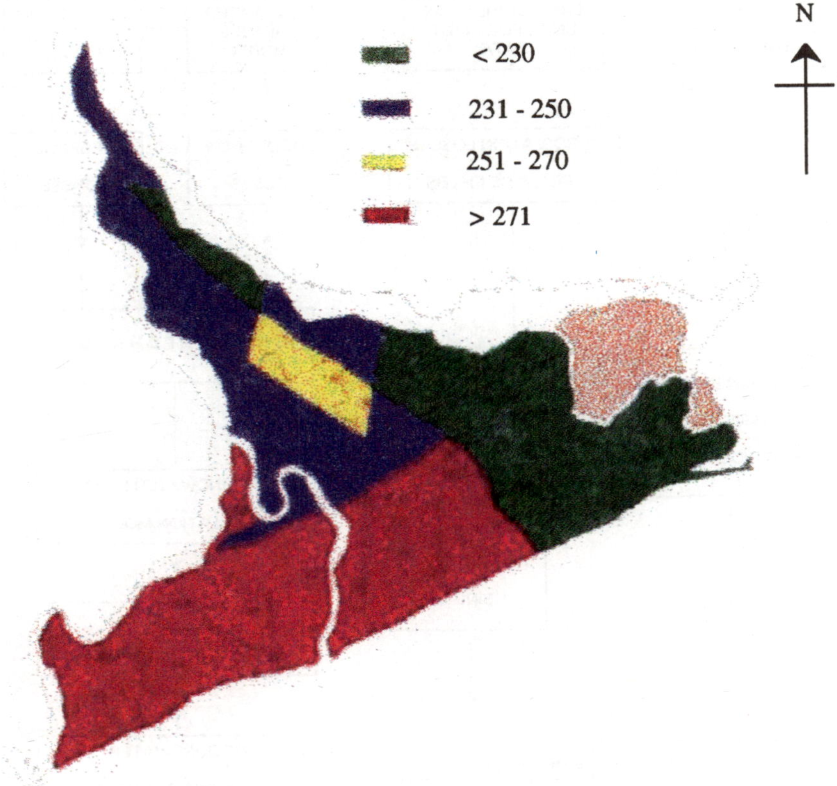

Figure 11. Spatial distribution of nitrate losses from rice fields in Godavari Delta Central Canal Irrigation project

3.4.1 GIS for Spatial Distribution of Recharge

The spatial distribution of recharge for variable weather, soil, land use and water supply conditions over the command area of an irrigation project was assessed using GIS. Field data of GCDC Irrigation Project were used as a case study to develop the procedures [33]. Daily data of 7 raingauge stations, discharges in 4 main canals, soil and command area maps, groundwater structures in each block/*mandal* were available. The soil map, raingauge Thiessen polygons, command area map and the block map were digitized to obtain the respective coverages using ARC/INFO (ver 3.4.2). By performing overlay operations sequentially with all the 4 coverages a fifth coverage was obtained. The polygons of the newly derived coverage are designated as ' basic simulation units'. Each polygon of this coverage is homogeneous with respect to the soil parameters, the daily rainfall and water supply, which are the inputs to the soil water balance model.

Seepage losses were assessed by using rating curves of canal discharges vs. seepage losses derived from a simple model of canal flows developed in earlier studies [52]. A generalised daily soil water balance model was developed to estimate the percolation losses for all crops. The model includes a separate subroutine for rice, which is grown under flooded conditions and under a distinct water management regime. The daily soil water balance model was run for each unit to obtain the percolation losses from various fields. The coupling between the GIS and model was achieved by developing suitable pre-processors for linking the input and output routines of the soil water balance model. The percolation losses and seepage losses were added to obtain the total annual recharge and displayed on the GIS. The recharge rates derived from GIS were verified by using a finite element based a groundwater basin simulation model. The groundwater levels predicted by the model were close to the observed levels confirming the recharge and its spatial distribution.

3.4.2 GIS for Nitrate Pollutant Loads to Groundwater

Nitrate losses from fertiliser applications vary spatially and in time with weather, soil, and water supply conditions over the command area of an irrigation project. The coupled soil water balance-nitrogen balance model developed above was used to assess spatially and temporally varying nitrate pollutant loads from rice fields over the GDCC project area using GIS [34]. Urea is the main fertiliser considered.

The coupled daily soil water balance - nitrogen balance model was run for each unit to obtain the annual percolation and nitrate leaching losses from various fields. The link between the GIS and model was achieved by modifying the input and output routines of the model. The annual nitrate loads in leached water were displayed on the GIS (Fig.10). The pollutant loads were input to a finite element groundwater transport to assess the nitrate concentrations in groundwater at the end of each season. The estimated pollutant loads and the transport model were validated by comparison with field data of nitrate concentrations in wells at the end of monsoon season 1995.

3.4.3 Development of Decision Support System for Real Time Irrigation Management

Decision support systems can be developed for real time management of irrigation systems. This is achieved by suitably combining the real time data with the decision support system scheme developed for planning the irrigation system management. This was developed for the Sone Irrigation Project in Bihar [185]. A GIS of the Canal network of the project area displaying the various main / branch canals and distributaries has been developed A set of modular soil water balance models for rice and other crops, and canal flow models, developed and tested in earlier studies were adopted for use in the Sone project area. From the soil water balance model, information on irrigation scheduling (depth and timings) of crops is derived and the canal flow model helps account for seepage losses. Based on this, the biweekly irrigation requirements at the head of each distributary were derived. The GIS of the Patna canal system component of the command area and the rice water balance model have been dynamically linked. This linkage facilitates:

i) selection of the distributary of interest from the canal network GIS,
ii) running the rice field water balance model in real time up to current date in any year after entering the date in response to screen queries,
iii) preparing a report of the current water status in rice fields in the command area of the distributary transplanted on up to five different dates, and
iv). preparing a water indent for the irrigation requirements at the head of the distributary for the next 14 days of the irrigation cycle after accounting for conveyance losses.

A convenient feature of the software is that Individual distributaries can be selected at random by clicking on the map from the GIS files and reports of water status in fields and indents for water for the next irrigation cycle on any given date can be prepared.

3.5 Main System Management (Scheduling of Flows in Irrigation Canals)

During the early 70s, considerable emphasis was laid on on-farm development (OFD) works and improving the efficiency of water distribution at the farm level through a major national progarmme of "Command Area Development [15]. However, it came to light that unless the main system, comprising the network of main canal, branch canals, and distributaries does not function properly, the benefits of OFD can not be achieved. Inadequate conveyance network and inappropriate water distribution policies are identified as major factors for poor operational efficiency of large irrigation systems.

Normally the canals are designed for supply based distribution of water. However to obtain maximum irrigation efficiency canal scheduling has to match the demands for water. Appropriate models are required to be developed to determine delivery schedule of canal network i) to meet the water requirement of its command area ii) to determine optimal level of rotation and iii) to determine appropriate supply level.

3.5.1 Integrated Canal Scheduling

Rajput and Michael [101, 102] developed a computer based Integrated Canal Scheduling Model for scheduling the canal water deliveries on the basis of the actual water requirements of its command area. The water requirements of different crops were estimated on the basis of the moisture balance of the root zone. Operational schedules of different components of the canal network were determined based on the simulation of flows in a canal network.

For the purpose of simulation, the daily average value of actual evapotranspiration of the command area for a multi-crop system was obtained based on the weighted average evapotranspiration values of different crops grown at a time. Effective rainfall was considered equal to the rainfall minus the surface run-off. Maximum value of the effective rainfall was limited to the soil moisture deficit on any day [85]. The surface run-off was computed using the Curve Number Technique. The quantum of water contributed from groundwater towards meeting the crop consumptive use requirements was estimated from the nearness of the water-table to the root zone and the capillary and conductive properties of the soil. For a multi-crop system, the average allowable maximum soil moisture depletion value was, estimated on the basis of weighted average allowable moisture deficits of different crops, their cultivated areas and field capacity, wilting point and bulk density of the soils of the area.

A conceptual canal system consisting of one main and four branch canals was considered. Depending upon the available supply rate at the head of the canal network and the carrying capacities of its different branches, the branches may operate at full supply level or less than that in different number of ways. All possible ways of operation of the four branches with respect to available supply rates at the head of the canal network were considered in the model.

The Integrated Canal Scheduling Model was applied [102] to the Sansad distributary system of MRBC, Gujarat, to determine the delivery schedules of its minors and sub-minors for a particular cropping pattern.

3.5.2 Optimum Operation Pattern of Canal System under Different Supply Rate Conditions

The four branches of the conceptual canal network could be operated in a number of ways depending upon the priority of operation. With each operation sequence, eight operation schedules could be adopted for the operation of the canal system, giving rise to 192 possible combinations of operating the canal system. This model, together with an exhaustive search model was applied to determine the optimum operation schedules for different minors and sub-minors of the Sansad distributary for a large range of available supply rates and supply levels [102].

3.5.3 Optimum Operation Pattern of a Canal System under Different Levels of Rotation

The *rotational* water distribution (RWD) system puts a constraint on continuous flow in canal system and instead calls for an " on" and "off " system. This is commonly followed in all the irrigation systems where any form of RWD is followed. The durations of "on" and 'off' periods need not be equal but depend upon the type of soils, crops and their consumptive use requirements. But different ratios of the 'on' and 'off 'durations would lead to different amounts of seepage losses and also would require different carrying capacities of the canal network which involves different amounts of earthwork and land areas and associated costs at the time of construction. An appropriate model could be developed to determine the optimum operation pattern for the Sansad Distributary system (MRBC), under different levels of rotation [104].

3.5.4 Water Management Interventions

In many instances, it is common to develop an "Optimal Cropping Pattern" using linear programming problem. Area constraints, water availability (normally seasonal), crop choice etc are used as constraints. However, it is common to notice that no such cropping pattern was ever adopted by the farmers. Intra-seaonal short duration (10 to 14-day period) shortages of water availability are the major limitation in adopting the optimal cropping patterns. The discrepancy between water supplies and water required by the crops could be as high as +85% to -85%. Appropriate water management interventions can bring down this disparity substantially. A simple agronomic intervention involving the minor shift in sowing / planting dates of crops could bring down the supply- requirement discrepancy to about +45% to - 45%. Further intervention in terms of appropriate timings and quantities of canal water releases, could bring down the discrepancy to as low as +15% to -15% [108].

4 Status of Research in Irrigation Water Management at Field Level

4.1 Development of On-Farm Water Management Technologies

Proper management of water at the field level plays a crucial role in improving the efficiency of irrigation and crop productivity. Fragmented land holdings with irregular field boundaries are a common feature of the canal command areas in the country. Lack of well-maintained media of conveyance of water from the outlet to property heads remains a major, bottleneck in the efficient functioning of the canal irrigation systems. The fields below an outlet need to be properly levelled graded and laid out for smooth flow of water. Appropriate water distribution and control structures need to be installed where ever needed. The watercourses need to be properly aligned and lined so that the transmission losses and the cost involved are minimal. Further, the water distribution schemes to be such that the losses are minimal and the crops do not suffer stress. Research efforts on these lines are briefly mentioned here.

4.1.1 Optimal Layout of Water Courses in Outlet Command Area

For the successful operation of any water distribution system, water courses should touch each holding with at least one delivery point. Since the seepage losses below a canal outlet are a direct function of the length of water course, minimisation of the total length of water courses is essential.

The 'minimal spanning tree ' model is useful in arriving at an optimal network of interconnected points (nodes) in space with a minimum possible total length. The minimal spanning tree model was suitably modified [99] and applied for developing the plan for optimally laying-out the water courses in the command area of an outlet, taking off from the Sansad distributary, of MRBC command area. The network connects the outlet with each field in the area and provides for at least one delivery point for each field ensuring minimum length of the watercourse.

4.1.2 Optimal Layout Have Fields in Outlet Command Areas

Scattered fields with irregular boundaries require longer lengths of water courses to convey water to different fields. The effects of different field rectangulation schemes on lengths of water courses, seepage losses and water availability were studied [103] and an attempt was made to develop a suitable criteria for field rectangulation in the command area of canal outlets. Different field lengths are considered optimal under different conditions of soil type, land slope, size of irrigation stream and the crops grown in the area. The type of soil (medium loam soils), general land slope (0.20%) and the available size of the irrigation stream (30 lps) suggest the optimal field lengths to be around 100 m for efficient water application.

4.1.3 On-Farm Irrigation Structures

On-farm irrigation structures are crucial for efficient distribution and control of water at the field level. Over years, a concerted effort was made at the Water Technology Centre, Indian Agricultural Research Institute, New Delhi, to develop different types of on-farm irrigation structures such as in-situ flow measuring devices, drop structures prefabricated concrete channels, distribution boxes, control gate structures, spillways etc. [78, 79, 24, 49, 95, 96, 98].

4.1.4 Selection of Appropriate Lining Material

The pre-fabricated concrete channel sections developed at the Water Technology Centre have become very popular in the alluvial tracts of north India From the combined consideration of reduction of seepage loss and the cost of construction and installation, the pre-fabricated concrete channel section was rated as most suitable [87]. With some local need based modification, this technology was adopted on a large scale in the Sone and the Gander Command areas.

4.1.5 Soil Hydrophysical Characteristics

Knowledge of soil water retention and transmission characteristics is basic for efficient water management [176]. These soil hydrophysical characteristics are mostly influenced by soil texture, organic matter content, bulk density, etc., and play crucial role in soil water storage and transmission. Hence attempts were made to obtain empirical estimates of these characteristics. Singh *et al.*, [194] proposed the following relation ship for saturated hydraulic conductivity (K_s):

$$K_s = a+b1*SI+b2*CL+b3*OM\ b4*CEC \quad (20)$$

where

SI = Silt content (%),
CL = Clay content (%),
OM = Organic matter content (%),
CEC = Cation exchange capacity, $(cmol(p^+)\ kg^{-1})$,
a is a constant and b1, b2, b3 and b4 are regression coefficients.

A procedure for estimating field capacity and permanent wilting point for alluvial, black and red soils of India was developed by grouping soils into different textural classes. The estimates of Filed capacity and permanent wilting point for Indian soils compared favourably with those of US soils. The values obtained were validated using a soil water balance model

4.1.6 Use of Marginal Quality Water for Irrigation

As mentioned in section 2.5.1, availability of fresh water will ultimately become the key constraint for food production. As the population continues to increase, the per capita availability of water dwindles The countries with the amount of water available is between 1000 - 1600 m^3 /year /per capita, are said to be " water stressed" while those whose water availability falls below 1000 m^3 /year/capita are identified as "water scarce "countries [136]. Although currently India is comfortably situated in respect of water availability, with the unabated population, the water demands in future, have to be met through, besides judicious planning and use, use of marginal quality water, particularly for agriculture.

Water is considered suitable for agriculture when it has no osmotic or specific toxic effect on crop or animal production, when it contains no solute affecting the chemical or hydraulic properties of soil and when it does not cause deterioration of ground or surface water [233]. These adverse conditions are caused primarily by accumulation of salts in the root zone of crops. Accordingly, use of marginal quality of water for agriculture may be based on ensuring no or minimum accumulation of salts in the root zone of plants, suitable changes in land use and cropping activity and other administrative measures such as pricing of water to prevent misuse.

4.1.6.1 Maintenance of salinity regime of the root zone

Accumulation of salts in the root zone can be prevented by regular supply of adequate water either by leaching or by adopting special methods of irrigation. In situations of inadequate water availability or if the water is saline, the drip and pitcher methods of irrigation are found to be appropriate. These methods would ensure that the salts do not accumulate near the roots, maintain low soil moisture tension, and thus protect the plants from adverse effects. Dubai *et al.*, [40] reported that with pitcher irrigation, vegetables like cabbage, brinjal, ridge gourd etc, could be successfully grown without loss of yield with saline water upto 9.8 ds/m.

4.1.6.2 Use of Multiple Quality of Irrigation Water

Multiple quality water resources can also be used in arid climates through either blending of marginal quality water with good quality water in the supply system to a predetermined quality to match with the salt tolerance of the crop, or through alternate irrigation with good and marginal quality waters from different sources (eg: canal water and saline groundwater) [80]. The cyclic strategy has the advantages of i. Limiting use of good quality water at stages of crop growth that are sensitive to salinity, such as germination and seedling, and ii. Vast choice of crops.

4.1.6.3 Crop selection

Crops are known to have different levels of tolerance to salinity. When salinity can not be maintained at acceptable levels through the above methods, it desirable to choose crops/ varieties that are tolerant to salinity eg. Vegetables, wheat, barley, sorghum, wheat, tomato etc. Crop productivity of saline water irrigated soils can be increased by adopting selection of suitable crops, proper soil and water management practices along with judicious use of fertiliser and water [86, 191,232].

4.2 Efficient Methods of Irrigation

In view of the fact that the operational efficiency of the traditional flow irrigation systems are low, there is an immense need for adopting modern, efficient methods of irrigation. Alternate furrow irrigation, surge flow irrigation, and pressurised irrigation systems (drip and sprinkler irrigation) are considered to be efficient technologies. Research developments in design and operation of these systems are briefly reviewed here.

4.2.1 Surge Irrigation

Surge flow irrigation is a recent method of surface irrigation [205,206]. This is accomplished by "surging" the water down the furrows at timed intervals until the water reaches the end of the furrow. In the past few years, field researchers have investigated some significant new roles for surge that rely on its ability to distribute water uniformly, economise water use, reduce infiltration and deep percolation losses, and control runoff and drainage [3, 83]. Some very promising new applications of surge irrigation include furrow rice cultivation, control non-point source pollution and to automate fertigation through surface systems. This method has the ability to apply water uniformly, to create a shallow, uniform water profile and keep it at the root zone cutting down deep percolation. Surge irrigation is reported to boost the surface irrigation application efficiencies by as much as 40 percent matching drip efficiencies of 85 percent in studies by Washington State University [218]. Surge fertigation saves up to 25 percent in fertilizer costs. A study in the Grand Valley, indicated substantial reduction of salt load in return flows to the Colorado River.

In India, this is still a new topic. The effects of stream sizes, cycle on-times, cycle off-times and number of surges on advance of water front, tail water recession, tail water run-off, deep percolation and application efficiency, storage and distribution efficiencies are to be studied for Indian conditions of soils, crops and cultural practices. Some of these were evaluated and the optimum number of surges and furrow lengths was identified for different stream sizes and cycle on [228].

4.2.2 Pressurised Irrigation Systems

4.2.2.1 Drip system of irrigation

Drip irrigation system is comparatively a modern method of water application. It is expensive on the basis on initial investment but is appropriate for situations that call for high water use efficiency and undulating terrain. Considerable experimental research was carried out in the past 30 years world over as well as in India, to investigate the water savings and yield increases, design of appropriate components and their materials, moisture distribution and irrigation, and fertigation under drip irrigation [13].

4.2.2.1.1 Selection of appropriate drip emitters

An important component of a drip irrigation system is its dripper. Selection of drippers for a field situation depends upon the requirement of the discharge rate, soil, and the expected operating pressure. Different types of drippers suitable for varying operating conditions were introduced by different manufactures in India and abroad. The relative performance of different type drippers in terms of their operating pressure- discharge relationships were studied at length and their suitability for specific situations were identified [72].

4.2.2.1.2 Design of lateral line and sub-main line of a drip irrigation system

Design of lateral line and sub-main line of drip irrigation system for various plant layout and topography is another important aspect of the drip irrigation system design. Studies indicated that for satisfactory operation of a drip irrigation system, it is necessary to keep the overall pressure variation as well as friction losses in the system within some specified range [73].

4.2.2.1.3 Moisture front advance under a point source of water application

The shape of the moisture front as it advances with respect to time and space under the drip (point source) system of irrigation can be characterised as a semi-elliptic curve [71]. This was found to be true in the case of saline water also [135]. The rates of advance of the moisture front in the horizontal and vertical directions were observed to decrease with the increase in the value of elapsed time and were dependent on the soil type. The studies provided generalised solutions of the hydraulics of the soil moisture front useful in the design of optimum rates of water application and durations of drip irrigation.

4.2.2.1.4 Salinity distribution in crop root zone in drip irrigation

Studies on the influence of salinity build-up in the crop root zone, as a function of discharge rate at varying levels of salinity in irrigation water, showed that the discharge rates had a profound influence on the level of soil salinity at different depths in the crop root zone [135].

4.2.2.1.5 Studies on water requirement of crops under drip irrigation

Considerable research effort has been spent at several State Agricultural Universities for developing estimates of water requirements for several vegetable, horticultural and orchard crops in different agro-ecological situations. The main objective was to obtain irrigation schedules for different crops. Drip irrigation is well known to save water upto 60%, and also increase productivity of water upto 243% (Table15). A detailed review of this work is reported by INCID [13].

Table 15. Water Productivity Gains from Shifting to Drip Irrigation from Traditional Surface Irrigation [13]

Crop	Increase in yield, %	Saving of water, %	Increase in productivity of water, %
Banana	52	45	173
Cabbage	2	60	150
Cotton	27	53	169
Grapes	23	48	134
Potato	46	0	46
Sugarcane	20 to 33	30 to 60	70 to 243
Sweet potato	39	60	243
Tomato	Upto 50	Upto 39	49-145

4.2.2.2 Sprinkler system of irrigation.

4.2.2.2.1 Sprinkler system

Sprinkler systems sprinkle water in a manner similar to rainfall, such that the run off and deep percolation losses are minimised and uniformity of application is close to that obtained under rainfall conditions. As the water is conveyed through a pipe system, the conveyance losses are eliminated resulting in higher irrigation efficiencies. The savings of water upto 30 to 50% are reported in the literature facilitating increase in irrigation area upto 50% additional area. This method of irrigation is well recognised to achieve high water use efficiency, improved crop productivity, less labour requirement, adaptability to hilly terrain, suitability for water scarce areas, ability to reduce frost attack and facility to apply water-soluble fertilisers. The system is very well suited to in canal, tank, and well-irrigated areas. All close grown crops such as cereals, pulses, oil seeds, sugarcane, cotton and all plantation crops can be grown with sprinkler method of irrigation. An advantage of this method of irrigation is that undulated lands and shallow soil areas can be irrigated without any land leveling. During the past 20 years, considerable research was carried out at various State Agricultural Universities a research Institutions on adapting this technology to various agro- climatic regions. A report on the status of Sprinkler irrigation in India was developed by the INCID [183]. A major draw back of the system, besides its high cost, is the wind effect that distorts the uniformity of distribution. Further research effort is required to develop efficient and economic packages of design of the system configuration for its adaptation.

4.2.2.2.2 Micro sprinkler system

Micro sprinklers facilitate spraying of water under the tree canopy around the root zone of the trees (about 30cm high) and work under low pressure. This method is least effected by wind. The required exact quantity of water can be delivered to each plant daily at the root zone. Water is given only to the root zone area as in the case of drip irrigation unlike to the entire ground surface as in case of sprinkler irrigation method. This method is highly suitable for tree / orchard crops and vegetable crops in combination with alternate sources of Energy for pumping the water.

4.2.2.2.3 Low energy precision application (LEPA) systems

Most of the sprinkler units used in India are "Linear move " type and are suitable for rectangular field configuration. Recent innovations in these systems are Low Energy Precision Application (LEPA) systems. In these systems, the laterals are equipped with "drop tubes" (instead of riser tubes) fitted with very low-pressure orifice emission devices called "socks". Water is discharged just above the ground surface into dead end furrows or micro basins thus preventing soil erosion and runoff. These systems are not affected by wind forces and besides saving considerable energy, also have very high application efficiencies (98% percent) and uniformity of application. A variant of the LEPA system is Low Elevation Sprinkler Application (LESA) system, which is equipped with "bubbler" type of discharging devices instead of orifice type of the LEPA system [218].

4.2.3 Use of Alternate Source of Energy for Drip and Sprinkler Irrigation Systems.

Renewable energy resources present a technically sound alternative for pumping groundwater for irrigation, particularly with drip and sprinkler systems of irrigation. Wind mills, solar photovoltaic panels, biogas are being tested for their efficiency and are under use to a limited extent by farmers [18].

4.3 Rotational Water Distribution Systems

Several rotational water disrtribution systems such as warabandi, *Shejpali, Phad*, Block system etc., are in practice in various parts of the country. In essence, all these systems aim at achieving equity of water distribution among the farms. Each one was evolved based on the considerations of the physical irrigation system, crops grown administrative convenience and acceptability by the farmers. Among these the warabandi system and its variants are popular in many states in northern India.

4.3.1 Modifications of Warabandi System

Several Modifications in the warabandi rostering are attempted to achieve i) distribution of volume of water, instead of duration of flow, in proportion to the sizes of land holdings, ii) distribution of water in proportion to the actual water requirement of the crops grown, iii) incorporation of allowances for seepage losses , iv) adoption to the tube well command areas, v) accounting for time loss due to electricity failure vi) minimising the interference of multiple tube wells in operation and vi) equity in water distribution [97, 100,104]. Many of these modifications remain academic for want of proper testing and fine tuning at the field level. A major requirement for successful implementation of these schemes is to ensure that the designed discharges are maintained in the distributaries, which is still proving to be an allusive phenomenon.

4.3.2 Participatory Irrigation Management

Participatory Irrigation Management (PIM) refers to the involvement of irrigation water users in "all or some aspects" of irrigation management at part or at "all levels". "All aspects" implies initial planning and design, construction, supervision, financing, operation, maintenance, monitoring, evaluation of the system. "All levels" refers to the full physical limits of the irrigation project comprising capture, conveyance, and distribution and application [68]. This involves a major shift of responsibility of irrigation from governmental agencies to the Water User Associations (WUAs) [67]. Govt.of India accorded priority to PIM since the VIII plan period. Such a shift is associated with complex social, technical and administrative considerations. These include [182]:

i) type and elements of modes of transfer to water user's groups (eg: decisions on cropping systems, water releases, collection of water rates, maintenance of infrastructure, restoration and management of village tanks, organizing inputs etc.) and providing adequate training of farmers organizations,

ii) Administrative changes needed in the role of Governmental agencies , with a shift from controlling the irrigation system to helping the user groups do the same ; namely to function as facilitators and not as administrators,

iii) Legal policy, infrastructural and institutional issues including the role of NGOs and other catalytic agencies etc; to be addressed for a smooth shift of managerial responsibilities.

Thus the transition is not simple and much research is needed to achieve this. The recent amendment to the constitution of India devolving powers to the people through the *Panchayat Raj* system could be advantageously used. Andhra Pradesh is the first state to enact the A. P. Farmers Management of Irrigation System (APFMIS Act, 1997) to provide the legal policy framework. Several other states are in different stages of formulating the legislation and implementing PIM at different levels. While this shift to PIM is in its initial stages, appropriate guidelines for evaluation of the functioning of the WUAs and the impact of PIM on the water use efficiency and crop yields, are necessary. These are to be developed through research efforts. Some of the issues to be addressed are: assessing the functioning of i) conveyance, distribution and application systems, ii) crop production system iii) collection of water fees, and its investment in maintenance of the water conveyance system, iv) proper crop planning and releases of water and management schedules, v) farmers' organization network and its linkages with the state departments, etc. The management in the partnership mode should be robust for sustainability of reliable, adequate, equitable and timely water supply and enhanced agricultural production.

4.3.3 Diagnostic Analysis and Performance Evaluation

It is necessary to periodically evaluate the performance of the irrigation projects to identify bottlenecks in achieving the full performance of the system to recommend appropriate corrective measures. This is done by examining the extent to which the objectives of the project are realised, cost benefit relationships, and progress in the programme implementation. However, such an evaluation is often done from diverse viewpoints depending upon the background of the investigator. For instance, an irrigation engineer is concerned with the efficiency of operation and maintenance of the main canal system while the agricultural engineer is concerned with efficient field application of irrigation water from the outlet through the water courses and field channels to the crop root zone. Likewise, the socio-economist is concerned with the impact of the irrigation system on the socio-economic aspects of the farmer's community. A systematic procedure for this exercise is known as "Diagnostic analysis" and is carried out for various physical segments of the system.

i) *Main System*: At the main system level (up to the outlet), the components studied are: the condition and functioning of the physical systems such as the storage/ diversion structures, cross regulators, escapes, outlets, canal cross sections etc. and the canal discharges and seepage losses, system operation etc.

ii) *Farm Level*: At the Farm level, conditions of the water courses, field channels, control structures, seepage losses, soil irrigability, cropping pattern, yields, equity, adequacy and reliability of water supplies, etc. are studied.

Evaluation of the performance helps to characterize and quantify the performance of the system so that the scope for improvements as well as the impacts of the corrective measures adopted on the i) irrigation and agricultural aspect ii) economic aspects and iii) social aspects. This is normally attempted through use of several indices that reflect features such as adequacy and timeliness, reliability, and equity of the water supplied. Other indicators include area of land irrigated in relation to water supply, crop yields in the command area, actual income to the farmers/ha, departure from planned cropping pattern, comparison between canal head and tail performance, responsiveness of Irrigation Departments to farmers' needs etc., Bos et.al, [28] were among the first to formally suggest some indices. Subsequently several investigators proposed a large number of indices [1, 81, 151, 167, 229, 181]. Some of them are location specific while others are generic in nature.

5 Perspectives, Future Research and Development Needs

The World Commission on Water l proposed that " Every human being, now and in the future should have access to safe water for drinking, sanitation, food, and energy at reasonable cost, in an equitable manner that works in harmony with the nature. To achieve this, a holistic, systematic approach through an Integrated Water Resources Management (IWRM*)* is needed *for* managing water. This calls for a paradigm shift from administrative and political boundaries based planning to that of natural hydrologic units of catchment, basin, and aquifer. Such a shift provides for attention to interaction of the complex and dynamic environment, hydrological cycle, and the human socioeconomic system. This applies to all sectors of water use. Since agriculture is the major user of water any comprehensive discussion for planning and utilisation of water shall have emphasis on this sector [231].

India is endowed with a strong institutional and human resource base in the field of agriculture and allied sciences and technologies. A vast irrigation network, rich natural resource base, high precipitation and abundant sunshine, diverse environment, emerging private sector, gradual shift to farmers management of irrigation systems, present low productivity levels and low input use provide future scope for improving agricultural productivity, value addition, diversification and trade. This scenario, together with the advances in space, agricultural sciences, social sciences, information and communication technologies, biotechnology, management and computer sciences, enables new opportunities to improve the efficiency of water management in agriculture and ensure sustainability of Indian agriculture [231]

5.1 Strengths and Opportunities of Irrigated Agriculture

Efficient water management in agriculture holds the promise to fulfil the future food needs of India because of several comparative advantages. There are several common and uncommon strengths and opportunities for increasing food grain production through improved water management in India. Some of the prominent ones are listed below:

5.1.1 Common Strengths and Opportunities

i) India has a high average annual rainfall of 112 cm as compared to many other countries,

ii) India's irrigation infrastructure is the largest in the world (Table 2),

iii) In terms of the total cropped area, India is the second largest in the entire world (Table 9)

iv) There are Widely varied climatic zones across the country that cover temperate regions suitable for horticultural crops, flowers etc., to Tropical regions suitable for cereals, oilseeds and pulses, with ample Sunshine for over 11 months of the year,

v) the vast alluvial tracts of the entire Gangetic plains, and the fertile deltaic regions of major rivers like Mahanadi, Godavari, Krishna, and Kaveri, Endowed with rich water resources, posses excellent potential for food production,

vi) India has the world's third largest fertiliser industry,

vii) the large National Agricultural Research System comprising about 87 National Institutes and National Research Centres, 81 All India Co-ordinated Research Projects and 29 Agricultural Universities, employing a large number of highly trained manpower to provide technological inputs needed.

viii) ample scope for realising the full yield potentials for almost all the crops (Table 10 & 11).

5.1.2. Uncommon Opportunities [19]

High Technology Innovations that are likely to be developed in near future are:

- Membrane technology: For waste water treatment, and desalinization at low cost (@<1$ per cum.) thus increasing water availability to agriculture,
- Biotechnology: Possibility of developing
  - low water requiring crops (with the recent completion of
  - mapping of gene for rice crop, this has high possibility in near future)
  - High yielding plant varieties that are most environmental friendly,
  - plants that are salt tolerant
  - plants that are drought tolerant
  - plants with pest resistance (reduces pesticide pollution)

  - Hyper toxin accumulating plants to remove soil toxins

- Microbial technologies for waste water treatment for Agricultural reuse
- Increase in yields of rain fed agriculture
- separating heavy metals and other toxins from soil and water

5.2 Research and Development (R & D) Needs

The present efforts in R & D for improving water use efficiency should be intensified. For this, both location specific field studies, and analytical studies for developing systematic decision support systems for planning and real time operation of Irrigation systems for growing multiple crops under limited water supply, using modern tools like medium range weather forecasts, Simulation, Optimisation and GIS are required.

Further, there is need to have a paradigm shift in adopting the modern methods of drip and sprinkler irrigation systems in irrigation command areas, as well as in the non- command areas. These technologies need not any more be confined to areas with problem soils and undulating terrain only. No doubt, this calls for huge investments, and technological and infrastructural back up. But that is precisely what it takes to ensure high water use efficiency

Generalized packages of practices adaptable to location specific situations are required to be developed. As the participatory management progressively comes in to operation, the research results must be user friendly to the farmer 's societies, with adequate interactive interfaces. Some R&D thrust areas that need attention for water conservation and sustainable use in agriculture are listed below. Among these, while items at numbers iv, v, vii, viii, ix, x, xiii, xv, xvi, xvii, xxiii, and xxiv may be given long term priority, all the others need immediate attention.

i) Even after full development of the identified ultimate irrigation potential, more than 50% of the cultivated area in India will remain rain fed. Rainwater management in these areas is vital for food security. A balanced combination of traditional conservation practices and scientific approach to watershed management, water harvesting, and efficient use of the rain water for life saving supplementary irrigation, needs to be adopted.

ii) Integrated management of all sources of water, namely river, rain, ground, soil, and sewage waters, in a conjunctive manner in appropriate combinations would not only increase the water availability to crops but also mitigates problems of waterlogging and salinity build up.

iii) With increasing concern for sustainability and environmental security, a paradigm shift in irrigation water management research is needed. A formal frame wok to identify, formulate and analyse the resource management problems in their totality, and to evolve procedures, criteria for evaluating alternatives, is needed to ensure accountability. Thus irrigation water management has to be more knowledge intensive.

iv) Development of policies for co-operative management of watersheds and Irrigation command areas for harnessing water in a sustainable manner.

v) Development of guidelines to regulate the expansion of water markets and Water lords, legal framework introduces proactive measures to avoid water conflicts, and ensure economic growth,

vi) Appropriate technologies to quantify and minimise the uncertainty and dependability of food production on the vagaries of weather, climate change, Elnino effects, etc. and thereby achieve food security,

vii) The river flow storage capacity of the country has to be increased to a total of 400 km^3 from the present 193 km^3. At the same time, all out efforts are needed to renovate and modernise the existing tanks in the country .The estimated storage capacity of these tanks is 21.4 km^3.

viii) Detailed water balance studies are required for sustained development and use of groundwater resources. The recharge to groundwater from the vast net work of canal systems and farm ponds need to be scientifically assessed. Artificial recharge structures like percolation tanks, check dams sub-surface dams etc., need to be designed scientifically based on long term hydrology and geology of the area.

ix) The estimated evaporation losses from storage reservoirs and irrigation tanks are about 27 km^3, which is likely to increase to 56 km^3 when the total storage developed becomes 400 km^3. Economic and effective methods for control of evaporation losses are required to be developed.

x) It is estimated that with 70-80% of current domestic and industrial water use, if recycled after proper treatment, for agricultural use, about 2 m.ha could be irrigated. Appropriate technologies need to be developed for this.

xi) With the availability of IRS satellite imageries with high degree of resolution there is a need to harness this facility intensively for day to day real time operation of irrigation systems for monitoring the depth of standing water in paddy fields, and the plant water stress, etc., besides the current use in land use classification, planning of OFD works, crop monitoring including spread of diseases, and production estimates.

xii) Information technology has immense applications in water management. Development of appropriate Database Management systems and software That can incorporate real time data collection through automatic weather stations and medium range weather forecasts, modern electronic communication systems for canal automation, centralised system of monitoring and control of large irrigation systems, water management of crops grown under controlled environment etc., are some of the possibilities. Appropriate research efforts are needed to develop the necessary technologies.

xiii) The Expert system technology and the neural network technology, are extremely useful in developing practical solutions to operate complex irrigation systems. Present efforts of water management scientists in harnessing these technologies are extremely limited and are in elementary stage. Further research efforts in combination with other emerging technologies are necessary.

xiv) Participatory institutional mechanisms must be put in place at the project, distributary and the outlet command levels to involve all sectors of society in decision making. In view of total lack of experience in this aspect, several pilot studies are needed to evolve and improve the guidelines for efficient functioning of these institutional mechanisms. The Irrigation Departments should adopt a new role of facilitator.

xv) Legal provisions to enforce Full cost pricing of water , with appropriately targeted subsidies to the poor, and the principle of " *polluter pays and User pays* ".

xvi) Private sector to be entrusted with providing cost effective services in an accountable and transparent manner, and mobilise investment.

xvii) There is need for conducting anticipatory research for assessing the adverse environmental impacts of long term canal irrigation in black soil areas with medium rainfall, and develop guidelines for judicial use of canal water to prevent waterlogging and salinisation in such areas.

xviii) Generic ,as well as location specific, procedures are to be developed to ensure adequacy, timeliness, reliability, and equity of irrigation water supplies so as to minimize the mismatch between crop water requirements and the supplies, and to suggest optimal allocation of water among competing crops,

xix) procedures for performance assessment to determine the impact of irrigation and participatory management of irrigation water need to be developed.

xx) Investigate the declining groundwater conditions of the rice-wheat belt to ensure sustainability,

xxi) development methods to operationalise conjunctive use plans through specification of timings and amounts of use of canal water and groundwater, as well as suitable legislative framework needed to implement the plans.

xxii) strengthen the basic research on the understanding of the complex surface water - groundwater interactions under different hydrogeological situations, for planning conjunctive use and artificial recharge.

xxiii) Existing worldwide experience in use of saline water in irrigation indicates high potential for using such waters. However, documentation of long term effects of such use is very limited. Research needs to be carried out on this aspect.

xxiv) Site specific management strategies need to be developed for using marginal quality irrigation water sustainable agriculture. Information base on the nutrition value, shelf life of fruits, vegetables as affected by the water quality also need to be developed. Procedures for assessment of pollution caused by intensive long-term use of fertilizers and pesticides need to be developed.

xxv) Guidelines for using drip irrigation system with marginal quality water for various crops under different agro-climatic conditions.

xxvi) Continued resource illiteracy leads to not only wastage of water but also to undesirable environmental consequences. Socio-economic research is required to evolve a mechanism to create awareness among the farmers of the precious value of water and the use of modern technologies of water management.

xxvii) There is need to develop new knowledge for water management at farm level to ensure:

- Significant reduction in the use of irrigation water per unit area,
- timely removal of excess water from agricultural lands,
- sustainable cropping system in relation to availability of water,
- multiple uses of water in agricultural production programmes to enhance their productivity, particularly in the Eastern India where pisciculture is popular.

References

1. Abernathy C. L., Performance Measurement In Canal Irrigation, *IIMI/ODI - Irrigation Management Network Paper* No.**86/2d** (1986)
2. Abrol I.P. and Khosla B. K., *Nature, London* **212** *(*1392)
3. Allen R. R and Schneider A.D., Furrow Water Intake Reduction with Surge Irrigation or Traffic Compaction, *Applied Engineering In Agriculture* **8** (1992) pp. 455-460.
4. Amir J and Sinclair T. R., A Model of the Temperature and Solar Radiation Effects on Spring Wheat Growth and Yield, *Field Crop Research* **28** (1991) pp. 47-58.
5. Anonymous, Hand Book of Hydrology (Central Unit of Soil Conservation-Hydrology and Sedimentation, Ministry of Agriculture, GOI, New Delhi)
6. Anonymous, Water Requirement and Irrigation Management of Crops in India, Monograph No.**4** (New Series), (Water Technology Centre, Indian Agricultural Research Institute, New Delhi, 1977)
7. Anonymous, Reports of the Project Co-Ordinator of *The All India Co-Ordinated Research Project on Water Management,* Indian Council of Agricultural Research, 1975-77, 1977-79, 1979-81, 1981-83 and 84-85.
8. Anonymous, Basic Agronomic Research, Chapter IV in *Scientific Basis of Indian Agriculture: Seventy Five Years of Service* (Indian Agricultural Research Institute, New Delhi, 1980)
9. Anonymous, Report of the Ground Water Estimation Committee-Ground Water Estimation Methodology (Ministry of Irrigation, Govt. of India, New Delhi, 1984)

10. Anonymous, Report of The Working Group on Problem Identification in Irrigated Areas with Suggested Remedial Measures (Ministry of Water Resources, Government of India, 1991a)
11. Anonymous, Water logging, Soil Salinity, and Alkalinity (Ministry of Water Resources, Government of India, New Delhi, 1991b)
12. Anonymous, Irrigation Scheduling under Limited Water Supplies-Publication No.**218** (1991c), Central Board of Irrigation and Power, New Delhi.
13. Anonymous, Drip Irrigation - A status Report (INCID, Ministry of Water Resources, Govt.of India, 1994), p. 176.
14. Anonymous, Water and Related Statistics (Central Water Commission, Ministry of Water Resources, GOI, 1995), p. 343.
15. Anonymous, Command Area Development - Report of the Working Group, IXth Five Year Plan (Ministry of Water Resources, Government of India, New Delhi, 1996a)
16. Anonymous, Agricultural Drainage Experiments in Case Studies, Report of the Project Co-ordinator (All India Co-ordinated Research Project on Agricultural Drainage, Indian Council of Agricultural Research, New Delhi, 1996b)
17. Anonymous, The Andhra Pradesh Farmers Management Irrigation System Act, Act 11 (1997), - Act and Rules, Department of Irrigation & Command Area Development, Govt.of Andhra Pradesh.
18. Anonymous, Report of the Project Co-ordinator (All India Network Project on Performance and Optimal Utilisation of Water Using Solar Photo Voltaic Pumps, Indian Council of Agricultural Research, 1998).
19. Anonymous, Biotechnology and Water Security in the 21st Century, Global Water Scenario, and Emerging Challenges (Madras Declaration, M. S. Swaminathan Research foundation, Chennai, 1999)
20. Anonymous, MOWR Website, Ministry of Water Resources, GOI (2000)
21. Babu Ram, Sarma P. B. S. and Pandey C. M., Simulation of Vertical Drainage in Irrigated Field, *Journal Of Agril.Engg.* **27** (1990), pp. 102-111.
22. Bhamrah P. J. S., Conjunctive Use of Surface and Groundwater Resources for Sustained Productivity in Irrigated Areas - Problems, Case Studies and Future Perspectives, *Proc. National Seminar on Water Management for Sustainable Agriculture, - Problems and Perspectives for the 21st Century*, IARI, **April 15-17** (1998) New Delhi, pp. 268-275.
23. Bhattacharya A. K. and Sarkar T. K., Proposed Combination of Surface And Subsurface Drainage Systems for Madanpur, Kahdar, Asola and Ali Areas, *CBIP Publication No.118*, **1**(1972) pp. 138-142.
24. Bhattacharya A. K., Hydraulics of Flow Through Rough Channels -Design, Pre-fabrication and Testing of Drop Structure for Small Fall Structures-*Technology for Water Use and Management at Farm Level* (1973), *Indian Farming.*

25. Bhattacharya A. K., Ray S. B., Sinha A. K. and Ghosh G., Drainage in Tea Plantations (Water Technology Centre, Indian Agricultural Research Institute, New Delhi, 1996) p. 23.
26. Blank D and Delleur J. W., A programme for Estimating Runoff From Indiana Watersheds, Part I, Linear System Analysis in Surface Water Hydrology, and its Application to Indiana Watersheds, Technical Report. No. **4**, Purdue University Water Resources Research Centre, Indiana (1968).
27. Borg H. and Grimes D. W., Depth Development of Roots with Time: an empirical Description, *Transactions* **29** (1986), pp. 194-197.
28. Bos M. G. *et al.*, Standards for the Calculation of Irrigation Efficiencies, *ICID Bulletin*, **27** (1978) pp. 21-101.
29. Box G. E. P. and Jenkins G. M., Time Series Analysis: Forecasting and Control (Holden-Day publications, Sanfransisco, Ca., 1970)
30. Bredero Th. J., Concepts and Guidelines for Crop-Water Management Research - A Case Study for India (Oxford-IBH Publishing Co. Pvt. Ltd., 1991), p. 156.
31. CGWB, Background note, Colloquium on Strategy for Groundwater Development, Central Groundwater Board, Ministry of Water Resources, GOI, (1997).
32. Choudhary T. N., Water Management in Rice for Efficient Production, (Published by Directorate of Water Management Research, ICAR, Patna 1997), p. 63.
33. Chowdary V. M., Rao N. H. and Sarma P. B. S., Estimating Recharge Rates in Irrigation Project Areas using GIS, *Proc. Second Annual ESRI/ERDAS '97 User Conference*, ESRI, India, Nov.1997 a, New Delhi.
34. Chowdary V. M., Rao N. H. and Sarma P. B. S., Estimation of Nitrate Pollution Loads to Groundwater in Irrigation Projects Using GIS, *Proc. Second Annual ESRI/ERDAS'97 User Conference,* ESRI India, Nov (1997 b), New Delhi.
35. Chowdary V. M., Study of Non-Point Source of Pollution to Groundwater, (Ph.D Thesis, Division of Agricultural Engineering, Indian Agricultural Research Institute, New Delhi, 1998)
36. Dariane A. B. and T. C. Hughes, Application of Crop Yield Functions in Reservoir operation, *Water Resources Bulletin* **27** (1991), pp. 649-656.
37. Doraisamy P., Development of an Integrated Expert Decision System for Irrigation Scheduling (Ph.D Thesis, Division of Agricultural Engineering, Indian Agricultural Research Institute, New Delhi, 1992)
38. Doorenbos J and Pruitt W. O, Crop Water Requirements, FAO, *Irrigation and Drainage Paper* No.**24** (1977), FAO, Rome.
39. Doorenbos J. and Kassam A. H., Yield Response to Water, FAO, *Irrigation and Drainage Paper* No. **33** (1979), FAO, Rome.
40. Dubey S. K., Mandal R. C. and Gupta S. K., Pitcher irrigation for arid and semi-arid zones, in *Dry land Resources and Techniques* (1991), pp. 137-177, ICAR, New Delhi.

41. FAI, Fertiliser Statistics, Fertiliser Association of India (1997).
42. FAO, Crop Evapotranspiration - Guidelines for Computing Crop Water Requirements, FAO *Irrigation & Drainage Paper No.* **56** (1998), Food And Agriculture Oragnisation, Rome.
43. FAO, Agricultural Statistics, FAOSTAT, Food and Agriculture Organisation's Website, Dec.2000.
44. Feigen Baum E. A., Knowledge Engineering for the 1980's (1982), Heuristic Programming Project, Computer Science Department, Stanford University, Stanford, Calif. USA.
45. Ganesh H. V., and Hakkali J., Influence of Different Irrigation Schedules on Yield and Water Use Efficiency of Paddy Crop, *Proc. National Seminar on Water Management for Sustainable Agriculture- Problems and Prospects for the 21st Century* (1998), New Delhi, pp. 183-185.
46. Garg S. P., A System Analysis Approach to Water Management in Waterlogged Areas (PhD Thesis, Division of Agricultural Engineering, Indian Agricultural Research Institute, New Delhi, 1996)
47. Garnayak L. M., Singh N. P. and Singh Subedar, Productivity and Water Use by Late sown, Brassicas Influenced by Irrigation and Nitrogen, *International Congress on Agronomy*, Environment and Food Security for 21st Century, Nov. (1998), New Delhi.
48. Griffin Jr. D. M and Warington R. O., Examination of 2-D Groundwater Recharge Solution (By Rao & Sarma), *Journal of Irrigation & Drainage Engineering* **14** (1988), pp. 691-702.
49. Hai Pham Ngoc, Bhattacharya A. K. and Trikha S., Design, Prefabrication and Testing of Drop Structure for Small Fall, *ISAE Journal of Agricultural Engineering* **27** (1990), pp. 126-139
50. Hajilal M. S., Rao N. H. and Sarma P. B. S., Can Medium Range Weather Forecasts Influence Irrigation Scheduling?, *Current Science* **66** (1994), pp. 60-63.
51. Hajilal M. S., Sarma P. B. S. and Rao N. H., Optimal Reservoir Release for Irrigation Management, (XX1 Annual Convention, Indian Society Of Agricultural Engineers, Dec.1995)
52. Hajilal M. S., Rao N. H. and Sarma P. B. S., Real Time Operation of Reservoir Based Irrigation Systems., *Central Board of Irrigation and Power,* New Delhi, Publication No. **256** (1997), pp. 370.
53. Hajilal M. S., Rao N. H. and Sarma P. B. S., Planning Intraseasonal Irrigation Requirements in Large Projects. *Agricultural Water Management* **37** (1998-a), pp. 163-182.
54. Hajilal M. S., Rao N. H. and Sarma P. B. S., Real Time Irrigation Reservoir Operation, *Agricultural Water Management* **38** (1998-b), pp. 103-122.
55. Hall W. A. and Butcher W., Optimal Timing of Irrigation, *Journal of Irrigation and Drainage Division, American Society of Civil Engineers* **94** (IR2) (1968), pp. 267-275.

56. Hantush M. S., Growth and Decay of Ground Water Mounds in Response to Uniform Percolation, *Water Resources Research* **3** (1967), pp. 227-234.
57. Hasan Z., Remarks at the Inaugural Function by the Chairman, Central water Commission, Ministry of Water Resources, GOI, *Proc. National Seminar on Water Management for Sustainable Agriculture - Problems and Perspectives For the 21st Century* (April 1998), pp. 10-11.
58. Hazra C. R., Management Of Rainwater Resources on Watershed Basis For Sustainable Agricultural production-An Experience of Tejpura Watershed (Jhansi) on Changing Scenario, *Proc. National Seminar On Water Management For Sustainable Agriculture-Problems And Prospects For The 21st Century* (1998), New Delhi, pp. 241-248.
59. Hukkeri S. B. and Pandey S. L., Water Management of Irrigated Rice, *Research Bulletin* **41** (1983), Indian Agricultural Research Institute, New Delhi.
60. ICID, The Hague Declaration, International Commission of Irrigation and Drainage, The Hague (1993).
61. ICWE, The Dublin Declaration, The International Conference on Water and Environment, Dublin, Ireland (1992).
62. Jalota S. K. and Prihar S. S., Bare Soil Evaporation in Relation to Tillage, *Adv. Soil. Sci.* **12** (1990), pp. 187-206.
63. Jameison P. D., Martin R. J., Francis G. S. and Wilson D. R., Drought Effects on Biomass Production and Radiation Use Efficiency in Barley, *Field Crop research* **43** (1995), pp. 77-86.
64. Jameison P. D., Porter J. R., Goudriaan J., Ritchie J. T., Van Keulin H. and Stol W., A Comparison of the Models AFRCWHEAT2, CERES-Wheat, Sirius, SUCROS2 and SWHEAT with Measurements From Wheat Grown Under Drought ", *Field Crop Research* **55** (1998), pp. 23-44.
65. Jameison P. D., Crop Responses to Water Shortages in *Water Use in Crop Production,* ed. by Kirkham M. B., Food Products Press, New York (1999)
66. Jayanthi M, Singh K. M. and Singh R. N., Differential Influence of Drip and Surface Irrigation on Pest Complex of Groundnut, *Indian Journal of Entomology* **55** (1993), pp. 124-131.
67. Joshi L. K. and Dinkar V. S. (eds.), Waterlogged, Saline and Alkaline lands- Prevention and Reclamation- Based on *National Workshop on Reclamation of Waterlogged Saline and Alkaline lands* (Ministry of Water Resources, GOI, New Delhi, Dec. 1996)
68. Joshi L. K. (ed.), Management of Irrigation, A New paradigm, Participatory Irrigation Management (Ministry of Water Resources, Govt.of India, New Delhi, 1997)
69. Kapoor A. S., Biodrainage to prevent pollution of Surface Water from Irrigation Drainage, *Proc. National Seminar on Water Management for Sustainable Agriculture - Problems and Perspectives for the 21st Century,* IARI, (1998), New Delhi, pp. 332-336.

70. Katyal J. C., Ramana Rao B. V., Shrinivas Sharma and Mishra P. K., Rain water Harvesting in India-Assets and Liabilities, *Proc. 1st Agricultural science Congress* (1992), National Academy of Agricultural Sciences, New Delhi, India, pp. 238-248.
71. Kaul R. K., Moisture Front Advance Under A Point Source Of Water Application, *J. Agricultural Engineering* **XIX** (1982), pp. 1-8.
72. Kaul R. K., Rajput T. B. S. and Kumar A., Criteria for Selection of Appropriate Drip Emitters, *Annals Of Agril. Research* **8** (1987), pp. 145-145.
73. Khanna M., Bhattacharya A. K. and Kumar A., Design of Lateral Line and Submain Line of Drip Irrigation System for Various Plant Layout and Topography, *J.Indian Water Resources Society* **10** (1990), pp. 32-35.
74. Kumar P., Food Demand and Supply Projections for India, *Agricultural Economics Policy Paper*, 98-01 (1998), IARI, New Delhi, p. 141.
75. Loucks D. P., Stedinger J. R. and Haith D. A., Water Resources Systems Planning and Analysis (Prentice-Hall, New Jersey, USA, 1981)
76. Maidment V. R. and Chow V. T., Stochastic State Variable Dynamic Programming for Reservoir System Analysis, *Water Resources Research* **17** (1981), pp. 1578-1584.
77. Mehto D. N., Singh K. M. and Singh R. N, Influence of Irrigation on Succession and Population Build up of Insect Pests in Chickpea (Cicer arietinum Linn), *Indian Journal of Entomology* **49** (1987), pp. 297-329.
78. Michael A. M., Arora D. R, Bhattacharya A. K., Mandal A. and Gupta A. K., Handbook of Farm Irrigation Structures (Division of Agricultural Engineering, Indian Agricultural Research Institute, New Delhi, 1970) p. 82.
79. Michael A. M., Sarma P. B. S and Sinha A. K., Technology for Water Use and Management at Farm level *Indian Farming* **XXX** (1980), pp. 95-98.
80. Minhas P. S., Saline Water Management for Irrigation in India, *Agricultural Water Management* **30** (1995), pp. 1-24.
81. Molden D. J. and Gates T. K., Performance Measures for Evaluation of Irrigation Water Delivery System, *Jl Irrig. & Drainage Engg.* **116** (1990), pp. 804-823.
82. MOWR, Water Resources Development Plan of India, Policy & Issues (National Commission for Integrated Water Resources Development Plan, Ministry of Water Resources, GOI, New Delhi, 1999)
83. Musick J. T., Walker J. D., Schneider A. D., and Pringle F. B., Seasonal Evaluation of Surge Flow Irrigation for Corn, *Applied Engineering in Agriculture* **3** (1987), pp. 247-251.
84. NAAS, Agricultural Scientists' Perception on Indian Agriculture: Scene, Scenario and Vision, National Academy of Agricultural Sciences (1999), p. 78.
85. Nema R. K., Rajput T. B. S. and Bhattacharya A. K., Water Balance Studies in a Canal Command Area in Jabalpur. Lining of Small Irrigation Channels, *J. Irrigation And Power.* **40** (1996), pp. 385-391.

86. Paliwal K. V., Irrigation with Saline Water, IARI Monograph No.2, (New Series), Water Technology Centre, Indian Agricultural Research Institute, New Delhi, 1972.
87. Panda R. K. and Bhattacharya A. K., Lining of Small Irrigation Channels, *Irrigation and Power* **40** (1983), pp. 385-391.
88. Pandey S. L. and Sinha A. K., Studies on Water Table Position, Soil Properties, Crop Growth and Yield under Drained and Undrained Conditions, *Indian Journal of Agronomy*, **XVI** (1971), pp. 494-501.
89. Pandey S. L., Sinha A. K. and Murari K., Effect of Levels of Nitrogen on the Yield and Water Use Efficiency of Mustard Under Shallow Water Table Conditions, *Indian Journal of Agronomy* **24** (1979), pp. 10-12.
90. Pandey S. R., Bhattacharya A. K., Singh O. P. and Gupta S. K., Transient drain design equation incorporating the effect of evaporation, *Proc. National Symposium on Management of Irrigation Systems,* (Feb. 1988), Central Soil Salinity Research Institute, Karnal.
91. Porter J. R., AFRCWHEAT2: A Model of the Growth and Development of Wheat Incorporating Responses to Water and Nitrogen, *European Journal of Agronomy* **2** (1993), pp. 69-82.
92. Prihar S. S., Khera K. L., Sandhu K. S. and Sandhu B. S., Comparison of Irrigation Schedules Based on Pan Evaporation and Growth Stages in Winter Wheat, *Agronomy Journal* **68** (1976), pp. 650-653.
93. Prihar S. S. and Sandhu B. S., Irrigation of Field Crops and Practices, *Indian Council of Agricultural Research* (1987), p. 142.
94. Pundarikanthan N. V. and Kallapiran S. N., Tank Irrigation, Its Potential Management and Policy Perspectives, 1997, *Proc. Round Table on National Water Policy - Agricultural Scientists' Perceptions*, National Academy of Agriculture, New Delhi, pp. 14-38.
95. Rajput T. B. S. and Kumar A., Development and Performance Evaluation of Wheel Flow Metre *J. Of Agril. Engg.* **17** (1980), pp. 541-545.
96. Rajput T. B. S., A Simplified Recorder for Monitoring Irrigation Water Flow, *Agril. Mechanisation in Asia* **XV** (1986), pp. 32-34.
97. Rajput T. B. S. and Michael A. M., An Equitable Water Distribution System for Tubewell Command Areas *J. Of Indian Water Resources Society* **7** (1987), pp. 11-20.
98. Rajput T. B. S., Development and Calibration of an Automatic Field Water Flow Recorder, *Invention Intelligence* **14** (1987), pp.172-176.
99. Rajput T. B. S. and Michael A. M., Optimal Layout of Water Courses in Outlet Command Areas, *Irrigation and Power Journal* **45** (1988 a), pp. 29-37.
100. Rajput T. B. S. and Michael A. M., Modifications in Warabandi Rostering for Equitable Water Distribution, *Irrigation and Power Journal* **45** (1988b), pp. 55-67.

101.Rajput T. B. S. and Michael A. M., Scheduling of Canal Deliveries I - Development of an Integrated Canal Scheduling Model, *Irrigation and Power Journal*, **46** (1989 a), pp. 23-39.
102. Rajput T. B. S. and Michael A. M., Scheduling of Canal Deliveries II Application of the Integrated Canal Scheduling Model, *Irrigation and Power Journal*, **47** (1989 b), pp. 17-39.
103. Rajput T. B. S., Layout of Fields and Water Courses for Efficient Utilisation of Irrigation Water, *J. of Agril. Engg*. **Special Issue** (1991), pp. 17-26.
104. Rajput T. B. S., Optimum Operation Pattern of a Canal System under Different Levels of Rotation, *J. of Indian Water Resources Society* **12** (1992), pp. 252-258.
105. Randhawa N. S. and Sarma P. B. S. (ed.)- A Policy Document on National Water Policy, Agricultural Scientists Perceptions, National Academy of Agricultural Sciences, New Delhi (1996), p. 41
106.Randhawa N. S. and Sarma P. B. S. (ed.) - National Water Policy - Agricultural Scientists' Perceptions, Proc. of Round Table held in 1994, National Academy of Agricultural Sciences, New Delhi (1997), p. 469.
107.Rao K. V. G. K., Man's Interference with the Environment in Water Use - Problems of Waterlogging and Soil Salinity, Proc. of Round Table held in 1994 on National Water Policy - Agricultural Scientists' Perceptions, National Academy of Agricultural Sciences, New Delhi (1997), pp. 68-80.
108. Rao M. S., Studies on optimum conjunctive use in agriculture (Ph.D. Thesis, Division of Agricultural Engineering, Indian Agricultural Research Institute, New Delhi, 1984)
109. Rao M. S. and Sarma P. B. S., Analysis of Hydraulics of Cavity Wells, *Journal Of Agricultural Engg*, **XVIII** (1981), pp.127-131.
110. Rao N. H. and Sarma P. B. S., Growth Of Groundwater Mound in Response to Recharge, *Groundwater* **18** (1980), pp. 587-595.
111. Rao N. H. and Sarma P. B. S., Groundwater Recharge from Rectangular Areas, *Groundwater* **19** (1981-a), pp. 272-274.
112.Rao N. H. and Sarma P. B. S., Recharge from Rectangular Areas to Finite Aquifers, *Journal of Hydrology* **53** (1981-b), pp. 269-275.
113.Rao N. H. and Sarma P. B. S., Mathematical Modelling of Groundwater Recharge to Finite Aquifers, *Proc. International Conference on Modelling And Simulation,* Paris, France (1982), pp. 46-51.
114. Rao N. H. and Sarma P. B. S., Recharge to Finite Aquifers from Strip Basins, *Journal of Hydrology* **66** (1983), pp. 245- 252.
115.Rao N. H. and Sarma P. B. S., Recharge to Aquifers with Mixed Boundaries, *Journal of Hydrology* **74** (1984), pp. 43-51.
116.Rao N. H. and Sarma P. B. S., Impact of Canal Irrigation on Ground water Utilisation, *Proc. First National Water Convention* (1987), New Delhi.

117.Rao N. H., Field Test of a Simple Model of Soil Water Balance, *Journal of Hydrology* **91** (1987-a), pp. 179-186.
118. Rao N. H., Estimating Probable Weekly Rainfall, *Journal of Indian Water Resources Society* 7 (1987-b), pp. 14-20.
119. Rao N. H. and Sarma P. B. S., Water Resources Utilisation in an Irrigation Project in India, *Water Resources Development* **4**(1988), pp. 200-207.
120.Rao N. H., Sarma P. B. S. and Subhash Chander, A Simple Dated Water Production Function for Use in Irrigated Agriculture, *Agricultural Water Management* **13**(1988-a), pp. 25-32.
121.Rao N. H., Sarma P. B. S. and Subhash Chander, Irrigation Scheduling From a Limited Water Supply, *Agricultural Water Management* **15** (1988-b), pp. 165-175
122.Rao N. H., Seepage Losses From Canal Irrigation Schemes- Influence of Surface and Subsurface Conditions and Implications for Conjunctive Use, *Water Resources Development* **6** (1990), pp. 55-62.
123. Rao N. H., Climatic Fluctuations and Water Resources Development in India, *J.Indian Water Resources Society* **11** (1991), 32-36.2
124.Rao N. H. and Rees D. H., Simulation model of Daily Rainfall in the Monsoon Season, *J. Indian Water Resources Society* **11**(1991), pp. 29-34.
125.Rao N. H. and Rees D. H., Irrigation Scheduling of Rice with a Crop Growth Simulation Model, *Agricultural Systems* **39** (1992), pp. 115-132.
126.Rao N. H., Simulation Model of Weekly Rainfall for Application in Real-Time Irrigation Management, *Journal Of Indian Water Resources Society* **12**(1992), pp. 49-53.
127.Rao N. H., Sarma P. B. S. and Subhash Chander, Optimal Multicrop Allocation of Seasonal and Intraseasonal Irrigation Water, *Water Resources Research* **26** (1990), pp. 551-559.
128.Rao N. H., Sarma P. B. S. and Subhash Chander, Real-time Adaptive Irrigation Scheduling under a Limited Water Supply, *Agricultural Water Management*, **20** (1992), pp. 267-79.
129.Rao K. V., Development of an IEDSS for conjunctive use in irrigation projects (Ph.D Thesis, Division of Agricultural Engineering, Indian Agricultural Research Institute, New Delhi, 1996)
130.Rao V. V., Water management in major irrigation projects (Ph.D. Thesis, Division of Agricultural Engineering, Indian Agricultural Research Institute, New Delhi, 1992)
131.Rao V. V. and Sarma P. B. S., Regional Groundwater Modelling Using Finite Element Method - A Case Study, *Proceedings, Session-GIII, International Conference on Hydrology and Water Resources* (Kluwer Publications, Netherlands, 1993).
132. Ray S. B., Gautam O. P. and Dastane N. G., Water Requirements of Crops on Soil Moisture Basis, *Proc. 5th NESA Seminar on Irrigation Practices* (1964), pp. 413-415.

133. Ray S. B., Water Management in Saline and Sodic Soils, *Seminar on Current Thrusts in soil and Fertiliser Research in Relation to Crop Production* (1981), Indian Agricultural Research Institute, New Delhi.
134. Ray S. B., Consumptive Water Use, Irrigation Requirement, Water Requirement of Crops Under Mono and Multi Crop Command, Crop Factors and Effective Rainfall, Lecture delivered at the Vth Training of Trainers Programme (1988), Water Technology Centre, Indian Agricultural Research Institute, New Delhi.
135. Rema Devi A. N. and Michael A. M., Soil Moisture and Salinity Distribution in Crop Root Zone in Drip Irrigation with Saline Water, *J. of Agril. Engg.*, **XIX** (1982), pp. 27-36 .
136. Rosegrant M. W., Water Resources in 21st Century: Challenges, implication for action ,Food, Agriculture and the environment. Discussion paper, 20, IFPRI (International Food Policy Research Institute), Washington D.C. pp. 136-139.
137. Sarkar T. K., Design Manual No.1-Drainage System Design of a Research Farm at Indian Agricultural Research Institute (1980), Water Technology Centre, IARI, New Delhi.
138. Sandhu B. S., Singh Baldev and Ahuja T. S., Effects of Irrigation , Nitrogen and Rice Straw Mulch on Edaphic Environment and Growth of Summer Forage Sorghum, *Journal of Indian Society of Soil Science,* **35** (1987).
139. Sarma P. B. S, Delleur J. W and Rao A. R., Comparison of rainfall-run-off models for urban basin, *Journal of Hydrology*, **18** (1973), pp.329-347.
140. Sarma P. B. S., Some Aspects of Water Resources System Analysis, *Journal of the Institute of Engineers (India)*, **156** (1975a), pp. 81-85.
141. Sarma P. B. S., Nathan K. K. and Rao G. G. S. N., Analysis of Maximum Precipitation over Coastal Andhra Pradesh, *Second World Congress, IWRA*, New Delhi **3** (1975a), pp. 471-487.
142. Sarma P. B. S., Groundwater Management, *Commerce* **131** (1975-b), pp. 129-135.
143. Sarma P. B. S., Babu Ram and Pandey S. L., A Study of the Groundwater Hydrology of an Agricultural Farm, *Proc. National Symposium of Hydrology*, Roorkee **I** (1975b), pp. 11-15.
144. Sarma P. B. S., Modelling in Agricultural Water Use, *Symposium on Mathe - matical Modelling For Simulation And Solution Of Water Resources Problems,* Patna, Paper No.21 (1976), pp 1-9.
145. Sarma P. B. S., Kaul R. K. and Dogra A. K., A case study of sick tubewells in Ambala and Naraingharh tehsils of Ambalà district, Harayana, using Three-dimensional bore hole camera System, Tech. Report No.HYD-2., (1976), Water Technology Centre, Indian Agricultural Research Institute, New Delhi, pp. 22.
146. Sarma P. B. S. and Babu Ram, Simulation Of The Groundwater Basin of Agricultural Farm, *Fifth National System Conference*, Ludhiana **I** (1978), pp. 36-139.

147. Sarma P. B. S., Influence of Drought on Water Regime, Key Note Paper, Tech. Session No.2 on Water Quality and Water Regime, *International Symposium on Hydrological Aspects of Droughts,* Delhi, **II** (1979a), pp. A1-A9.

148. Sarma P. B. S., General Report for the Tech Session No.2 *International Symposium on Hydrological Aspects of Droughts,* Delhi, **II** (1979b), pp. 83-87.

149. Sarma P. B. S., Rao N. H. and Rao K. V. P., Calculation of Water Balance in the Crop Root Zone by Computer, *Journal of Hydrology* **45** (1980), pp. 123-131.

150. Sarma P. B. S. and Rao N. H., Systems Approach for Conjunctive Utilisation of Irrigation Water, *Proc. Third Afro Asian Regional Conference of International Commission on Irrigation and Drainage*, New Delhi, India (1980), pp.41-53.

151. Sarma P. B. S., Rao N. H., Sharma R. K. and Michael A. M., Water Management in Canal Command Areas - A Case Study of Paladugu Major, Nagarjunasagar Right Canal Command Area, Andhra Pradesh, Tech. Report No. HYD-7 (1982), Water Technology Centre, Indian Agricultural Research Institute, New Delhi.

152. Sarma P. B. S. and Rao M. S., Simulation Studies in the Design of Cavity Wells, *Journal of Agril. Engg.* **XIX** (1982a), pp.37-42.

153. Sarma P. B. S. and Rao M. S., Design of Cavity Wells - Stress Analysis Approach, *Journal of Agril. Engg.* **XIX** (1982b), pp. 21-26.

154. Sarma P. B. S., Rao N. H. and Rao K. V. P., A Computer Based Procedure for Calculating Water Balance in the Crop Root Zone, Tech. Report No. HYD-6 (1982), Water Technology Centre, Indian Agricultural Research Institute, New Delhi, p.43.

155. Sarma P. B. S., Assessment of Water Availability in Command Areas in Relation to Rainfall, Groundwater, Soil Moisture, Canal Releases and Crop Water Requirements, *National Seminar on Integrated Approach to Water Management in the Command Areas,* Indian Agricultural Research Institute (1983 a), pp. 43-53.

156. Sarma P. B. S. (ed.), *Resources Analysis and Plan for Efficient Water Management - A Case Study of Mahi Right Bank Canal Command Area, Gujarat, Research Bulletin* **42** (1983a), Indian Agricultural Research Institute, New Delhi, p. 360.

157. Sarma P. B. S., Rao N. H. and Michael A. M., Water Resources Utilisation and Management in Resources Analysis and Plan for Efficient Water Management - A Case Study of Mahi Right Bank Canal Command Area, Gujarat, *Research Bulletin* **42** (1983b), Indian Agricultural Research Institute, New Delhi, pp. 322-360.

158. Sarma P. B. S., Dogra A. K., Purshottam Chandra and Talwar S. C., A Case Study of Sick Tubewells in Vadodra, Kaira, Ahmedabad, Gandhinagar, Mehsana, Banaskantha and Kutch Districts of Gujarat State using Bore-Hole Camera System, WTC Tech. Report No. HYD-8. (1983 b), p. 24.

159. Sarma P. B. S., Purshottam Chandra and Talwar S. C., Photographic Survey of Sick Tubewells of South Arcot, Tanjavur and Ramanad Districts of Tamilnadu, (1984a) WTC Tech. Report No. HYD - P32.
160. Sarma P. B. S., Purshottam Chandra and Talwar S. C., Investigations of Sick Tubewells in Allahabad, Jaunpur Districts of Uttar Pradesh, WTC. Tech. Report No.HYD-1 (1984b), p. 26.
161. Sarma P. B. S., Estimation of Runoff, General Report Technical Session No. III, *Proc. Symposium on Estimation of Runoff for Surface and Subsurface Drainage,* New Delhi, **2** (1985), pp. 27-32.
162. Sarma P. B. S., Purshottam Chandra and Talwar S. C., Survey of Sick Tubewells in Khammam Warangal, West and East Godavari Districts of AP, WTC Tech. Report No.HYD-11, (1985), p. 32.
163. Sarma P. B. S., Rao N. H., Tiwari P. N. and Sharma P. K., Water Resources Utilisation and Management, *Resources Analysis for Integrated Development, Sultanpur District*, Water Technology Centre, Indian Agricultural Research Institute, New Delhi (1986 a) pp. 240-249.
164. Sarma P. B. S., Rao N. H, Tiwari P. N. and Sharma P. K., Water Resources and Irrigation, *Resources Analysis for Integrated Development, Sultanpur District*, Water Technology Centre, Indian Agricultural Research Institute, New Delhi (1986 b), pp. 240-249.
165. Sarma P. B. S., Ashwani Kumar, Rao N. H., Tiwari P. N. and Sharma P. K., Water Resources And Irrigation, *Resources Analysis for Integrated Development, Rae Bareli District*, Water Technology Centre, Indian Agricultural Research Institute, New Delhi (1986 c), pp. 252-306.
166. Sarma P. B. S., Ashwani Kumar, Rao N. H., Tiwari P. N. and Sharma P. K., Water Resources Utilisation and Management, Resources Analysis for Integrated Development, Rae Bareli District, Water Technology Centre, Indian Agricultural Research Institute, New Delhi (1986 d), pp. 337-377.
167. Sarma P. B. S., Rao N. H., Sharma R. K. and Michael A. M., Rotational Water Distribution in Nagarjunasagar Canal Command Area, *Indian Journal Of Agricultural Sciences* **56** (1986e), pp. 520-31.
168. Sarma P. B. S., Rao N. H., Tiwari P. N. and Sharma P. K., Ground Water, Resources Analysis for Integrated Development, Sultanpur District, Water Technology Centre, Indian Agricultural Research Institute, New Delhi, (1986 f), pp. 250-277.
169. Sarma P. B. S., Ashwani Kumar, Rao N. H., Tiwari P. N and Sharma P. K., Groundwater, *Resources Analysis for Integrated Development, Rae Bareli District,* Water Technology Centre, Indian Agricultural Research Institute, New Delhi, (1986g), pp. 307-323.
170. Sarma P. B. S., Rao N. H. and Tiwari K. N., Upconing of Salt Water-Fresh Water Interface beneath a Skimming Well, *Journal Of The Institution Of Engineers (India)*, **67** (1987), pp. 335-342.

171. Sarma P. B. S., Purshottam Chandra and Subash Talwar, Photographic Investigations and Suggestions for Rehabilitation of Sick Tubewells, *Vth Groundwater Congress,* Hyderabad, **I** (1987).
172. Sarma P. B. S. and Abel Immaraj., Estimation of Aquifer Parameters through Identification Problem, *Hydrology Journal* **1** (1990), pp. 13-24.
173. Sarma P. B. S. and Rao N. H., Hydraulics of Flow to Partially Penetrating Wells, *Journal of Applied Hydrology*, **3** (1990), pp. 19-28.
174. Sarma P. B. S., Computer Applications in Irrigation Water Management, *National Seminar of Use of Computers in Hydrology and Water Resources,* New Delhi, Lead paper Session III on Crop Water Demands and Irrigation Water Management Proc. **I** (1991), pp. 3(I)-3(Xiv).
175. Sarma P. B. S. and Mani A., A Simple Soil Water Balance Model, *Hydrology Journal* **3&4** (1992), pp.13-20.
176. Sarma P. B. S., Factors Affecting Retention and Movement of Water in Soils, *National Workshop on Soil & Environment* (1993), National Academy of Sciences, Bangalore Session.
177. Sarma P. B. S., Rao N. H. and Doraisamy P., An Integrated Expert Decision Support System for Irrigation Scheduling, *Proc. Workshop on Integrated Development of Irrigated Agriculture,* Central Board of Irrigation and Power, April 28-29 (1994), New Delhi, pp.262-270.
178. Sarma P. B. S., Issues of Irrigation System Management, *Proc. Special Workshop on Irrigation Water management,* Water Technology Centre, Indian Agricultural Research Institute, New Delhi (1995), pp. 12-16.
179. Sarma P. B. S. and Rao V. V., Identification of Critical Dry Spells for Irrigation Planning, *Water and Energy International* **54** (1997 a), pp. 29-36.
180. Sarma P. B. S. and Rao N. H., Use of Simulation and Optimization Models in Water Resources Planning and Management, *Proc. National Water Policy - Scientists Perceptions* eds. by Randhawa N. S. and Sarma P. B. S. (1997b) pp. 234-243.
181. Sarma P. B. S. and Rao V. V., Evaluation of an Integrated Water Management Scheme - A Case Study of Paladugu Major, *Agricultural Water Management,* **32** (1997c), pp. 181-195.
182. Sarma P. B. S., Participatory Irrigation Management, Ch.No.30, *Agricultural Scientists' Perceptions of National Water Policy* eds. by Randhawa N. S. and Sarma P. B. S. (1997), pp. 419-423.
183. Sarma P. B. S. and Kumar A., Sprinkler Irrigation in India (INCID, Ministry of Water Resources, Govt.of India, New Delhi, 1998)
184. Sarma P. B. S., Garg S. P. and Rao N. H., System Analysis Approach for Biodrainage of waterlogged regions, *8th International Drainage Workshop, ICID.,* New Delhi, vi, Group 5 (2000), pp.47-60.
185. Sarma P. B. S., Rao N. H., Nathan K. K., Development of a Decision Support System for Real time Operation of an Irrigation System - Patna Canal

Command Area- Sone Command Irrigation Project, Bihar - Project Completion Report (2000)

186. Sarma P. B. S. and Chowdary V. M., An Approach for Assessment of Non-Point Source Pollution to Groundwater in Irrigated Areas, *International Conference on Water Resources management*, April 10-12 (2000), Univ. of California, Davis, USA (Abstract).

187. Serageldin Ismail, ICRISAT's achievements and future challenges, ICRISAT Silver Jubilee Celebrations (1997), Hyderabad, A.P.

188. Shakya S. K., Gupta P. K. and Kumar D., Innovative Drainage Techniques for Water Logged Sodic Soils, *Technology Bulletin* **3** (1995), Dept. of Soil and Water Engineering, Punjab Agricultural University, Ludhiana.

189. Sharma S. D. and Michael A. M., Chapter in Resource Analysis and Plan for Efficient Water management - a case study of Mahi Right bank Canal Command Area, Gujarat, Water Technology Centre, Indian Agricultural Research Institute, New Delhi (1983).

190. Sharpley A. N. and Williams J. R., EPIC - Erosion Productivity Impact Calculator- (1) Model Documentation , pp. 235, & (2) *User's Manual* , USDA, Technical Bulletin,No.1768 (1990), p. 127

191. Singh A. K., Effect of Saline Water on yield and water use efficiency of crops, *Annals of Agriculture Research* **10** (1989), pp. 39-48.

192. Singh M., Bhattacharya A. K., A software (RIMTRAIN) for estimating total rainfall at desired recurrence intervals, Water Technology Centre, Indian Agricultural Research, Institute, New Delhi (1996), p. 19.

193. Singh N. P., Singh Subedar, Bahl D. K. and Jha D., Response of Safflower (Carthamus tinctorious L.) to Irrigation and Nitrogen Application, *Annals of Agricultural Research* **1** (1980), pp. 40-52.

194. Singh R, Das D. K. and Singh A. K., Prediction of hydrological Characteristics from basic properties of Alluvial soils, *J. Indian Soc. Soil Science* **40** (1992), pp.180-183.

195. Singh Subedar and Singh N. P. and Singh Mahatim, Influence of Irrigation and Phosphorus on Growth and Yield of Lintel, *Indian Journal of Agricultural Sciences* **53** (1983), pp. 225-229.

196. Singh Subedar, Singh N. P. and Bandyopadhyay S. K., Effect of Limited Irrigation on Seed Production , Oil Yield, and Water Use by Indian Mustard, *Annals of Agricultural Research* **18** (1997), pp. 265-269.

197. Singh R. N., Singh Subedar and Singh K. M, Effect of Sowing Time and Irrigation on Aphid Incidence and Yield of Mustard, Abstract in *the National Seminar on Strategies on Pest Management,* Indian Agricultural Research Institute, New Delhi (1981)

198. Sinha A. K. and Pandey S. L., Effect of Depth to Water Table on Performance of Rice Crop, *Indian Journal of Agronomy* **XXVIII** (1983), pp. 65-69.

199. Sondhi S. K. and Sarma P. B. S., Surface And Groundwater Development Alternatives for Canal Command Areas Using Simulation Model, *Proc. of Regional Workshop on Groundwater Modelling,* Roorkee (1986), pp. 107-119.
200. Sondhi S. K., Sarma P. B. S. and Rao N. H., Assessment of Groundwater Resources of A Canal Command Area, *XXIII Annual Convention of Indian Society of Agricultural Engineers,* Jabalpur (1987)
201. Sondhi S. K. and Sarma P. B. S., Spatial Distribution of Net Groundwater Recharge, *First Indian Water Congress* **1** (1987), New Delhi.
202. Sondhi S. K., Rao N. H. and Sarma P. B. S., Assessment of Groundwater Potential for Conjunctive Water Use in a Large Irrigation Project in India, *Journal Of Hydrology* **107** (1988), pp. 283-295.
203. Sondhi S. K. and Sarma P. B. S., Groundwater Basin Simulation of Mahi Right Bank Canal Command Area, Gujarat India, *International Workshop on Appropriate Methodologies for Development and Management of Groundwater Resources in Developing Countries, IGW-89,* Hyderabad, **II** (1989a), pp. 581-593.
204. Sondhi S. K. and Sarma P. B. S., Sensitivity Analysis in Groundwater Modelling. *International Workshop on Appropriate Methodologies for Development and Management of Groundwater Resources in Developing Countries, IGW-89,* Hyderabad, **II** (1989b).
205. Stringham G. E. and Keller J., Surge Flow for Automatic Irrigation, ASCE, *Irrigation And Drainage Division Speciality Conference,* Albuquerque, **NM** (1979), pp. 131-142.
206. Stringham G. E., Surge Flow Irrigation, *Utah Agricultural Experiment Station Research Bulletin* (1988), p. 515.
207. Sudar R. A., Saxton K. B. and Spomer R. G., A predictive Model of Water Stress in Corn and Soybean, *Trans.* ASAE, pp. 97-102.
208. Surajbhan, *Phaslon Mei Jal Prabhand* (Hindi), ICAR Publication. 1983.
209. Surajbhan, Soil And Water Conservation Practices for Sustainable Agriculture in Rainfed areas with Particular Reference to Northern Plain Zone of India, *J. Soil Water Conservation* **40** (1996), pp. 44-56.
210. Tanwar B. S., Conjunctive Use Strategy Imperative in Controlling Water Table, Energy Loss and Rural Economy, *Proc. National Seminar on Water Management for Sustainable Agriculture, - Problems and Perspectives for the 21st Century,* IARI, April 15-17 (1998), New Delhi, pp. 284-291.
211. Tiwari K. N, Studies on Hydraulic Design of Skimming Wells (M.Sc. Thesis, Division of Agricultural Engineering, Indian Agricultural Research Institute, New Delhi, 1980)
212. Tiwari K. N. and Sarma P. B. S., Reducing Infiltration Capacity of Soils through Surface Treatments for Runoff Harvesting, *Ind. Journal of Power and River Valley Development* (1987a), pp. 138-141.
213. Tiwari K. N. and Sarma P. B. S., Water Harvesting by Runoff Inducement for Irrigation, *Journal of Institution of Engineers (India)(AG)* (1987b) pp. 52-56.

214. Tiwari K. N. and Sarma P. B. S., Water Harvesting Technology for Life Saving Irrigation, *Ind. Journal of Power & River Valley Dev.* (1988), pp. 269-274.
215. Tiwari K. N. and Sarma P. B. S., Prediction of Runoff from Water Harvesting Catchments, *Journal of Irrigation & Power* **48** (1991), pp. 109-123.
216. Tsakiris G and Kiountouzis E., Optimal Intraseasonal Irrigation Water Distribution, *Advances in Water Resources* **7** (1984), pp. 89-92.
217. USDA, Hydrology, National Engineering hand Book, Section 4 (Soil Conservation Service, US Department of Agriculture, Beltsville, Maryland, USA, 1964)
218. USDA, Agricultural Research Service, Web site, visited in 1999.
219. UNCED, United Nations Conference on Environment and Development, The Earth Summit, Rio De Jenero (1992)
220. Vaux H. J. and Pruitt W. O., Crop Water Production Functions ed. by Hillel D., *Advances in Irrigation* **2** (1983), pp. 224-255.
221. Vedula S. and Mohan S., Real-time multipurpose Reservoir Operation: A Case Study, *Hydrological Sciences Journal* **35** (1990), pp. 447-461.
222. Vedula S. and Mujumdar P. P., Optimal reservoir operation for Irrigation of Multiple Crops, *Water Resources Research* **28** (1992), pp. 1-9.
223. Verma H. N. and Sarma P. B. S., Analysis of Short Duration Rainfall for Planning Rainfed Crops, *Journal of Agril. Engg.* **25** (1989 a), pp. 14-19.
224. Verma H. N. and Sarma P. B. S., Critical Dry Spells and Supplemental Irrigation to Rainfed Crops, *Journal of Indian Water Resources Society* **9** (1989 b), pp. 12-16.
225. Verma H. N. and Sarma P. B. S., Criteria for Identifying Effective Monsoon for Sowing in Rainfed Agriculture, *Journal of Irrigation and Power* **47** (1990a), pp. 177-184.
226. Verma H. N. and Sarma P. B. S., Design of Storage Tanks for Water Harvesting in Rainfed Areas, *Agricultural Water Management* **18** (1990 b), pp. 195-207.
227. Verma H. N. and Sarma P. B. S., Hydrologic Design of Runoff Storage Tanks in Rainfed Areas, *Hydrology Journal* **1** (1990 c), pp. 45-54.
228. Visalakshi K. P. and Rajput T. B. S., Development of Design Procedures for Surge Flow Irrigation in Furrow, *J. of Indian Water Resources Society* **17** (1997), pp. 13-17.
229. Vyas S. K., Evaluation of performance of canal irrigation system - a case study (M.Sc. Thesis, Division of Agricultural Engineering, Indian Agricultural Research Institute, New Delhi, 1988)
230. Vyas S. K. and Sarma P. B. S., An Analytical Model for Estimation of Flows at Various Locations in a Canal Command Area, *Journal of Irrigation and Power* **49** (1992), pp. 125-137.
231. WCW, Report of the World Commission on Water, World Commission on Water, The Hague Conference, (March 2000), *Water Resources Development*, **16** (2000), pp. 289- 320.

232. Yadav B. R. and Paliwal K. V., Growing vegetables with saline irrigation water, *Indian Horticulture* **35**, pp.11-13.
233. Yarn B. and Frenkel H., Water Suitability for Agriculture in Management of Water Use in Agriculture ed. by Tanji K. K. and Yaron B., Springer Verlog (1994), pp. 25-46.
234. Yeh W. W. G., Reservoir Management and Operation Models: A State of the Art Review, *Water Resources Research* **21** (1985), pp. 1797-1818.

REMOTE SENSING APPLICATIONS TO WATER RESOURCES

D. NAGESH KUMAR

Civil Engineering Department
Indian Institute of Technology, Kharagpur – 721 302,
INDIA
E-mail: nagesh@civil.iitkgp.ernet.in

Remote sensing has emerged as a major tool in studying and analyzing the complex water resource systems. With the advent of earth observation satellites, better understanding of the land surface conditions has become possible which is essential for several applications in hydrology, meteorology and agronomy. This article highlights various applications of remote sensing to water resources such as hydrology, watershed management, flood plain management, drought monitoring, irrigation management, irrigated crop yield assessment among others. It also includes micro-wave remote sensing applications to water resources with the availability of Synthetic Aperture Radar (SAR) data of European Satellites. Geographic Information System (GIS) has further improved the utility of remotely sensed data by linking the spatial database with temporal database at different locations. A discussion on GIS applications to water resources is also presented. Global Positioning System (GPS) has helped in collecting precise 'ground truth' for better utilization of remotely sensed data (Precision agriculture etc.) A brief introduction to GPS with its adaptability to water resources management is presented. Various remote sensing missions and their capabilities are briefly presented with special emphasis on Indian remote sensing (IRS) satellites and the future plans with specific missions for better utilization of earth resources. A state-of-the-art review of research in remote sensing applications to water resources is presented.

1 Introduction

Remote sensing is the science and art of obtaining information about an object, area or phenomenon through the analysis of data acquired by a sensor that is not in direct contact with the target of investigation. Remote sensing (RS) studies include data collected with measurements on the ground (hand-held, truck-mounted, etc.) and from a variety of airborne and spaceborne platforms (e.g. satellites) Satellites may be in geo-synchronous orbit or sun- synchronous orbit. A variety of sensors, which measure reflective, thermal, and dielectric properties of the earth's surface are available [28]. Both active sensors, which send a pulse (in microwave, thermal range etc.) and measure the return pulse and passive sensors, which measure emissions or reflectance from natural sources (e.g. solar energy), are used. Sensors used for water resources applications cover a broad range of the electromagnetic spectrum. RS techniques indirectly measure hydrological variables. So the electromagnetic variables measured by RS techniques have to be related to the hydrological variables empirically or with transfer functions.

Remote sensing can play useful role in harnessing available water resources wealth. There are several areas in the field of water resources wherein RS proves useful for effective applications – particularly in surveying and inventorying. The International Satellite Land-Surface Climatology Project (ISLSCP) encourages scientists to use remote sensing data to better understand natural processes on the global land surface [88]. Large-scale field experiments have been conducted to validate remote sensing algorithms for estimating surface parameters and fluxes such as evapotranspiration [112]. There is ample scope for the application of remote sensing in the assessment of various components of hydrologic cycle, quantification of these components in various environs and the fluxes of water through these environs. Some of the fields include hydrologic studies, river morphology, reservoir dynamics and sedimentation, watershed conservation, command area planning and management, flood estimation and forecasting, ground water studies, water quality and environmental protection. Remote sensing application in hydrology is briefly reviewed by Schultz [84]. A review of remote sensing applications to ground water studies is presented by Meijerink [52]. An exhaustive list of various applications to water resources, wherein RS may substitute or complement or supplement the conventional methods, is given by Balakrishnan [2]. A state-of-the-art review on satellite remote sensing applications in irrigation management is given by Bastiaanssen [5]. This article presents an overview of the applications of remote sensing to various aspects of water resources. Narayanan [62] presented number of articles popularizing the use of remote sensing in various fields. Joseph [42] and Kalyanaraman [43] presented the past, present and future of Indian remote sensing missions and their capabilities. More information about Indian space program can be found at http://www.isro.org/.

2 Hydrological Studies

Hydrological processes are highly dynamic phenomena which vary both in time as well as space. Conventional measurements of these hydrological processes are accomplished by in-situ or point measurements. These are then interpolated or extrapolated to get the aerial estimates. The advent of RS technology has opened new vistas for the study of various components of hydrologic cycle.

Some examples of parameters, used in hydrological modeling, which have been derived from satellite data with sample references are presented in Table 1.

Table 1. Parameters in hydrology and water resources currently obtainable from satellite remote sensing data (modified from [47])

Parameter	Satellite/ Sensor	Wavelength or Frequency	Spatial Resol-ution	Coverage	Sample references
Snow area	NOAA	0.62, 10.8 mm Bands 1, 4	1 km	2 per day	[45, 12]
Snow Depth	GOES	0.65 mm	2 km	2 per hour	[22, 14]
	Nimbus 7	37 GHz	30 km	-do-	
Snow water equivalent	SSM/I	19.3, 37 GHz	25 km	2 per day	[92, 107]
	MOS-1-MSR	23, 31 GHz	23-32 km	-do-	
Changes in snowmelt	ERS-1	5.3 GHz C-band SAR	30 m	35 days	[21]
Surface Temperature	NOAA	10.80 mm Band 4	1 km	2 per day	[23]
Evapotrans-piration	NOAA	0.62, 0.91, 10.8,12.0 mm	1 km	2 per day	[71]
	GOES	0.64,11.5 mm	2-8 km	2 per hour	[11]
Precipitation	Meteosat	0.65 mm	3 km	2 per hour	[70]
Land cover/ Land use	Landsat 5 MSS	0.55, 0.65, 0.75, 0.95 μm	80 m	8-16 days	[46, 111]
Vegetation	NOAA	0.62, 0.91 μm	1 km	2 per day	[1]
	IRS 1C/1D (WIFS)	0.92, 0.67 μm	188 m	5 days	[48]
Suspended Sediment/	Landsat 5 MSS	0.55, 0.65, 0.75, 0.95 μm	80 m	8-16 days	[78]
Algal growth	IRS P4 OCM		360 m	2 days	[44]
Spring runoff	Nimbus 5	19 GHz	30 km	2 per hour	[109]
Changes in soil moisture	JERS-1	1 GHz L-band SAR	30 m	35 days	[69]
Groundwater	Landsat	0.95 μm	80 m	8-16 days	[10]
Water depth	Landsat	0.48, 0.56, 0.66 μm	30 m	8-16 days	[33]
Oceanography	IRS P4 OCM	412,443,490,520 555,670,765 nm	360 m	2 days	[43]
	MSMR	6.6, 10.65, 18, 21 GHz		2 days	

There are three broad categories for using remote sensing in hydrological studies [80].

- Simple qualitative observations/ assessments are made. e.g. A visual observation of a photo that water from industrial effluent into a stream has a different colour than the stream water, suggesting a site for collection of a sample.
- Information on geometric form, dimensions, patterns, geographic location and distribution are derived for features such as land cover categories that influence runoff, evapotranspiration and soil moisture.
- Development of correlation between the remotely sensed observations and the corresponding point measurements on the ground for estimation of a hydrologic parameter. Examples include the estimation of rainfall, soil moisture, snow depth, sediment load etc.

2.1 Rainfall Estimation and Monitoring

Conventional means of rainfall monitoring using a network of rain gauges has limitations due to their sparse density more so in remote areas as well as over oceans. Radar is an active microwave remote sensing system operating in the 1 mm – 1 m region of electromagnetic spectrum (EMS) In a radar system, a pulse of electromagnetic energy is transmitted as a beam, which is partially reflected by cloud or rainfall, back to the radar. Radar has the unique capability to observe the areal distributions of rainfall and provide real-time estimates of rainfall intensities. Operational use of ground based radar for rainfall monitoring is limited due to smaller range and being costly. With the advent of satellites, it has become possible to obtain spatially continuous and homogeneous data over large areas including oceans in real-time. The use of satellite data for estimating rainfall is mainly based on relating brightness of clouds observed in imagery to rainfall intensities.

Various aspects of rainfall hydrometeorology amenable to improved analysis using satellite data are [3]: (i) Delineating the boundaries of areas likely to get rain, (ii) Assessing basin rainfall totals over time, (iii) Assessing extreme events of rainfall, (iv) Assessing the climatology of rainfall distributions and (v) forecasting of rainfall especially in regions with sparse data.

The wavelengths most commonly used for rainfall studies are [4]:

- Visible (VIS) : 0.5 – 0.7 μm
- Infrared (IR) : 3.5 – 4.2 μm and 10.5 – 12.5 μm and
- Microwave (MW): 0.81 to 1.55 cm

Most of the meteorological satellites currently used for precipitation estimation are either geostationary or polar-orbiting satellites. Geostationary satellites (e.g. NOAA, GOES, GMS, Meteosat, INSAT) typically carry infrared (IR) and visible (VIS) imagers with spatial resolution from 1-4 km. A geostationary satellite positioned over equator can provide high frequency (hourly or better) images of a portion of the tropics and middle latitudes, while a polar orbiting satellite provides roughly twice-daily coverage of the entire globe. Polar orbiters also fly in a low earth orbit, which is more suitable for the deployment of microwave imagers on account of the later's coarse resolution. High quality microwave imagery was made available after the launch of SSM/I (Special Sensor Microwave Imagery) with spatial resolution in the order of 10 km. SSM/I data is found more reliable for rainfall estimation. However it suffers from two limitations viz., coarse temporal resolution and coarse spatial resolution. The geometrical effects of three-dimensional rain clouds on radiances observed by an oblique viewing radiometer such as SSM/I have been examined and were found to be considerable [31].

There are a good number of satellite rainfall estimation algorithms documented in the literature. For example, 55 algorithms were evaluated in the Third Algorithm Intercomparison Project (AIP-3) of the Global Precipitation Climatology Project (GPCP) including 16 VIS/IR, 29 MW and 10 mixed IR/MW algorithms. Petty and Krajewski [68] presented a review of some of the important algorithms.

Cloud indexing method [30] is normally used to estimate daily rainfall over an area through statistical averaging of cloud-rainfall relationships. In this approach three most important types of clouds viz., cumulonimbus, cumulocongestus and nimbostratus are considered. This method is widely used in support of broad scale hydrology.

To meet the growing need for stably calibrated, long, time series data sets for global climate change, a number of organized scientific programs or projects are actively promoting the development, validation and/or access to the user community of satellite-derived precipitation products. Pathfinder precipitation products produced from SSM/I data using GSCAT algorithm are available freely from Distributed Active Archive Center (DAAC) of NAA, USA (http://daac.gsfc.nasa.gov) National Climate Data Center (NCDC) is currently producing rainfall products and distributing for both 1^0 and 2.5^0 grids (http://www.ncdc.noaa.gov) from a combination of archived microwave and infrared satellite data, surface rain gauge data and numerical forecast model output.

2.2 Snow Mechanics, Mapping and Monitoring

Snow in hydrology is a renewable resource and is one of the most complicated parameters to be measured. In India during summer months rivers rising in the Himalayas are substantially fed by snowmelt runoff. Periodic snow cover monitoring is essential for assessing the snowmelt runoff likely to occur. Conventional methods have serious limitations in the study of dynamic processes like snow and its monitoring complicated by inaccessibility. Considering the vastness of snow clad watersheds, aerial surveys are expensive and of little use. Satellite remote sensing has a vital role to play to obtain near real-time snow cover area (SCA) maps with good accuracy. Depletion curves of SCAs indicate the gradual areal diminishment of the seasonal snow cover during the snowmelt season. The typical shape of depletion curves can be approximated by the equation [2].

$$S = 100/\,(1+e^{bn})$$

where S is SCA in % obtained from satellite imagery, b is a coefficient and n is the number of days before (-) and after (+) the date at which S=50%.

The snow extent over the globe has become available on daily basis since 1972 from National Oceanic and Atmospheric Administration – Advanced Very High Resolution Radiometer (NOAA-AVHRR) satellite, assuming no cloud interference at 1.1 km spatial resolution. Passive microwave observations of snow cover extent are assured every 1-2 days from SSM/I due to cloud penetration capability but the spatial resolution is coarse (about 25 km)

US National Weather Service (NWS) distributes products of periodic river basin snow cover extent maps (http://www.nohrsc.nws.gov) from NOAA-AVHRR and Geostationary Operational Environmental Satellite (GOES) Some good examples of operational applications of such snow extent data are already there in India. Initially Ramamoorthi [73, 74] started to use NOAA-AVHRR data in a simple regression approach for empirical forecasts of seasonal snow melt runoff in the Sutlej river basin (43,230 km^2) and both the forecasts were extended to other basins. Ramamoorthi [74] also used satellite data as input to a snowmelt runoff model (SRM) for short term forecasts. Kumar *et al.* [49] further developed the idea for use in operational forecasts of daily and weekly snowmelt runoff in the Beas (5,144 km^2) and Prabati (1,154 km^2) river basins in India. Another type of application to operational snow hydrology is production of monthly hemispherical snow maps from GOES and NOAA-AVHRR data for climatological applications. EOS Multifrequency Passive Microwave Radiometer (MPMR) scheduled to be launched in 2000 can obtain 8 km resolution by using a much larger antenna than currently used for SSM/I. A detailed review of applications space borne remote sensing for snow hydrology is given by Rango [76].

2.3 *Rainfall-Runoff Studies*

Runoff in the form of water volume flowing through a river cross-section during a specified time interval cannot be measured on the basis of remote sensing data alone. This may evolve in future, if it becomes possible to measure the width and depth of a river cross section with the aid of remotely sensed data and flow velocities by ultrasonic or laser methods. At present RS data is indirectly used to determine runoff with the aid of hydrologic models. RS data is used either as model input or for the determination of model parameters or both. Most existing models do not incorporate the use of RS capabilities. Therefore it is necessary to develop structures of hydrological models which are amenable to the spatial and temporal resolution provided by RS data. Areal distribution of information from RS data can be better utilized by the application of distributed deterministic models. The spatial resolution of a model structure and the spatial resolution of its input data should have some correlation. For example, it makes no sense to structure a model in space according to the IRS 1-C PAN pixel size of 5m and use as its input precipitation data from one gauge for an area of several thousand square kilometers. Further, there has to be a reasonable correlation between the applied resolution to time and to space. For example, if only monthly runoff values are being used, there is no need to seek for a high spatial resolution data (say IRS LISS II with a spatial resolution of 36.25 m) Peck et al [66] identified 13 variables which can be obtained using remotely sensed data with some degree of success. They also presented a review of the existing hydrologic models with regard to their adaptability to remote sensing data. Similar studies were also presented in Balakrishnan [2] and Schultz [85].

Non-linear regression models based on RS and ground truth data were used for runoff modeling. A detailed sensitivity analysis for river basins in tropical West Africa was carried out by Papadakis *et al.* [65] to obtain information on the resolution required for the modeling effort in time, space and spectral channels. Meteosat data from European geostationary satellite with a high temporal resolution of 30 min and spatial resolution of 5 km was used for modeling monthly runoff for a large catchment area. The model functions in two consecutive steps: (i) Estimation of monthly precipitation values with the aid of Meteosat IR data and (ii) transformation of the monthly rainfall volumes into the corresponding runoff volumes with the aid of rainfall/ runoff model. For the three available spectral channels of Meteosat (IR, Visible and Water Vapor) it was shown that only IR information is relevant. Following a modified Arkin method [24], a monthly cloud cover index (CCI) is estimated indicating all pixels having a temperature in the IR channel below a specified threshold value. This threshold value varies from region to region. Monthly precipitation values are then obtained from daily CCI values as a nonlinear function of CCI [65]. Monthly runoff computed on the basis of rainfall estimated with the help of satellite imagery compared well with the observed runoff.

There have been many efforts to improve the performance of existing hydrological models by the use of RS data for improved parameter estimation. A very well received approach is for the improvement of the Soil Conservation Service (SCS) runoff Curve Number (CN) model [103]. Various SCS models allow computation of direct runoff volumes Q, in terms of land use and soil type without requiring hydrological data for calibration. The volume of direct runoff is given by the equation:

$$Q = \frac{(P - I_a)^2}{(P - I_a) + S}$$

where P is volume of rainfall, I_a is initial abstraction and S is retention parameter.

In SCS model, $I_a=0.2\ S$. Thus Q becomes a function of rainfall and S only. In practice a runoff curve number, CN, is defined as transformation of S according to the relationship, CN = 1000/ (S+10) The CN depends on the hydrological soil group and land use description. Conventionally, the curve numbers are given in tables. The introduction of RS data allowed a better estimation of the land use and thus a more reliable estimation of the relevant curve number. Two approaches have been used for estimating CN using remotely sensed data. In one approach land cover information was obtained using RS data which was then combined with general soil data to estimate the CN. Ragan and Jackson [72] used Landsat multi spectral data and this approach for runoff estimation. The other approach made use of Landsat average watershed reflectance ratios of spectral bands 0.5-0.6 μm and 0.8-1.1 μm to directly compute the CN without using ancillary soil data [9]. Remote sensing data can be a valuable tool in runoff prediction especially in areas experiencing rapid land use changes. Satish Chandra *et al.* [82] used Landsat imagery and aerial photographs to derive land use and vegetal cover data for Upper Yamuna basin and then obtained morphometric and relief characteristics of the basin. These were in turn used to derive runoff coefficients for various land uses for use in simulation of runoff, employing rational formula.

3 Watershed planning and management

Proper planning and management is essential for conservation of water and land resources for optimum productivity. Remote sensing via aerial and space borne sensors can be effectively used for watershed characterization and assessing watershed priority evaluation problems, potentials and management requirements and periodic monitoring. Various physiographic measurements which could be obtained from remotely sensed data include watershed area, size and shape, topography, drainage pattern and landforms. Spectral bands in the wavelengths 0.6-0.7 μm and 0.8-1.1 μm have been found to be useful for physiographic mapping of drainage basins. Stereoscopic attribute of aerial photographs permits quantitative assessment of landforms and evaluation of basin topography that can be used to develop or update the topo maps. Remote sensing due to its more current nature can provide better and more reliable information for quantitative analysis of drainage networks. Near Infra Red (NIR - 0.8-1.1 μm) band reveals the contrast between land and water features and is best suited for mapping large streams where as the visible red band (0.6-0.7 μm) facilitates differentiation among land, vegetated and non-vegetated features and is most useful for drainage network delineation [81]. The information on drainage network/ pattern is largely a reflection of the lithology and structure of the basin, stream orders, stream length, stream frequency, bifurcation ratio, stream sinuosity, drainage density and linear aspects of channel systems and can be obtained using RS data.

The most complex hydrological models are water balance models since they simulate all components in a water balance on and below the earth's surface and also in the atmosphere to a certain extent. Hydrological processes to be modeled in time and space in a water balance model include (1) precipitation, which usually serves as model input, (2) evapotranspiration, (3) runoff (surface and sub-surface), (4) storage change in the unsaturated zone and (5) ground water flow. Remote sensing data come into the water balance model at several levels [85]:

- Precipitation may be estimated from weather radar or from satellite imagery.
- Evapotranspiration models use leaf area index (LAI) data or NDVI (Normalized Differential Vegetation Index) for consideration of the transpiration of plants. These may be derived from Landsat, IRS or NOAA-AVHRR data [19, 114].
- Interception model also uses LAI or NDVI [20].
- Soil storage component uses land use classification from RS data [86].

Airborne laser altimeters provide quick and accurate measurements for evaluating changes in land surface features and is an effective tool to understand watershed properties. Airborne laser measurements can be used to measure directly topography, stream channel cross sections, gully cross sections, soil surface roughness and vegetation canopy height, cover and distribution [77].

Watershed degradation of soil and land resources could be mapped and monitored via remote sensing for reclamative measures. The mapping of soil degradation involving salinity/ alkalinity, water logging, erosion, desertification, shifting cultivation, excessive permeability, wetlands etc are successfully done in various regions using satellite imagery. Multitemporality (repetitive coverage) of satellite data is very much useful for change detection studies such as growth of desertification, flood damage area and encroachment of ravines on agricultural lands.

Remote sensor data also facilitates identification and classification of existing and potential erosion-prone areas for taking reclamation/ preventive measures. Effect of various watershed characteristics on soil erosion can be evaluated using aerial and satellite data. Many methods of erosion detection and measurement can be based on RS. Tueller and Booth [102] listed the following as erosion features identifiable from RS data:

- Erosion potential associated with changes in vegetation and litter
- Changes in soil type and soil colour
- Occurrence of dendritic soil patterns
- Occurrence of sand dunes
- Definition between bare soil or rock and
- Vegetal cover.

Spanner *et al.* [94] demonstrated the potential of using various data sources, through a data base approach, to map potential soil loss using Universal Soil Loss Equation (USLE) The data sources among others include Landsat MSS data, digitized USDA

SCS soil maps, digitized NOAA precipitation isopluvial maps and USGS digital elevation model topographic data.

4 Irrigation management

Irrigation is the largest consumer of fresh water. Seckler *et al.* [87] estimated that 70% of all water used each year produces 30 to 40% of world's food crops on 17% of all arable land. As water scarcity becomes more acute and competition for fresh water intensifies, better irrigation management will be required to achieve greater efficiency in the use of this valuable resource. Achieving food security through irrigation while combating water logging and salinization, to ensure sustainable agriculture requires quantitative and repetitive analysis of the irrigation processes. Distinct spatial variations in soil properties, soil moisture status, cropping conditions and micro-meteorology occur within an irrigation scheme. Although information on irrigation practices can be acquired by conventional survey methods, they have subjective character and usually differ from survey to survey. Consequently, remote sensing from space, which can regularly provide objective information on the agricultural and hydrological conditions of irrigated area, has a great potential for enhancing the management of irrigation systems. Studies have established that the presence of crops can be determined and several biophysical parameters can be measured with an accuracy of over 80% using RS data. In addition, RS data facilitates the construction of long time series (covering as much as 20 years) for investigation of changes in irrigation conditions. Many publications [5, 54, 99, 105] present the state-of-the-art of remote sensing applications in irrigation management. Molden et al [56] presented various indicators for comparing performance of irrigated agricultural systems. Bastiaanssen [5] categorized remote sensing publications, which explicitly refer to irrigation management. These categories include irrigated area, land use, crop water needs, use and stress, crop yield, soil salinity, water logging and reservoir mapping. Earlier works mainly included soil salinity mapping, land use and the quantification of area under irrigation. Recent works included crop classifications and crop yield estimation due to improved sensors onboard Landsat, SPOT and IRS yielding higher spatial resolution [59].

Thematic land classes can be derived digitally by grouping pixels, having similar spectral signatures, from measurements of individual bands throughout the EMR spectrum. Usually this classification is made using the visible, near-infrared and middle infrared parts of the spectrum. *Supervised classification* approach is the most common methodology for forming classes based on similar spectral reflectance [89]. In this approach, pixels are assigned to classes (i.e., training classes) verified on the ground (ground truth) in selected areas. These training sets represent a small percentage of an image. Then maximum likelihood criterion based on *a priori* probabilities [97], is used for classifying the entire image. Alternative criteria for classifying the images are based on minimum Euclidean distance to mean and Mahalanobis distance. *Unsupervised classification* algorithm clusters pixels multispectrally into classes through the use of standard statistical approaches such as the centroid and ward methods and does not rely on assigned classes from field visits. Cluster analysis technique also has some advantages. Huumeman and Broekema [38] presented an approach for classification of multi sensor data using a combination of image analysis techniques. A well-known phenomenon in pixel based image classification is the 'mixed' pixel, which comprise of a heterogeneous land surface i.e., two or more objects occur within a single pixel [5]. In agricultural areas, crop variety, seeding date and supply of fertilizer are often unevenly distributed among and within the crop fields. Due to this, the spectral reflectance of a given crop may contain a wide scatter [51]. *Fuzzy classification*, with more continuous land cover classes, has been developed to account for variations in natural conditions [108]. Fuzzy classifications are based on the relative strength of a class membership that a pixel has, relative to all defined classes and it does not require any assumption about the statistical distribution of the data. Neural network classifier [8, 34] also helps in thematic classification of an image for different crops.

4.1 Spectral Vegetation Indices

A vegetation index is a common spectral index that identifies the presence of chlorophyll. The index is based on reflectance in the red spectral region, R (0.62 to 0.70 μm) and a portion (0.7-1.1 μm) of the near infrared spectral region, NIR. Chlorophyll has a relative low reflectance in the red band (strong absorption) and a relatively high reflectance in the NIR band. Multiple combinations of red and NIR band data have been formulated and various crop parameters identified from vegetation indices using biophysics. Normalized Difference Vegetation Index (NDVI) is given by (NIR-R)/(NIR+R) SAVI (Soil adjusted vegetation index), TSAVI (Transformed soil adjusted vegetation index) and WDVI (Weighted difference vegetation index) indices are the improved indices by reformulation to minimize the back scattering of canopy-transmitted, soil-reflected radiation in partial canopies. Introduction of GEMI (Global environment monitoring index) and ARVI (Atmospherically resistant vegetation index) made it easier to quantify the time series of vegetation indices, despite continual change in atmospheric conditions and aerosol effects. Rondeaux *et al.* [79] reviewed the merits of most of the classical and upgraded vegetation indices recommended for application in agronomy.

4.2 Irrigated Area Assessment

Identification of irrigated area and separation from non-irrigated areas using remote sensing has started from early 80s. Thiruvengadachari *et al.* [101] estimated the irrigated area from IRS LISS II data and it was noticed that the estimated area has far exceeded the authorized area for irrigation, highlighting violations both in cropped area and cropping pattern. Visser [106] tested different methods to distinguish irrigated from non-irrigated areas. He found that supervised classification procedures performed (average accuracy of 92.5%) better than a multiplier of reflectance ratios in Landsat Thematic Mapper (TM) bands 3, 4 and 5 (90% accuracy) and than principal component analysis (85%) Nageswara Rao and Mohan Kumar [61] described the classification of irrigated crops through various vegetation indices (NDWI, NDVI and GVI) derived from Landsat TM and suggested that NDWI is better suited for identifying irrigated crops. Hussin and Shaker [37] classified the main land-use types in Sumatra, Indonesia using Landsat MSS and TM data combined with radar measurements from the European Remote Sensing Satellite (ERS-1 SAR) The spatial resolution of ERS-1 SAR (Synthetic Aperture Radar) was 12.5 m with a swath width of 100 km and a wavelength of 5.3 GHz or 5.6 cm (C-band) The multisensor spectral classification of TM and SAR data gave less satisfactory results, due to the speckle noise of SAR data, although paddy fields could be identified better. Paddy fields appeared very dark on the radar image, which is quite useful for delineating irrigated rice basins when the skies are cloudy. Texture labeling and texture spectrum analysis of ERS-1 SAR data was carried out by Nagesh Kumar *et al.* [60] to delineate irrigated area when the data from optical sensors was not available due to cloud cover.

Identification of different crop types by pattern recognition from multispectral high resolution satellite images is a classic remote sensing research problem in agriculture. Research on this issue seems to be continuously improving with better spectral and spatial resolution capabilities and sophisticated classification algorithms. Menenti *et al.* [53] adopted a multi-index, mutli temporal crop classification procedure, based on time composites of vegetation index and designed a transformed vegetation index (TVI), which expresses the density of full ground cover. They then combined TVI with the sample ratio to identify the cropping patterns in the irrigated plains of Italy. Use of TVI values allowed them to correct for variations in crop development due to differences in planting dates, irrigation water supply and farmers' practices. Jayasekera and Walker [41] applied a hybrid classification procedure in the densely vegetated Gal Oya irrigation project in Sri Lanka. They used principal component analysis and unsupervised classification of Landsat MSS bands and identified five categories of paddy practices, which differed in plant vigor, canopy density, and depth of standing water. Pedley and Curran [67] used SPOT-XS data for a field-by-field classification in South Yorkshire, U.K. The accuracy for 12 land cover classes was 46% at pixel scale and 55% at field scale. The best accuracy (62%) was achieved by using measures of prior probabilities and texture within a pre-field format. Thiruvengadachari *et al.* [101] used IRS LISS I image for irrigated crop classification in Bhadra reservoir command, Karnataka, India both by visual and digital interpretations and also evaluated the performance of the irrigation system.

In Netherlands, Schotten *et al.* [83] used eight ERS-1 SAR precision images with a 12.5 m resolution to identify potatoes, sugarbeats, winter wheat, maize, spring barley, winter rape, beans, onions, peas, grass, lucerne and orchards with a field based maximum likelihood classifier. The images were captured during the growing season between May 12 and November 3. Field size varied from 1 to 20 hectares and an overall classification accuracy of 80% could be obtained after an extensive ground survey. A SPOT-XS image with a 20 m resolution was used to fix the geometry of the agricultural fields prior to the classification procedure.

RADARSAT Scan SAR (Synthetic Aperture Radar) data of different dates were used to analyze the signature of rice crop in West Bengal, India [64]. The analysis showed that the lowland practice of cultivation gives distinct signature to rice due to the initial water background with relatively stable back scatter. Around 94% classification accuracy was achieved using two date data. The 300 km swath with 50 m resolution of ScanSAR was found cost effective for large area crop monitoring.

IRS-P3 Modular Optoelectronic Scanner (MOS-B) spectrometer data over parts of Northern India was evaluated by Singh *et al.* [91] for Wheat crop monitoring involving (a) sub pixel Wheat fractional area estimation using spectral mixing approach and (b) growth assessment by red edge shift at different phenological stages. Results obtained were compared with those obtained using IRS WiFS (Wide Field Scanner)

Accuracy levels attained in crop classification with the help of different sets of satellite images was summarized by [5].

Biophysical parameters are important for irrigation management because they reflect water and production issues. They include fractional vegetation cover and leaf area index (LAI) for vegetation development, surface roughness, surface emissivity, surface temperature, surface resistance, crop coefficients and transpiration coefficients for crop evapotranspiration and crop yield estimation. Bastiaanssen [5] described these parameters in detail along with their estimation based on remotely sensed data. Remote sensing applications for evapotranspiration monitoring over land surface is presented by Kustas and Norman [50].

5 Drought management

Agricultural drought assessment can be made with the help of spectral vegetation indices explained in the previous section. In a well executed project (ongoing) viz., 'National Agricultural Drought Assessment and Monitoring System', Dept of Space, India, is providing biweekly drought bulletins (June to December) for 246 districts in ten different states of India. The drought assessment is based on a comparative evaluation of satellite observed green vegetation cover (both area and greenness) of a district in any specific time period, with that of similar period in previous years. NDVI images derived from NOAA AVHRR data are used for this purpose. This nation wide early warning service has been found to be useful for providing first alert of drought conditions [98]. At present IRS 1D WiFS data with higher spatial resolution is being used for the same purpose.

6 Soil moisture estimation

As soil moisture controls the water balance in the crop root zone and also in the estimation of surface runoff, many attempts were made to retrieve soil moisture data from airborne and space borne multi spectral remote sensors. Bastiaanssen [5] presented a review of studies on soil moisture estimation based on remote sensing. Recent attempts to convert thermal infrared measurements into soil moisture maps are based on surface roughness [6] or use of the evaporative fraction to assimilate soil moisture in soil-vegetaion-atmosphere-transfer (SVAT) models [104]. Major contrast between the dielectric constants of water (~80) and dry soil (~3.5) produces very different propagation characteristics of the electromagnetic wave in soils having different moisture levels. For describing the soil moisture under given conditions, wave lengths between 0.3 cm (100 GHz) and 30 cm (1 GHz) are the most effective. Microwave radiometry reflects the moisture conditions in the top 10 cm of soil. The use of microwave emission is perturbed by surface roughness, attenuation and microwave emission by canopies and to a lesser degree by soil texture. Njoku and Entekhabi [63] presented the potentials of passive microwave technology for soil moisture estimation. Airborne microwave radiometry with spatial resolutions of a few meters, has shown greater success [15] compared to space borne passive microwave with very course spatial resolution. Radar instruments, which are active microwave sensors, are being used more frequently due to their improved spatial resolution [40]. Musiake *et al.* [58] conducted comparative study between microwave radars on board ERS (C-band) and JERS (L-band) for determining the soil moisture in agricultural fields and found good correlation (0.93) between the two radiometers. Engman and Chauhan [27] prepared a summary of the applications using microwave remote sensing, including radiometry, to detect soil moisture.

7 Soil salinity

Rising water tables, due to recharge from irrigation canals and watered fields, due to naturally poor groundwater quality or due to rock weathering, may cause soil salinity problems. In irrigated areas, the features, indicating salinity in increasing order, are stunted crop growth, poor and patchy germination, crop stress, death of crop, encroachment of halophytic species, bare soils with efflorescence and salt crust development. Visible reflectance of leaves from plants growing on such salt-affected soils is lower prior to plant maturation and higher there after. Steven *et al.* [96] showed that indices nearer to middle infrared indices are proper indicators of chlorosis occurring in stressed crops (normalized difference of TM bands 4 and 5) This index is immune to color variations and provides an indication of leaf water potential. Bastiaanssen [5] presented a review of different sensor applications for soil salinity detection and concluded that Landsat TM bands 5 and 7 are frequently used to detect soil salinity and drainage anomalies. Chaturvedi *et al.* [16] and Singh and Srivastav [90] used microwave brightness and thermal infrared temperatures synergistically for the identification of salt affected areas. Interpretation of microwave data was done by means of a two layer model with fresh and saline groundwater. Larger wavelengths (L-band, P-band) are capable of penetrating the soil and retrieving information from a soil layer rather than just from the soil surface. Combined remote sensing and geographic information system (GIS) with ancillary information on soil types, digital elevation data etc will yield better results for mapping soil salinity.

A general guideline for the use of remote sensing in irrigation management in terms of classifying cropping patterns, crop communities, land cover and land use cannot be given. However, a set of potentially successful components can be suggested [5].

- Ground truth on crop types across the entire classification domain
- Multitemporal, high resolution images
- At least 15 pixels per field
- Hybrid classifications based on both unsupervised and supervised classifications
- GIS-based aerial and contextual classifiers
- Spectral bands in the near-infrared, middle-infrared and microwave spectral range
- Definition of fuzzy classes for conditions with variable cultivation practices.

The potential usefulness of selected remote sensing applications for day-to-day and season-to season management of water resources in large irrigation schemes is summarized in Table 2, adopted from Bastiaanssen [5].

Wolters *et al.* [113] mentioned that satellite remote sensing cannot be effectively utilized for the day-to-day operation of irrigation system due to low frequency of high resolution images which are not compatible with the flexibility that canal operations require. But with the recent satellite missions (e.g. IRS 1C, WIFS), images can be acquired at higher frequency. Sriramany and Murthy [95] demonstrated the capability of remote sensing for mitigating crop stress or water logging in Greater Mae Klong System in Thailand.

Table 2. Irrigation Management Information that can be derived from Remote sensing

Parameter	Need for field data	Preferred Principle
Precipitation	High	Cold cloud duration and rain gauge network
Surface runoff	High	Curve number method
River discharge	High	Polarization differences, altimetry
Potential evapotranspiration	Low	1. Two-step Penman-Monteith equation with crop coefficient 2. Two-step Penman-Monteith equation 3. Radiation type of expression
Potential transpiration	Low	Transpiration coefficients
Potential evaporation	Low	Difference between Potential evapotranspiration and Potential transpiration
Actual evapotranspiration	Low	SEBAL algorithm
Actual transpiration	Moderate	LAI, transpiration coefficient and SAR
Actual evaporation	Moderate	Difference between actual evapotranspiration and actual transpiration
Crop stress indicators	Low	Water deficit index
Crop yield	Moderate	NDVI at heading stage
Relative yield	Low	Time-integrated relative transpiration
Topsoil moisture	Moderate	C-band, L-band microwave
Root-zone moisture	High	Inverse surface resistance and LAI
Soil salinity	High	Visible, near infrared, thermal infrared, microwave synergy
Salt minerals	High	Hyperspectrometry

In Crop Acreage and Production Estimation (CAPE) project under Remote Sensing Applications Mission (RSAM) in India, the acreage and production estimates [93] for wheat, rice, groundnut and sorghum were studied over a large area of 70.3 million hectares covering 21.4% of the geographical area of the country.

Application of remote sensing in irrigation management should be worked out through a two-way structure of demonstration and implementation. Demonstration projects should be initiated by international organizations (such as FAO, UN) The aim should be to reduce the gap between the information which remote sensing is technically capable of producing and the information needed for managing irrigation water for improving crop productivity. Elements that are successfully evaluated in the demonstration phase should be transferred to the implementation phase in close coordination with professionals. Also training of technicians and researchers on water related remote sensing issues is essential to achieve tangible results.

8 Reservoir sedimentation

Generally, suspended sediment causes the most serious pollution of water bodies. This not only reduces the reservoir storage and its life but also restricts the use of water for the intended multiple purposes. Remote sensing data is an important source in monitoring sedimentation of lakes and reservoirs through repetitive coverage. Quantifiable relationships between suspended sediment and remotely sensed data have been established. Remote sensing can be used in the following ways to monitor reservoir sedimentation [2].

- Inventorying the watershed runoff potential for taking steps to minimize sediment in the runoff and maximize clear water recharge.
- Identification and delineation of flood front during a storm event and establishing correlation between flood front extent and storm magnitude to predict future impact
- Mathematical modeling of sedimentation by correlating reflectance values with sedimentation rate.

Some of the variables used in hydrodynamic models of sedimentation can be measured using air and space borne sensors. They include near surface turbidity, Sechhi disc transparency, suspended solids concentration, surface temperature, flow velocity and surface area. Remotely sensed data generally represent fine materials, as large particles get deposited in the delta. These fine particles play an important role in biological and chemical dynamics of water bodies. Landsat MSS digital data has provided significant insight into analysis and understanding of near surface turbidity characteristics [110]. According to Moore [57], (i) turbid water is more reflective than clear water at all visible and NIR wavelengths, (ii) the remote signal from turbid water represents only near surface conditions and (iii) measured reflectance is dependent on the wave length used, size and shape of the particles present and their reflectance, absorption and refraction characteristics. Ratios of MSS bands, band 4/ band 1, band 3/ band 1 and band 2/ band 1 were found suitable for estimation of suspended sediments of high, high to low and low concentrations respectively. Spectral mixture analysis was used to estimate the concentration of suspended sediments in surface waters in Amazon basin wetlands by Mertes *et al.*, [55]. Surface concentration of suspended sediment load was estimated for many reservoirs and lakes in India by Chakraborti et al [13].

A comparative study of suspended sediment concentration potential derived from four band (100-300 nm) Landsat MSS, five broad band (40-300 nm) Landsat TM and eight narrow band (20-40 nm) IRS-P4 OCM spectral bands with that of the conventional (NIR-Red and NIR+Red) indices was made by Kaur and Rabindranathan [44]. A specific numerical index is proposed based on broad/ narrow n-wave band data from Landsat MSS/TM or IRS-P4 OCM spectral data. Prediction accuracies were observed to be the highest with the proposed index calculated from narrow OCM-P4 spectral data.

9 Flood monitoring and flood plain monitoring

Floods are regular phenomena in many parts of the world including India. Flood damage surveys are essential not only to assess the extent and severity of damage caused by the floods periodically in river valleys but also for economic evaluation of flood control measures. Moreover timeliness of information is crucial for tackling events like flood. Conventional methods have serious limitations in this aspect. Remote sensing facilitates the flood surveys by providing the much needed information on flood inundated areas, river course and its spill channels, spurs and embankments affected/ threatened etc so that appropriate flood relief and mitigation measures can be planned and executed in time. Through proper selection of platform and sensor, remote sensing can offer a quick and reasonably accurate means of surveying the spatial extent and assessment of flood damage. The effectiveness of existing flood control works in containing the flood can be assessed and vulnerable reaches identified for strengthening. New structures can be planned wherever necessary. Near real time monitoring of number of flood events of Indian rivers viz., Brahmaputra, Ganga, Kosi, Jhelum and Godavari has demonstrated that valuable information can be provided regarding flood affected areas for planning flood relief activities [75]. Availability of microwave data (ERS-1 SAR) has helped mapping flood inundation even during cloud covered periods [32].

10 Geographic information system

Geographic Information System (GIS) technology provides tools for effective and efficient storage and manipulation of remotely sensed information and other spatial and non-spatial information [29]. The strength of GIS comes from its ability to analyze data representing a particular point, line, or polygon. All features of the landscape can be reduced to one of these three spatial data categories. Water bodies, Soil type and cropped areas are all examples of features that are represented as polygons (areas) in a GIS database. Canal networks, roads and rivers are all lines, and features such as point elevation, precipitation data from a rain gauge are points in GIS. Some possible layers in GIS are shown in Figure 1.

Remote sensing images, effectively integrated within a GIS, can be used to facilitate measurement, mapping, monitoring and modeling activities [25]. Building a GIS database includes import or entry of data from different sources and digitizing data from different source documents. Data themes can range from hydrological (e.g. runoff) and climatological (e.g. temperature, precipitation) to data from both point and areal measurements.

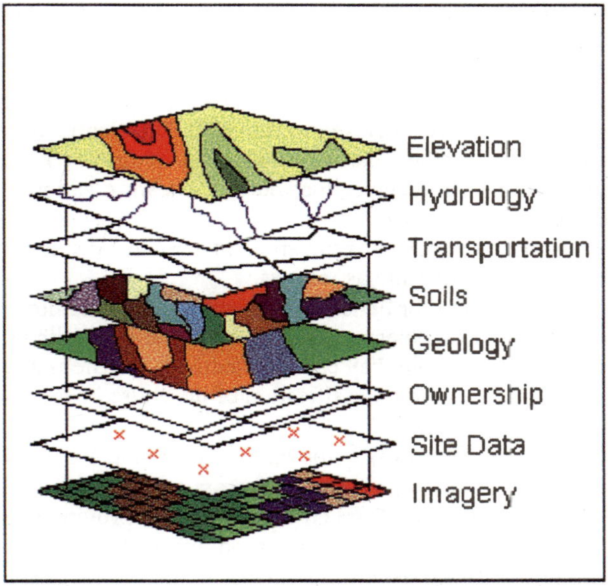

Figure 1. Typical GIS Data Layers

GIS analyses allow the user to perform a wide variety of investigations such as

- Proximity analyses, neighborhood operations (e.g. identifying objects within a certain neighborhood fulfilling specific criteria)
- Determine the relationships between data sets within such a neighborhood
- Temporal operations and analyses
- Generation of new information by combining several data layers and attributes (e.g. by splitting or aggregating etc)

DeVantier and Feldman [18] presented a review of use of GIS, Digital Terrain Modelling (DTM) and remote sensing data in hydrologic modeling. Baumgartner *et al.* [7] showed the advantages of integrated use of remote sensing, GIS, DBMS (Database Management Systems) and hydrological models.

11 Global positioning system

Global Positioning System (GPS) is a Satellite Navigation System funded by and controlled by the U. S. Department of Defense [17, 35]. The nominal GPS Operational Constellation consists of 24 satellites, which orbit the earth in 12 hours [36]. This constellation provides the user with five to eight Satellites visible from any particular point on the earth at any given time. GPS provides specially coded satellite signals, which can be processed in a GPS receiver, enabling the receiver to compute position (latitude and longitude), velocity and time. Four GPS satellite signals are used to compute positions in three dimensions. Earlier precise signal (encrypted signal) from GPS was available only for U.S. defense purposes. Recently GPS signal is made public for precise location [39]. Differential positioning (DGPS) involves determination of the relative coordinates of one or more receivers with respect to the position of a receiver located at a known position referred to as the Base (Monitor reference station)

GPS has an effective role in utilization of remote sensing and GIS in water resources. Some of its utilities are as follows.

- Ground truth collection for better classification of the satellite imagery. For example, different crop types can be located on the field with GPS during ground truth collection and they can be effectively transferred to the satellite imagery as training tests for classification of the entire image of an irrigated area into different crop types.
- Ground truth collection to verify the interpreted results from a remote sensing imagery.
- Point elevation information from GPS survey for improving the existing contour map, DEM model or any other topo information.
- Transfer of any point source information (e.g. rain gauge data) to the precise location in a satellite imagery or GIS layer.

12 Future scope

Impact of remote sensing on hydrology will be significant in future for several reasons.

- It has the ability to provide spatial data rather than point data
- It has the potential to provide measurements of hydrological variables which are not available through traditional techniques, such as soil moisture, snow water content
- It has the ability to provide long-term globe-wide data for remote and generally inaccessible regions of the Earth at regular intervals.

Most of the advances in using remote sensing for hydrology have come from new areas of hydrological analysis where existing methods were unsatisfactory and areas where sufficient data was non existent. These areas include General Circulation Model (GCM), land parameterizations, snow hydrology and measurement of soil moisture [26].

In the near future, more complex multi spectral, multi temporal and multi variate data sets of different origins will be available from different satellite platforms. The needs for long term monitoring of the hydrologic cycle are a high temporal (12 hours) and a moderate spatial (100-250 m) resolution as well as consistent data systems (sensors) and data sets. The imaging spectrometer – especially the moderate (MODIS by NASA) or medium (MERIS by ESA) resolutions systems – will improve the collection of remote sensing data. In addition to a high temporal and moderate spatial resolution, these systems offer an excellent spectral resolution, which improves the utility in hydrological applications.

Future Indian remote sensing (IRS) missions include [43]:

- CARTOSAT-1 which is meant as a mapping mission for cartographic applications, will carry two panchromatic cameras, one looking in the aft direction and the other in Fore direction in order to provide stereo pair images during the same satellite pass. The spatial resolution contemplated is 2.5 m with a height resolution more than 5 m.
- OCEANSAT satellite carrying a scatterometer, thermal infrared sensor, an altimeter and an ocean color monitor (OCM similar to that in IRS-P4)
- CLIMATSAT satellite with a payload of microwave radiometer, humidity sounder and a radiation budget monitor. The satellite will provide scientific data pertaining to the tropical regions viz., data on water vapor, ice in clouds, cloud liquid water using microwave radiometers at frequencies 10, 18, 23, 37, 85 and 157 GHz and humidity profiler around 183 GHz in 6 channels to get the vertical humidity profile. In addition, a radiation budget monitor operating from wave length of 0.2 μm will provide data for total radiation budgeting.

All these missions will provide vital information for hydrologic budgeting, analysis and various applications in water resources.

For comparison of global data sets, where satellite remote sensing plays a vital role, procedures for geo-coding (including digital elevation models for a three dimensional correction), calibration, normalization and atmospheric correction, must be standardized.

The processing algorithms have to be adapted for more synthetic, integrative and automated analyses including contextual pixel analysis. In addition, digital photogrammetry, computer science and digital cartography should be integrated. Future interpretation systems (expert systems and Fuzzy systems) based on artificial intelligence with *a priori* expert knowledge stored in a database can then utilise the integrated findings to get at more enlightened decisions.

13 Conclusions

Remote sensing applications in different fields of water resources discussed in detail reveal the strong potential of use of remote sensing for water resources planning and management.

Studies have suggested that the use of remote sensing data in hydrology and water resources could yield very high benefit/cost ratios from savings in flood damage and in improved planning of irrigation and hydro electric production. Such applications are well suited for hydrologic modeling.

Active microwave remote sensing data from satellites offers the potential of (almost) all-weather application due to their penetration capabilities through clouds and shadows when compared to the optical sensors. However, the necessary algorithms are not universally applicable.

The current usage of remote sensed data in hydrologic modeling is relatively low. One reason is, most hydrological models in operational use are not designed to use spatially distributed data which is a prerequisite to the sensible use of remotely sensed data. Research is needed into the development of generalized algorithms and into the design of hydrological models more suited to the routine use of remotely sensed data. Another reason for the low usage of remotely sensed data in hydrological modeling is the lack of appropriate education and training. Operational agencies and consultants predominantly use traditional techniques. Potential users should be properly trained and appraised about the advantages of use of remote sensing.

Satellite remote sensing can also be an appropriate tool to help alleviate some of the hydrometric data collection and management problems facing many developing countries.

Progress in water resources research depends on the availability of adequate data for model development and validation. Remote sensing plays a vital role in this process and can address successfully some of the previously intractable problems. For the water resources applications, proper image processing of remotely sensed data and spatio-temporal analyses with GIS and a complete integration of image processing, GIS and GPS are necessary.

References

1. Allison E. W., Brown R. J., Press H. and Gairns J., Monitoring drought affected vegetation with AVHRR, In: *Quantitative Remote Sensing: An Economic Tool for the Nineties*, *Proc. IGRASS'89, IEEE*, (1989) pp.1961-1965.
2. Balakrishnan P., Issues in water resources development and management and the role of remote sensing, **Report # ISRO-NNRMS-TR-07-86** (Indian Space Research Organization, Bangalore, India, 1987) p. 208.
3. Barret E. C. and Curtis L. F., Introduction to environmental remote sensing (Chapman and Hall, New York, USA 1982).
4. Barret E. C. and Martin D. W., The use of satellite data in rainfall monitoring (Academic press, New York, USA 1981).

5. Bastiaanssen W. G. M., Remote sensing in water resources management: The state of the art (International Water Management Institute, Colombo, Sri Lanka 1998) p. 118.
6. Bastiaanssen W. G. M., Pelgrum H., Droogers P., de Bruin H.A.R. and Menenti M., Area-average estimates of evaporation, wetness indicators and top soil moisture during golden days in EFEDA, *Agricultural and Forest Meteorology*, **87** (1997) pp. 119-137.
7. Baumgartner M. F. and Apfl G. M., Remote sensing and geographic information systems, *J. Hydrological Sciences*, **41** (1996) pp. 593-608.
8. Benediktsson J. A., Swain P. H. and Ersoy O. K., Neural network approaches versus statistical methods in classification of multi source remote sensing data, *IEEE Transactions on Geoscience and Remote Sensing*, **28** (1990) pp. 540
9. -551.
10. Blanchard B. J., Third ERTS 1 Symposium, NASA SP 351, **1**(1974) pp.1089-1098.
11. Bobba A. G., Bukata R. P. and Jerome J. H., Digitally processed satellite data as a tool in detecting potential groundwater flow systems, *J. Hydrology*, **131** (1992) pp. 25-62.
12. Bussieres N., Louie P. Y. T. and Hogg W., Implementation of an algorithm to estimate regional evapotranspiration using satellite data (Canadian Climate Centre, Downsview 1989).
13. Carroll T. and Baglio J., Techniques for near real-time snow cover mapping using AVHRR satellite data, *Poster paper at 57th Western Snow Conference*, NOAA/NWS (1989), Minneapolis, USA.
14. Chakraborty A. K., Rao V. V., Raju P. L. M. and Prasad V. H., Remote sensing in water resources, *Proc. of the Silver Jubilee Seminar*, IIRS (1992), Dehradun, India.
15. Chang A. T. C., Foster J. L., Hall D. K., Rango A. and Hartline B. K., Snow water equivalent estimation by microwave radiometry, *Cold Regions Research Technology*, **5** (1982) pp. 259-267.
16. Chanzy A., Schmugge T. J., Calvet J. -C., van Oevelen P., Grosjean O. and Wang J. R., Airborne microwave radiometry on a semi-arid area during HAPEX-Sahel, *Journal of Hydrology*, **188-189** (1997) pp. 285-209.
17. Chaturvedi L., Carver K. R., Harlan J. C., Hancock G. D., Small F. V. and Dalstead K. J., Multispectral remote sensing of saline seeps, *IEEE Transactions on Geoscience and Remote Sensing*, **21** (1983) pp. 239-250.
18. Dana P. H., The Geographer's Craft Project (Department of Geography, The University of Colorado at Boulder, USA, 2000).
19. DeVantier B. A. and Feldman A. D., Review of GIS applications in hydrologic modelling, *Journal of Water Resources Planning and Management*, **119** (1993) pp. 246-261.
20. Dicknson R. E., Henderson-Sellers A., Kennedy P. J. and Wilson M. F., Biosphere-Atmosphere Transfer Scheme,(BATS) for the NCAR community

climate model, (Tech. Note NCAR/TN-275+STR, National Center for Atmospheric research. Colorado, USA, 1986).

21. Dicknson R. E., Henderson-Sellers A., Rosenzweig C. and Sellers P. J., Evaportanspiration models with canopy resistance for use in climate models, a review, *Agricultural Meteorology*, **54** (1991).
22. Donald J. R., Seglenieks F. R., Soulis E. D., Kouwen N. and Mullins D. W., Mapping of partial snow cover during melt season using c-band SAR imagery, *Proc. of 15th Canadian Symposium on Remote Sensing*, Toronto, Canada (1992) pp. 170-175.
23. Donald J. R., Soulis E. D., Thompson N. and Malla S. B., Using GOES visible data to extend snow course in southern Ontario. In *Application of Remote Sensing in Hydrology*, ed. by Kite G. W. and Wankiewicz A., (Proc. Symp. No.5, NHRI, Saskatoon, Canada, 1990) pp. 69-78.
24. Dousset B., AVHRR derived cloudiness and surface temperature patterns over the Los Angeles area and their relationship to land use. In *Quantitative Remote Sensing: An Economic Tool for the Nineties*, *Proc. IGRASS'89, IEEE* (1989) pp. 2132-2137.
25. Dugdale G., Hardy S. and Milford J. R., Daily catchment rainfall estimated from Meteosat, *Hydrological Process*, **5** (1991).
26. Ehlers M., Greenlee D., Smith T. and Star J., Integration of remote sensing and GIS: Data and data access, *Photogrammetric Engineering and Remote Sensing*, **57** (1991) pp. 669-675.
27. Engman E. T., Remote sensing applications to hydrology: future impact, *J. Hydrological Sciences*, **41** (1996) pp. 637-648.
28. Engman E. T. and Chauhan N., Status of microwave soil moisture measurements with remote sensing, *Remote Sensing of Environment*, **51** (1995) pp. 189-198.
29. Engman E. T. and Gurney R. J., Remote sensing in Hydrology (Chapman and Hall, London, 1991).
30. Estes J. E., Remote sensing and GIS integration: Research needs, status and trends, *ITC Journal*, **1** (1992) pp.2-10.
31. Follansbee W. A., NOAA Technical Memo NESS 44 (NOAA, Washington, USA, 1973).
32. Haferman J. L., Krajewski W. F., Smith T. F. and Sanchez A., Three dimensional aspects of radiative transfer in remote sensing of precipitation: Application to the 1986-COHMEX storm, *J. Applied Meteorology*, **33** (1994) pp. 1609-1622.
33. Hall D. K., Remote sensing applications to hydrology: imaging radar, *J. Hydrological Sciences*, **41** (1996) pp. 609-624.
34. Hallada W. A., Mapping bathymetry with Landsat 4 thematic mapper, preliminary findings. In *Proc. of 9th Canadian Symposium on Remote Sensing* (Ottawa, Canada, 1984) pp. 277-285.

35. Hepner G. F., Logan T., Ritter N. and Bryant N., Artificial neural network classification using a minimal training set: Comparison to a conventional supervised classification, *Photogrammetric Engineering and Remote Sensing*, **56** (1990) pp. 469-473.
36. Hoffmann-Wellenhof, Lichtenegger B. H. and Collins J., *GPS: Theory and Practice*. (3rd ed., Springer-Verlag, New York, USA , 1994).
37. http://www.colorado.edu/geography/gcraft/notes/gps/gps_f.html
38. Hussin Y. A. and Shaker S. R., Optical and radar satellite image fusion techniques and their applications in monitoring natural resources and land use changes, *AEU(Archiv fur Elektronik und Ubertragunstech)*, **50** (1996) pp. 169-176.
39. Huurneman G. and Broekema L., Classification of multi-sensor data using a combination of image analysis techniques, Paper E-2 in *Proceedings of the 17th Asian Conference on Remote Sensing* (Survey Dept, Colombo, 1996).
40. Interagency GPS, End of selective availability of GPS, http://www.igeb.gov/sa/ (2000).
41. Jackson T. J., Schmugge J. and Engman E. T., Remote sensing applications to hydrology: soil moisture, *J. Hydrological Sciences*, **41** (1996) pp. 517-530.
42. Jayasekera A. A. and Walker W. R., Remotely sensed data and geographic information systems: For management and appraisal of large scale irrigation projects in the developing countries. In *Advances in planning, design and management of irrigation systems as related to sustainable land use*, ed. by Feyen J., Mwendera E. and Badji M., (Center for Irrigation Engineering, Leuven, Belgium, 1990) pp. 453-461.
43. Joseph G., Retrospective and Prospective of Remote Sensing in India, *Journal of Indian Society of Remote Sensing*, **24** (1996) pp. 133-143.
44. Kalyanaraman S., Remote sensing data from IRS Satellites- past, present and future, *Journal of Indian Society of Remote Sensing*, **27** (1999) pp. 59-70.
45. Kaur R. and Rabindranathan S., Ground validation of an algorithm for estimating surface suspended sediment concentrations from multi-spectral reflectance data, *Journal of Indian Society of Remote Sensing*, **27** (1999) pp. 235-251.
46. Kite G. W., Using NOAA data for hydrological modelling, In: *Quantitative Remote Sensing: An Economic Tool for the Nineties, Proc. IGRASS'89, IEEE* (1989) pp. 553-558.
47. Kite G. W. and Kouwen N., Watershed modelling using land classification, *Water Resources Research*, **28** (1992) pp. 3193-3200.
48. Kite G. W. and Pietroniro A., Remote sensing applications in hydrological modelling, *J. Hydrological Sciences*, **41** (1996) pp. 563-592.
49. Krishna Prasad V., Yogesh Kant and Badarinath K. V. S., Vegetation discrimination using IRS-P3 WiFS temporal dataset - A case study from Rampa forests, Eastern Ghats, A.P., *Journal of Indian Society of Remote Sensing*, **27** (1999) pp. 149-154.

50. Kumar V. S., Haefner H and Seidel K., Satellite snow cover mapping and snowmelt runoff modelling in Beas basin. In: *Snow, Hydrology and Forests in High Alpine Areas* (IAHS publication no. 205, 1991) pp. 101-109.
51. Kustas W. P. and Norman J. M., Use of remote sensing for evapotranspiration monitoring over land surfaces, *J. Hydrological Sciences*, **41** (1996) pp. 495-516.
52. Leguizamon S., Pelgrum H., Azzali S. and Menenti M., Unsupervised classification of remotely sensed data by means of the Fuzzy C-means approach. In *Fourier analysis of temporal NDVI in the Southern African and American Continents*, ed. by Azzali S. and Menenti M. (Netherlands, Report 108, 1996) pp. 25-36.
53. Meijerink A. M. J., Remote sensing applications to hydrology: groundwater, *J. Hydrological Sciences*, **41** (1996) pp. 549-562.
54. Menenti M., Azzali S., Collado D. A. and Leguizamon S., Multitemporal analysis of Landsat multispectral scanner (MSS) and Thematic Mapper (TM) data to map crops in the Po valley (Italy) and Mendoza (Argentina). In *Remote sensing for resources development and environmental management*, ed. by Damen M. C. J., Sicco Smit G. and Verstappen H. Th. (Balkema, Rotterdam, 1986) pp. 293-299.
55. Menenti M. (ed.), Remote sensing in evaluation and management of irrigation, *Instituto Nacional de Ciencia Tecnicas Hidricas* (Mendoza, Argentina, 1990).
56. Mertes L. A. K., Smith M. O. and Adams J. B., Estimating suspended sediment concentrations in surface waters of Amazon basin wetlands from Landsat images, *Remote Sensing of Environment*, **43** (1993) pp. 281-301.
57. Molden D. J., Sakthivadivel R., Perry C. J. and de Fraiture C., Indicators for Comparing Performance of Irrigated Agricultural Systems (Research Report 20, International Irrigation management Institute, Colombo, 1998).
58. Moore G. K., Satellite surveillance of physical water quality characteristics, *Proc. of 12th International Symposium of Environment* (Ann Arbor, USA, 1978) pp. 445-462.
59. Musiake K., Oki T., Nakaegawa T. and Wakasa K., Verification experiment of extraction of soil moisture information using SAR mounted on JERS-1/ ERS-1, In Col. 2 of *Final report of JERS-1/ ERS-1 systems verification program* (Ministry of International Trade and Industry and National Space Development Agency, Tokyo, Japan, 1995) pp. 617-624.
60. Nagesh Kumar D., Satellite Remote Sensing in Irrigation Management. In: '*Remote Sensing Applications in Applied Geosciences*', ed. by Saumitra Mukherjee (Manak Publications, New Delhi, India, 1999) pp. 79-97.
61. Nagesh Kumar D., Jonna S. and Thiruvengadachari S., Texture labelling and texture spectrum analysis of ERS-1 SAR data, *Proc. of National symposium on microwave remote sensing* (Ahmedabad, India, 1994) pp. 174-181.

62. Nageswara Rao P. P. and Mohankumar A., Cropland inventory in the command area of Krishnarajasagar project using satellite data, *International Journal of Remote Sensing*, **15** (1994) pp. 1295-1305.
63. Narayanan L. R. A., Remote sensing and its applications (University Press India limited, Hyderabad, India, 1999).
64. Njoku E. G. and Entekhabi D., Passive microwave remote sensing of soil moisture, *Journal of Hydrology*, **184** (1996) pp. 101-129.
65. Panigrahy S., Chakraborty M., Manjunath K. R., Kundu N., Parihar J. S., Evaluation of RADARSAT ScanSAR Synthetic Aperture Radar data for Rice crop inventory and monitoring, *Journal of Indian Society of Remote Sensing*, **28** (2000) pp. 59-65.
66. Papadakis I., Napiorkowski J. and Schultz G. A., Monthly runoff generation by nonlinear model using multi spectral and multi temporal satellite imagery, *Adv. Space Research*, **13** (1993).
67. Peck *et al.*, NASA CR 166674, (prepared for Goddard Space Flight Center, Meryland, USA, 1981).
68. Pedley M. I. and Curran P. J., Per-field classification: An example using SPOT HRV imagery, *International Journal of Remote Sensing*, **12** (1991) pp. 2181-2192.
69. Petty G. W. and Krajewski W. F., Satellite estimation of precipitation over land, *J. Hydrological Sciences*, **41** (1996) pp. 433-452.
70. Pietroniro A., Soulis E. D., Kouwen N., Rotunno O. and Mullins D.W., Using wide swath C-band SAR imagery for basin soil moisture mapping, *Canadian J. Remote Sensing*, **Special issue** (1993) pp. 77-82.
71. Pietroniro A., Wishart W. and Solomon S. I., Use of remote sensing data for investigating water resources in Africa. In *Quantitative Remote Sensing: An Economic Tool for the Nineties, Proc. IGRASS'89 IEEE* (1989) pp. 2169-2172.
72. Price J. C., The potential of remotely sensed thermal infrared data to infer surface soil moisture and evaporation, *Water Resources Research*, **16** (1980) pp. 787-795.
73. Ragan R. M. and Jackson T. J., Runoff synthesis using Landsat and SCS model, *J. Hydraulics Division,* **106** (1980) pp. 667-678.
74. Ramamoorthi A. S., Snowmelt runoff studies using remote sensing data, *Proc. Indian Academy of Sciences*, **6** (1983) pp. 279-286.
75. Ramamoorthi A. S., Snow cover area(SCA) is the main factor in forecasting snowmelt runoff from major basins. In *Large scale effects of seasonal snow cover* (IAHS publication no. 166, 1987) pp. 187-198.
76. Ramamoorthi A. S., IRS-1A data utilization in Flood studies, In *Natural Resources Management: A New Perspective*, ed. by Karale R. L. (NNRMS, Bangalore, India, 1992) pp. 196-202.
77. Rango A., Space-borne remote sensing for snow hydrology applications, *J. Hydrological Sciences*, **41** (1996) pp. 477-494.

78. Ritchie J. C., Remote sensing applications to hydrology: air-borne laser altimeters, *J. Hydrological Sciences*, **41** (1996) pp. 625-636.
79. Ritchie J. C. and Schiebe F. R., Monitoring suspended sediment with remote sensing techniques. In *Hydrologic Application of Space Technology* (IAHS publ. No. 160, 1986) pp. 233-243.
80. Rondeaux G., Steven M. and Baret F., Optimization of soil-adjusted vegetation indices, *Remote Sensing of Environment*, **55** (1996) pp. 95-107.
81. Salmonson V. V., Water resources assessment. In *Manual of remote sensing*, ed. by Colwell J. (American Society of Photogrametry and Remote Sensing, 1983) pp. 1497-1570.
82. Salmonson V. V. and Rango A., Chapter 18. In *Remote Sensing in Geology*, ed. by Seigal B. S. and Gillespie A. R. (John Wiley & Sons, New York, 1980).
83. Satish Chandra, Sharma K. P. and Kashyap O., Watershed studies using simulation models for Upper Yamuna catchment. In *Application remote sensing methods to hydrology* (Univ. of Roorkee, India, 1984).
84. Schotten C. G. J., Van Rooy W. W. L. and Janssen L. L. F., Assessment of the capabilities of multi-temporal ERS-1 SAR data to discriminate between agricultural crops, *International Journal of Remote Sensing*, **16** (1995) pp. 2619-2637.
85. Schultz G. A., Remote sensing in hydrology, *J. Hydrology*, **100** (1988) pp. 239-265.
86. Schultz G. A., Remote sensing applications to hydrology: runoff, *J. Hydrological Sciences*, **41** (1996) pp. 453-476.
87. Schumann A. H. and Funke R., GIS-based components for rainfall-runoff models. In *Proceedings of HydroGIS'96 Symposium* (Vienna, Austria, IAHS Publ. No. 235, 1996).
88. Seckler D., Amarasinghe D., Molden R., De Slilva and Barker R., World water demand and supply. 1990 to 2025: Scenarios and issues (Research Report 19, International Irrigation management Institute, Colombo, 1998).
89. Sellers P. J., Meeson B. W., Hall F. G., Asrar G., Murphy R. E., Schiffer, R. A., Bretherton F. P., Dickinson R. E., Ellingson R. G., Field C. B., Huemmrich K. F., Justice C. O., Melack J. M., Roulet N. T., Schimel D. S. and Try P. D., Remote sensing of the land surface for studies of global change: Models-algorithms-experiments, *Remote Sensing of Environment*, **51** (1995) pp. 3-26.
90. Settle J. J. and Briggs S. A., Fast maximum likelihood classification of remotely sensed imagery, *International Journal of Remote Sensing*, **8** (1987) pp. 723-734.
91. Singh R. P. and Srivastav S. K., Mapping water logged and salt affected soils using microwave radiometers, *International Journal of Remote Sensing*, **11** (1990) pp. 1879-1887.
92. Singh R. P., Dadhwal V. K. and Navalgund R. R., Wheat crop inventory using high spectral resolution IRS-P3 MOS-B Spectrometer data, *Journal of Indian Society of Remote Sensing*, **27** (1999) pp. 167-173.

93. Slough K. and Kite G. W., Remote sensing estimates of snow water equivalent for hydrologic modelling, *Candian Water Resources Journal*, **17** (1992) pp. 323-330.
94. Space Application Centre, Crop Acreage and Production Estimation (Status Report, ISRO, Dept. of Space, Govt. of India, 1990).
95. Spanner M. A., Strahler A. H. and Estes J. S., *Proc. of 16th Symposium on Remote Sensing of Environment* (1982).
96. Sriramany S. and Murthy V. V. N., A real-time water allocation model for large irrigation systems, *Irrigation and Drainage Systems*, **10** (1996) pp. 109-129.
97. Steven M. D., Malthus T. J., Jaggard F. M. and Andrieu B., Monitoring responses of vegetation to stress. In *Remote Sensing from research to operation: Proceedings of the 18th Annual Conference of the Remote Sensing Society* (University of Dundee, Nottingham, U.K., 1992) pp. 369-377.
98. Strahler A. H., The use of prior probabilities in maximum likelihood classification of remotely sensed data, *Remote Sensing of Environment*, **10** (1980) pp. 135-163.
99. Thiruvengadachari S., Space technology for disaster warning and management (Fourth session of Scientific and Technical Committee for IDNDR, 1-5 February, Delhi, India, 1993) pp. 1-14.
100. Thiruvengadachari S. and Sakthivadivel R., Satellite remote sensing techniques to aid irrigation system performance assessment: A case study in India (Research Report 9, International Irrigation management Institute, Colombo, 1997).
101. Thiruvengadachari S., Harikishan J. and Nagesh Kumar D., Satellite Evaluation of Current status of Irrigation management in Rajolibanda diversion scheme command, Mahaboobnagar District, AP, *NNRMS Bulletin*, (Dept. of Space, Bangalore, NNRMS, (B)-17, 1993) pp. 39-41.
102. Thiruvengadachari S., Jonna S., Raju P. V., Murthy C. S. and Nagesh Kumar D., Evaluation of Irrigation water management in Bhadra Project command area through Satellite Remote Sensing techniques (Research Report, Water Resources Group, National Remote Sensing Agency, Hyderabad, India, 1993).
103. Tueller P. T. and Booth D. T., *Proc. of American Society of Photogrammetry*, Arizona, (1975) pp.708-753.
104. US Department of Agriculture, Soil Conservation Service, National Engineering Handbook, Section 4, Hydrology (US Govt. Printing Office, Washington DC, USA, 1972).
105. Van den Hurk B. J. J. M., Bastiaansen W. G. M., Pelgrum H and van der Meygaarden E., Soil moisture assimilation for numerical weather prediction using evaporative fraction from remote sensing, *Journal of Applied Meteorology*, **36** (1997) pp. 1271-1283.
106. Vidal A. and Sagardoy J. A. (eds.), Use of remote sensing techniques in irrigation and drainage, *Water Reports* **4**, (FAO, Rome, 1995).

107. Visser T. N. M., Appraisal of the implementation of water allocation policies, (ICW Note 1963, Wageningen, Netherlands, 1989).
108. Walker A. E., Gray S. A., Goodison B. E. and O'Neill R. A., Analysis of MOS-1 microwave scanning radiometer data for Canadian Prairie snow cover. In *Application of Remote Sensing in Hydrology*, ed. by Kite G. W. and Wankiewicz A., (Proc. Symp. No. 5, NHRI, Saskatoon, Canada, 1990) pp. 319-330.
109. Wang F., Fuzzy supervised classification of remote sensing images, *IEEE Transactions on Geoscience and Remote Sensing*, **28** (1990) pp. 194-201.
110. Wankiewicz A., Microwave satellite forecasting of snowmelt runoff. In *Quantitative Remote Sensing: An Economic Tool for the Nineties* (Proc. IGRASS'89, IEEE, 1989) pp. 1235-1238.
111. Wertz D. L., Mealor W. T., Steele M. C. and Pinson J. W., Correlation between multispectral and near surface turbidities, *Photogrammetric Engineering and Remote Sensing*, **42** (1973) pp. 695-701.
112. Whiting J., Determination of characteristics for hydrologic modeling using remote sensing. In *Application of Remote Sensing in Hydrology*, ed. Kite G. W. and A. Wankiewicz (Proc. Symp. No. 5, NHRI, Saskatoon, Canada, 1990) pp. 79-91.
113. WMO (World Meteorological Organization), Concept of the Global Energy and Water Experiments (GEWEX), *Report WCRP 5*, Geneva, Switzerland (1988).
114. Wolters W., Zevenbergen A. W. and Bos M. G., Satellite remote sensing in irrigation, *Irrigation and Drainage Systems*, **5** (1991) pp. 307-323.
115. Xinmei H., Lyons T. J., Smith R. C. G. and Hacker J. M., Estimation of land surface parameters using satellite data, *Hydrologic Processes*, **9** (1995).

MODELLING RESERVOIR OPERATION FOR IRRIGATION

S. VEDULA

Department of Civil Engineering, Indian Institute of Science, Bangalore – 560 012
INDIA
E-mail: svedula@civil.iisc.ernet.in

A brief overview of the state of the art in modelling reservoir operation for irrigation is presented. Because of the rapid growth of modelling techniques and applications in the recent past, only those publications, by and large, which appeared in the literature in the last decade dealing with a reservoir for irrigation and using some kind of mathematical modelling technique for optimizing the system, are considered in this paper. There has been a tremendous development, during this period, in the model formulations increasingly reflecting practical situations, and in the articulation of modelling techniques to enable solutions to these. It is hoped that this trend continues to grow and the systems analysts would be able to bring in enough motivation among water managers to applying them through innovative approaches to enable the irrigation sector to benefit from them.

1 Introduction

Publications on systems and simulation proliferated in the recent past in water resources systems, signaling the recognition of their potential for application in planning and management of reservoir water for various uses such as municipal and industrial use, irrigation, hydropower production, and conjunctive use. In developing countries like India, however, irrigation continues to be the largest consumer of water, because of its importance in food production. Water has indeed become an important economic commodity in the irrigation sector, dictating the need to policy planning to harness this scarce resource, optimally.

This paper has been prepared in the backdrop of three broad guidelines in presenting a brief review of the state of the art on modelling reservoir operation for irrigation. First, because of the increasing rate of development in model formulation and solution techniques being reported in reputed journals in the recent past, it is proposed to base this review on the works that appeared in the literature primarily in the last ten years. Secondly, it is aimed to limit the scope of the review to modeling reservoir irrigation, and therefore, only those papers, which by and large, deal with a reservoir for irrigation, and involve some kind of optimization/simulation technique, are proposed for inclusion.

The third guideline has to do with the extent of coverage. Two developments in the recent past, both of which are challenging, primarily facilitated the nature of publications in this field. First is the articulation in the model formulation of specific problems with reasonable assumptions and approximations, yet without losing the basic ground realities, and making them amenable to solution with known systems techniques, and secondly, the modifications for innovative use of the most popular mathematical modelling techniques such as linear programming (LP), dynamic programming (DP) and simulation. In this process, even minor modifications in the model formulation to incorporate additional features (or relaxation of some earlier assumptions) are likely to have significant effect in motivating model applications. It is realized that this is going to be a continuous and evolving process as demonstrated by the recent advances in the development of modelling in this area.

The guidelines mentioned above also form the limitations of the text of this paper. A few publications, which are related to the topic of the paper, but which are strictly outside of these guidelines, also find a mention in this paper. The aim of the paper is to give a brief overview of the type of problems addressed and techniques used in this context, with a pointer to the perspectives for the future. Omission of any significant work or publication in the area is purely inadvertent.

1.1 Reservoir Operation

The temporal patterns of demand and supply of water, on a regional scale, are most usually out of phase with each other necessitating the need for storage in times of surplus for use in times of deficit. The management of the stored water holds the key in an environment of ever increasing water demands with limited supply. Strategies for optimal management of the existing storage reservoirs, and planning in phases for capacity expansion for the future are essential for sustained growth of water utilization.

Reservoir operation, in the context of this paper, involves decision making in respect of reservoir releases to be made in time. It is the determination of the specific policy (reservoir operating policy) to be followed in each intraseasonal period, which when followed over a long period of time, would result in the best of the specified objective for the system. The objective for a reservoir system used for irrigation could be to maximize the efficiency of water use, to maximize the annual physical crop output, to maximize the net benefits from irrigated agriculture, and so on. The policy, when determined by the model solution, is a sequence of decisions on reservoir releases among the decision periods within the year. The release to be made at the beginning of a given period is a function of the state of the system at the beginning of the period, and the expected inputs (e.g., inflow to the reservoir, rainfall in the command area, evaporation, crop growth) during the period. The (optimal) reservoir release, in each period, is linked to the irrigation allocation to different crops in the area commanded by the reservoir, as dictated by the objective function. These allocations depend on the crop water requirements for optimum growth the cropping season. The degree of detail (of the crops, crop growth, crop water requirements, soil properties of the cropped areas, and other field requirements) to be accounted in the model formulation depends on the need, context and the nature of the system being modelled.

2 General - Reviews

A few publications, which summarized the developments in reservoir systems modelling from time to time, are cited below.

Mujumdar and Narulkar [21] present a survey of the optimization models used in multireservoir planning and operation problems. A critical review of models and applications is not attempted, however, but most of the collected information is presented in a concise form through tables.

Ramesh and Mujumdar [36] present an overview of stochastic optimization models used in reservoir planning and operation. Stochastic optimization methodologies are discussed in considerable detail applicable to real time reservoir operation.

Datta [7] presents a review of operation models for single and multiple reservoirs. Some issues concerning real time operational models such as smaller time steps, uncertainties possible with using larger time steps, nonlinearities due to actual operational conditions, the importance of real time forecasts and criteria for performance evaluation of an operational model are discussed. In this process, models already developed, some of which are being actually used for operation are briefly described.

Banerjee [2] outlines the operational strategies and procedures for developing operational rules and discusses the operational features of the DVC water resources system.

Raman and Sunil Kumar [34] deal with the development and application of expert systems for flood management and drought management. For the drought management expert system, a LP model was used to generate optimal cropping patterns from the past drought experiences and from synthetic drought sequences.

Mohan and Armugam [20] present an overview of expert system applications in irrigation management. The paper reviews the adaptability and suitability of expert systems applications in the domain of irrigation management. A detailed review of the existing applications of expert systems is presented under three classes: expert systems proper, intelligent front-ends, and hybrid systems. Additional research on expert systems application to domains such a real -time irrigation scheduling, reservoir operation involving stochastic nature of inflows and evapotranspiration demand, and integrated operation of irrigation system components is needed to evolve guidelines for optimal water use. The same authors, Armugam and Mohan [1], developed an integrated decision support system (DSS) that aids the operation of a tank irrigation system in south India. The knowledge base of the DSS was developed form the knowledge derived from field experts and the results of an optimization model. Shortages in irrigation water supply simulated from the DSS were less than those occurring in the actual operation practiced by water authorities.

Vedula and Mujumdar [44] present an overview of the developments in modelling for reservoir operation for irrigation. The models in this field progressively improved from time to time to take into account more and more aspects in the complex decision making mechanism with uncertainty in such factors as water supply, rainfall and crop water demands. Recent works in this area have promised a great hope for meaningful real time applications of the developed models.

3 Models for Reservoir Operating Policy

The primary challenge in the reservoir operation models lay with the extent to which uncertainty in the hydrologic variables affecting the system has been accounted in the model formulation. Some of the most important variables, which are pertinent in this context are reservoir inflow, rainfall in the irrigated (reservoir command) area, and crop water demands.

Dariane and Hughes [6] developed a model for real time operation of an irrigation reservoir with the objective of maximizing the values of multiple crop yields during a season. The model employs monthly additive and product forms of crop yield functions for dry matter and grain crops, respectively. The nonlinear optimization model uses a log transform to reduce nonlinearities in the model, uses deterministic demands, rainfall, and predicted values of inflow, which can be updated at the beginning of each month. Model results were compared with those from a standard operating policy simulation.

Dudley and Scott [8], extended the previous models (developed by Dudley and colleagues) to develop the most comprehensive model integrating irrigation water demand, supply and delivery management in a stochastic environment with application to the case study of Gwydir Valley of northern New South Wales, Australia, where most farmers have on-farm storages in which to store water in excess of immediate needs. Previously developed suites of models integrate irrigation water supply and demand management for simplified surface reservoir supply systems. A central stochastic dynamic programming (SDP) model is supported by simulation models, including a soil water-plant growth model. That modeling is extended in this paper to include decisions about timing and quantity of reservoir water releases into the delivery system, resulting in the integration of supply demand and delivery management.

Ravikumar and Venugopal [37] developed a three phase optimal operation model for a typical South Indian irrigation system. First is a simulation model to simulate the command area of the reservoir, drawing water both from the reservoir canals and the tanks in the command area, to determine the expected demand sequences. Second is a SDP model to obtain optimal release policy, treating both demand and inflow as stochastic, and the third is simulation using the optimal release policy from the SDP model to study the degree of failure for different reservoir storages at the beginning of the crop seasons.

Peng and Buras [29] developed a nonlinear programming (NLP) solution, to overcome the curse of dimensionality associated with the use of DP, which also integrated the objectives of several stakeholders, objectives which were noncommensurate. The NLP model employed a time series model to generate synthetic inflows into the storage facilities in order to simulate future operations of the system. Two sets of simulations were run, and a statistical test was used to terminate them. The model allows decision makers to adjust operation targets for different alternative operation scenarios.

4 Multicrop Models

Sunantara and Ramirez [41] carried optimal seasonal multicrop irrigation water allocation and optimal stochastic intraseasonal (daily) irrigation scheduling using a two-stage decomposition approach based on a stochastic dynamic programming (SDP) methodology. In the first stage, the optimal seasonal water and acreage allocation among the crops is defined using deterministic DP with the objective of maximizing total benefits from all crops. Seasonal crop production functions are obtained using single crop SDP incorporating the physics of soil moisture depletion and stochastic properties of precipitation. In the second stage, optimal intraseasonal irrigation scheduling is performed using a single crop SDP conditional on the optimal seasonal water allocation of stage one. Optimal daily irrigation decision functions are obtained as function of root-zone soil moisture content and the currently available irrigation water.

A short-term real-time reservoir operation model is developed by Mujumdar and Ramesh [23] for multiple crops. The model differs from earlier models in that it considers interdependence of crop water allocation across time periods and its ability to provide an adaptive release policy. Integration of reservoir operation with on-farm utilization of water by crops is achieved through the use of two models: an operating policy model, which optimizes reservoir releases across time periods in a year, and an allocation model, which optimizes irrigation allocations across crops within a time period. The model application is illustrated with a case study of a reservoir in India. The authors subsequently proposed [Mujumdar and Theegavarapu, (22)], a short term reservoir operation model for multicrop irrigation, which overcomes some of the limitations (especially of one soil moisture state variable for all crops) of the earlier model by replacing the two DP formulations with a single LP formulation.

Vedula and Mujumdar [43] developed a model for the optimal reservoir operating policy for irrigation of multiple crops using SDP. Intraseasonal periods (of ten days) smaller than the crop growth stage form the decision intervals of the model to facilitate irrigation decisions in real situations. Reservoir storage, inflow to the reservoir and the soil moisture in the irrigated area are treated as state variables. Optimal crop water allocations to individual crops are determined taking into account crop competition for water, impact of crop yield to water deficit, and effect of soil moisture dynamics on crop water requirements. A linear growth of crop root is assumed until the root grows to the full depth by the end of the vegetative stage.

Rainfall in the irrigated area is considered deterministic. Model application is illustrated to an existing reservoir case in India.

Vedula and Nagesh Kumar [45] made some improvements to the previous model and developed an integrated model, based on seasonal inputs of reservoir inflow and rainfall in the irrigated area, to determine the optimal reservoir release policies and irrigation allocations to multiple crops, in two modules. The first is an intraseasonal allocation model to maximize the sum of relative yields of all crops, for a given state of the system, using LP. Reservoir storage continuity, soil moisture balance, and crop root growth are considered. The second is a seasonal allocation model to derive the steady state reservoir operating policy using SDP. Seasonal inflow and rainfall are considered stochastic variables. Model application to the case study of an existing reservoir is illustrated.

Vedula and Sreenivasan [47] developed a chance constrained reservoir irrigation model integrating the reservoir release decisions with the crop water allocation decisions, taking into account soil moisture balance and crop root growth with time, for a specified reliability of meeting the irrigation requirements of multiple crops. The model can be used to determine the maximum possible reliability level for a given project and to find whether it is the inflow or the reservoir capacity that limits this in a specific case. The model has the capability to take heterogeneous soil types. Application of the model is demonstrated taking an existing reservoir as a case.

Hajilal *et al.* [11] propose that the real-time adaptive operation of reservoir based canal irrigation systems be carried to in two phases: a plan phase and operation phase. In the plan phase, biweekly reservoir releases are optimized for a set of target releases and mean inflows by minimizing a water supply deficit function using DP. In the operation phase, real time releases are made according to plan for the current interval. The four steps namely, releasing supplies from reservoir according to seasonal plan in the current interval, determining the fraction of the releases available to crops at field level, allocating the field level supply to competing crops, and modifying the reservoir release plan for the remaining intervals of the year based on inflows and forecasts, operate sequentially in real-time for each interval. As the season advances, this cycle repeats successively from the first to last intervals of the year. Model application is demonstrated using a case study irrigation project in India.

The problem of finding an optimal, closed loop control strategy for water distribution among crops grown in an irrigation system was addressed by Hiessl and Plate [12]. The control strategy is then optimized using a mixed simulation optimization method. The simulation module is used to represent the system and the basic control structure mathematically, while the optimization module is used to find an optimal control strategy for the system. The simulation module contains two submodules: one which describes the physical components in the system (reservoir irrigation area) and one which describes the controller which connects the reservoir and the irrigated area. The optimization algorithm uses a orthogonal line search similar to Rosenbrock's method. A multiplicative type production function is assumed. Inflow to the reservoir is considered stochastic and rainfall is considered negligible.

Mujumdar and Vedula [24] used reliability, resiliency and productivity indices to study the performance of a single purpose irrigation reservoir in Karnataka, India. Three different optimal operating policies are derived, having increasing mathematical complexity using SDP. The first policy ignores soil moisture contribution to crop demand. The last two policies incorporate a detailed soil moisture dynamics model as an integral part of the SDP. The last policy considers, in addition, optimal allocation of water among the irrigated crops when there is competition for water. The reservoir releases are simulated under each optimal operating policy using synthetically generated inflows. It was observed that inclusion of soil moisture dynamics into the optimization model that derives the operating policy for the reservoir enhances the performance of the system,

Mayya and Rama Prasad [17] made a comprehensive study of modelling tank irrigation systems typical of those in peninsular India. They investigated a number of issues relating to water availability and agricultural technology in these. The method involves developing a LP model to optimize the net profit from the system and to determine the optimal cropping pattern under the influence of various parameters, e.g., animal power, labor, fodder production, the resources of farmers, and the nutritional energy requirement of the system, in addition to water availability. The crucial nature of these factors and as well as the irrigation efficiency is analyzed. The solution reveals the effectiveness of prevailing agricultural practices consistent with the availability of water resources in the initial crop season. In the subsequent paper they [Rama Prasad and Mayya, (35)], critically discuss the issues of a delayed start and water deficit in tank irrigation systems vis-a-vis the energy requirements. To optimize the grain yield of rice, it is essential to start the agricultural operations in the second week of July so that favorable climatic conditions will prevail during flowering and yield formation stages. Because of low inflow during the initial few weeks of the crop season, often farmers are forced to delay planting until sufficient rain and inflow have occurred or to adopt deficit irrigation during this period. The delayed start affects the grain yield, but will lead to an improved efficiency resulting in the energy resources becoming critical during peak requirement time.

A methodology for optimizing design of integrated tank irrigation system was developed by Srivastava [40] taking into account the constraints, objectives, options of the location, conveyance and application methods, crops and their irrigation levels, productivity, and demand of the population.

5 Multiobjective / Multipurpose Systems

Vedula [46] studied the optimal irrigation planning of the Cauvery river basin in peninsular India, which is extensively developed in the downstream reaches and has a high potential for development in the upper reaches. A four reservoir system is modelled using LP to find optimum cropping patterns, subject to land, water and downstream release constraints. Maximizing net economic benefits and maximizing irrigated crop area are examined as the conflicting objectives in multiobjective planning analysis to determine the tradeoff between the two. This study is a more elaborate presentation for different scenarios of an earlier study by Vedula and Rogers [48].

Vedula and Mohan [42] developed a real-time operational methodology for multipurpose reservoir operation for irrigation and hydropower generation with application to the Bhadra reservoir system in Karnataka, India. The three phases of modelling methodology consist of determining the optimal release policy using SDP, streamflow forecasting using an adaptive ARIMA model and a real time simulation model using the inputs of the earlier two phases. A comparison of the optimal monthly real-time operation with the historical operation demonstrates the relevance, applicability, and the relative advantage of the proposed methodology.

A decision support system is developed by Michalland *et al.* [18] to get the set of non-inferior solutions using the constraint, penalty and weighting methods in reservoir management for irrigation and hydropower generation. The method is based on SDP with daily time steps using 50 annual values of daily historical inflows and marginal electricity production costs. The model yields trade-off curves of electrical production losses versus, respectively, irrigated surface area, crop yield, and agricultural benefits. A numerical application to the French case of the Bort reservoir illustrates the methodology in analyzing the trade-off between hydropower and irrigation. They feel that the current economic justification of single purpose reservoir management for hydropower ought to be changed.

Sinha *et al.* [39] present a screening model for selecting and sizing potential reservoirs and hydroplants on a river basin. The system is designed to meet the annual irrigation and hydropower demands at prescribed levels of reliability. A linked simulation optimization framework is used for formulation. For sizing of reservoirs, a new sequent trough algorithm is used that considers evaporation automatically. Generator capacities are determined by maximizing the local net revenues produced from hydropower generation. Application to some river basins in India demonstrates that the formulation leads to a reduction in the system storage and cost of development compared to the state agency operation.

Raju and Kumar [32] examine the selection of the best compromise irrigation plan in the multiobjective context. Net benefits, agricultural production, and labour employment are the three conflicting objectives studied. Multiobjective optimisation, cluster analysis and multiobjective decision making methods are combined in a three-stage procedure. Two MCDM methods, PROMETHEE-2 and a newly developed method EXPROM-2 are employed in the evaluation. The methodology is applied to a case study of Sri Ram Sagar Project, Andhra Pradesh, India. Sensitivity analysis indicated that ranking pattern is quite robust to parameter changes as far as the first two positions are concerned.

6 Conjunctive Use Models

Lall [15] developed an optimization model for selecting between candidate surface-water reservoirs and ground water development. The model may be used to perform a preliminary screening of alternatives for water-supply development, and to identify storage capacities at reservoir sites, as well as aquifer pumping at candidate locations. A hybrid monthly simulation-optimization model is used. A modified sequent peak algorithm is used for reservoir sizing, and a unit response matrix approach is used to model the ground water subsystem.

Peralta *et al.* [30] present optimal sustained ground-water yield and conjunctive water-use strategies for northeastern Arkansas, based on water demands projected for the five decades of 1990-2039. Each strategy consists of spatially and temporally distributed values of ground-water and surface-water use. Ground-water simulation/optimization (S/O) models are used to attempt to satisfy temporally increasing water needs for alternative future management scenarios. The S/O models employ a sequential steady-state embedding approach and contain over 1600 embedded ground-water and river-volume balance constraints per decade (stage).

An integrated approach to reservoir, irrigation, and cropping management which links four different models- a hydrologic model, a crop growth simulation model, an economic model based on LP, and a DP model – is developed by Evers *et al.* [9] and applied to an irrigation district in a subhumid climate with an irrigation reservoir large enough for over-year storage. Two different types of results are presented based on repeated simulations for various planning horizons. The first provides the probability that each of the various farm plans (land/crop/water allocation) will be chosen as the optimum in the first year of the planning horizon. The second approach provides probability distributions of accumulated initial reservoir level and a particular farm plan in the first year of the planning horizon. Based on families of probability-revenue curves, an irrigation manager can simultaneously evaluate crop, irrigation, and reservoir management options.

Belaineh *et al.* [3] present a simulation / optimization model that integrates linear reservoir decision rules, detailed simulations of stream/aquifer system flows, conjunctive use of surface and ground water, and delivery via branching canals to water users. State variables, including aquifer hydraulic head, streamflow and surface water/aquifer interflow, are represented through discretized convolution integrals and influence coefficients. They find the more detailed the model represents the physical system, the better is the conjunctive use management.

Chandra Sekhar *et al.* [5] developed a conjunctive use model to maximize an integrated measure of the relative yields of multiple crops in an irrigated area served by a surface reservoir taking into account the nature of crop and its timing, the type of soil on which the crop is grown, crop competition for water, reservoir release constraints, soil moisture balance in the root zone of the crop, the effect of deficit water supplies on crop yield and ground water balance. The proposed LP model integrates reservoir modelling and ground water flow modelling using finite element technique with appropriate linkages. The model is applied to Vanivilasa Sagar Reservoir command in Karnataka State, India.

7 Large Systems

The complex problem of irrigation management in a large heterogeneous basin is solved by Paudyal and Das Gupta [28] by using a multilevel optimization technique. The objective is to obtain a high level of economic efficiency in the irrigation development and water use system within a hydrologically feasible policy domain. The solution strategy is based on the physical decomposition of a large system into interconnected subsystems. A computationally efficient algorithm that can be implemented on a microcomputer is developed to solve the multilevel linear programming model by an iterative procedure. A case study is presented for illustrating the model developed.

Mizyed *et al.* [19] demonstrate the utility of optimal control theory for the deterministic operation of very large multireservoir systems for the Mahaveli system in Sri Lanka. The model is designed to minimize the hydropower shortage subject to meeting specified irrigation demands. Two alternative approaches are explored for optimal operation of the system. The first is a monthly application of the optimal control algorithm to find an optimal policy for the next year, based on the current storage and forecasted or historical inflows and demands. The second is an implicit stochastic approach, in which linear operating rules are derived using deterministic optimal control and historical data. Both alternatives are shown to give reasonable and comparable results, the later approach being useful in terms of saving computer time and storage requirements.

Vijaya Kumar *et al.* [49] used a simulation-optimization procedure to evaluate the extent of interbasin transfer of water in the Peninsular Indian river system consisting of 15 reservoirs on four river basins. The system-dependent simulation model developed incorporates the concept of reservoir zoning to facilitate releases and transfers. A large number of solutions generated by simulation are screened by Box complex nonlinear programming algorithm. System performance is evaluated with respect to reliability, resiliency, vulnerability and deficit ratio. The results indicate that operation of the system of 15 reservoirs as a single unit increases the existing utilization significantly.

Jain *et al.* [14] present a study of the operation of the Sabarmati system, consisting of four reservoirs and three diversion structures to satisfy municipal and industrial water supply, irrigation and flood control. Rule curves are derived for the reservoirs for conservation regulation of the system. Simulation is used to fine tune the rule curves to achieve the targets to the best possible extent.

An integrated optimization model is developed by Malek-Mohammadi [16] for planning irrigation systems. Surface reservoir capacity, ground-water and spring withdrawal, delivery system capabilities, area of land to be developed for irrigation and cropping pattern are considered as interacting parts of the system. Chance constraint optimization is used with mixed integer linear programming to maximize the net benefit associated with the development. Nonlinear cost functions are linearized using a new generalized technique. Model results with sensitivity analysis aid in the selection of optimum design for different scenarios.

Ponnambalam and Adams [31] developed a general algorithm for stochastic optimization of mutireservoir systems and used a heuristic algorithm to determine the optimal operational policies for five of the major reservoirs of the Parambikulam-Aliyar irrigation and power project in India. A closed loop suboptimal policy was determined for the stochastic inflow, deterministic demand problem, subject to storage and cumulative release constraints imposed by an interstate water-sharing agreement. The closed loop policies were further used to determine optimal rule curves for the major reservoirs. A comparison of the simulated optimal policies and the past performance of the project demonstrates the utility of the derived policies determined with the proposed algorithm.

8 Neural Networks / Fuzzy Models

Application of artificial neural networks and fuzzy set theory to reservoir optimization has been in vogue for a decade now, with a view to tackle the nonlinearities and uncertainties in the inherent hydrological processes, which are difficult to model by the classical methods of optimization. The experience and success of their application have been, however, varied.

Neural networks and multiple linear regression was used along with dynamic programming algorithm by Raman and Chandramouli [33] to derive the general reservoir operating policies to improve efficient management of the available waters of the Aliyar Dam in Tamil Nadu, India, based on 20 years of historical data. The objective is to minimize the squared deficit of the release from the irrigation demand. The performance is compared with SDP and standard operating policy (SOP) models to conclude that the policy from the neural network procedure coupled with the DP algorithm provided better performance than the other models.

Shrestha *et al.* [38] derived operation rules for a multipurpose reservoir based on a fuzzy rule-based model. The case study of the Tenkiller Lake in Oklahoma illustrates the methodology. Operation rules are generated on the basis of economic development criteria such as hydropower, municipal and industrial and irrigation demands. The fuzzy rule based model operates on an 'if-then' principle, where the 'if' is a vector of fuzzy explanatory variables or premises and 'then' of fuzzy consequences. The reservoir storage level and demands are used as the premises and release from the reservoir is taken as the consequence. Different performance indices are calculated and two figures of merit, engineering sustainability and engineering risk are developed for evaluating the rules generated by the model.

Panigrahi and Mujumdar [27] developed a fuzzy rule based model for the operation of a single purpose reservoir using the 'if - then' principle similar to the previous study. The steps in modelling include construction of membership functions, for inflow, storage, demand and release, formulation of fuzzy rules, implication and defuzzification. Simulated reservoir operation with a steady state policy provides the knowledge base necessary for the formulation of the fuzzy rules. The methodology is illustrated through the case study of the Malaprabha reservoir in Karnataka, India.

Bouchart and Goulter [4] argue that the existing metrics of performance used to identify optimal reservoir management policies may in fact only be nominally rational and in fact at odds with the true attitudes and perceptions of the decision makers and users affected by the water releases. They present a model capable of replicating the manner in which risks associated with release decisions are perceived, interpreted, and compared by a decision maker. The model is based upon neural network theory and enables the more complete representation of the risk of a particular decision to be considered in making decisions on reservoir releases.

Jain and Srivastava [13] study the application of Artificial Neural Networks (ANNs) for reservoir inflow prediction and operation. The Upper Indravati multipurpose project, in the state of Orissa, India, is selected as the focus area. The project provides irrigation to 128,000 ha of agricultural land and generates 600 MW of electric power. An ARIMA model and an ANN-based model were fitted to the monthly inflow data series and their performance compared. Reservoir operation policies were formulated through DP. The results of intercomparison indicate that the ANN is a powerful tool for input output mapping and can be effectively used for reservoir inflow forecasting and operation.

Naresh and Sharma [25] present a two-phase artificial neural network approach for finding optimal scheduling of interconnected hydropower plants. The objective is to maximize hydropower generation and to satisfy the irrigation requirement as far as possible. Scheduling of Bhakra-Beas interconnected reservoir system is presented. The proposed technique is compared with augmented penalty function methodology and demonstrates the potential of achieving better results.

Gupta *et al.* [10] present a study dealing with the real world problem of irrigation water management of evolving suitable cropping pattern, which should be in harmony with optimal operation of the multi-reservoir system in the basin. The Narmada river basin system is simulated on a monthly basis based on a 30 year historical data. A multiobjective fuzzy linear programming (MOFLP) area allocation module has been formulated to cope with the diverse/conflicting interests of different decision makers. Variable irrigation demand has been incorporated into the model considering high variation in precipitation. Varying cropping patterns in the command area, one for each year, have been analyzed based on the historical record.

Neelakantan and Pundarikanthan [26] used a back propagation neural network and trained to approximate the simulation model developed for the Chennai city water supply problem. The neural network is used in a submodel in a Hooke and

Jeeves nonliner programming model to find 'near optimum policies'. The results are further refined using conventional simulation-optimization model.

9 General Remarks

The review has been necessarily brief keeping in mind the scope stated at the beginning of the paper. The applicability obviously depends on the features that a model supports, and its limitations. The field conditions can be very widely different from place to place. For example, the Indian scenario in irrigation is somewhat different from that of the west. There are marked differences in both the supply and demand sectors. On the supply side, the availability of river water is controlled by monsoon climatology and is practically restricted to a few months in a year, while the demand is spread all year round, except perhaps for a few weeks in the summer season. The demand side is dominated by the need to sustain irrigated agriculture by providing subsidies to the country's large farming community backed up by acceptable water pricing policies. Small land holdings, seasonal water supplies, subsidies to agricultural inputs, suitable water pricing and tax policies are but a few of the issues that must be borne in mind in the development of appropriate modelling methodologies to aid water resources planning to suit to Indian conditions.

9.1 Performance Evaluation and Reliability

Periodic review and performance evaluation (against well defined criteria) are essential elements of sound management of water resource systems. Systems techniques provide ample opportunities to aid this process. Whatever be the goals of a project, one essential economic requirement is that it shall operate at maximum efficiency. The efficiency maximization can be achieved by operating the physical system as per the optimal policies, to result in the best combination of outputs. Every existing system can be put to this evaluation to examine the possible gap between efficient and existing operations, and identify the required improvements in the actual operation.

One of the popular concepts of operating storage reservoirs for irrigation is the reliability of meeting irrigation requirements. In a single reservoir system, for example, one can determine the maximum reliability with which irrigation demands can be met from an existing reservoir for given hydrology, or one can determine the minimum required capacity of a new reservoir for different levels of reliability. In either case, using optimization techniques ensures maximum resource utilization. Naturally, the maximum reliability that can be attained is limited, by either the inflows or the existing capacity. In the former case, the reservoir is oversized, and in the later the reservoir capacity is inadequate (to realize the maximum possible reliability), for given inflow record. Systems analysis for reservoir operation is extremely useful in exploring whether or not the maximum possible reliability is in fact being achieved. General criteria, on similar lines, for such an exploratory study for multi-reservoir systems would be very useful in practice.

9.2 Model Articulation

As one can see, there are models in the literature, which deal with reservoir operation with an aim of determining the steady state operating policy for the reservoir, which gives the maximum benefits in the long run. These models are based on lumped irrigation demands. On the other hand, there are models, which deal with optimal irrigation allocation to multiple crops for known or given surface water availability at different times of the year. These models take into account details of the crops and crop growth in the field, but do not address the issues connected with the reservoir operation for the withdrawal of water, which forms the supply to the irrigated area. There are then integrated models, which integrate the features of reservoir operation with those of multicrop irrigation. It is primarily these models that are the subject of this paper. These models not only give the optimal steady state operating policies, but also give the optimal crop water allocations in the intraseasonal periods of the crop season. These models therefore are necessarily complex in nature and have to be formulated under assumptions, which may not represent the true conditions in the field everywhere. But then the real challenge in these integrated models lies in relaxing the assumptions one by one, and thereby adding additional features, which are close to the ground, into the modelling process, yet not falling out of the purview of solution techniques. That continues to remain a challenge, however.

10 Perspectives for the future

Irrigation continues to be the most promising potential economic activity in the developing countries in the years to come. Resource scarcity and technology development would certainly provide incentives to look for alternate solutions and strategies for harnessing the available resources to maximize the physical / economic outputs with the best use of the available water coupled with suitable cultural practices. In this scenario, the need for modelling techniques for resource-economics optimization with potential for practical application will even be more relevant and perhaps critical in the future years than now.

Modelling techniques, revolutionarily new, are not necessarily what one should be looking for, but articulation in the model formulations, which take into account newer features and which can still be solved with the existing techniques remains the challenge facing the systems analyst today. Accounting for variable uncertainties to aid in the reliability analysis of the system performance and real time operation would further enhance the motivation to apply the developed models in the field. It is hoped that future research will be directed towards this, and the irrigation sector benefited even better through its use.

References

1. Arumugam N. and Mohan, S., Integrated decision support system for tank irrigation system operation, *Journal of Water Resources Planning and Management* **123** (1997) pp. 266-273.
2. Banerjee, B. K., Operation of surface water reservoirs, *Jal Vigyan Sameeksha*, Indian National Committee on Hydrology Publication on Reservoir Operations, **8** (1993) pp. 53-62.
3. Belaineh, Getachew, Peralta R. C. and Hughes T. C., Simulation/optimization modelling for water resources management, *Journal of Water Resources Planning and Management*, **125** (1999) pp. 154-161.
4. Bouchart, Francois J.-C. and Goulter I. C., Is rational decision making appropriate for management of irrigation reservoirs?, *Journal of Water Resources Planning and Management*, **124** (1998) pp. 301-309.
5. Chandra Sekhar, G., Mujumdar, P. P. and Vedula S., A conjunctive use model for multi-crop irrigation, Proceedings of International Conference on Large Scale Water Resources Development in Developing Countries - New Dimensions of Prospects and Problems, Khatmandu, October 20-23, (1997) MM111-MM116.
6. Dariane, A. B. and Hughes T. C., Application of crop yield functions in reservoir operation, *Water Resources Bulletin*, **27** (1991) pp. 649-656.
7. Datta, Bithin, Operation models for single and multipurpose reservoirs - A review, *Jal Vigyan Sameeksha*, Indian National Committee on Hydrology Publication on Reservoir Operations, **8** (1993) pp. 1-12.
8. Dudley, Norman J. and Scott B. W., Integrating irrigation water demand, supply, and delivery management in a stochastic environment, *Water Resources Research*, **29** (1993) pp. 3093-3101.
9. Evers, A. J. M., Elliot R. L. and E. W. Stevens, Integrated decision making for reservoir, irrigation, and crop management, *Agricultural Systems*, **58** (1998) pp. 529-554.
10. Gupta, A. P., Harboe R. and Tabucanon M. T., Fuzzy multiple criteria decision making for crop area planning in Narmada river basin, *Agricultural Systems*, **63** (2000) pp. 1-18.
11. Hajilal, M. S., Rao, N. H. and Sarma P. B. S., Real time operation of reservoir based canal irrigation systems, *Agricultural Water Management*, **38** (1998) pp. 103-122.
12. Hiessl, H. and Plate E. J., A heuristic closed-loop controller for water distribution in complex irrigation systems, *Water Resources Research*, **26** (1990) pp. 1323-1333.

13. Jain, S. K. and Srivastava D. K., Application of ANN for reservoir inflow prediction and operation, *Journal of Water Resources Planning and Management*, **125** (1999) pp. 263-271.
14. Jain, Sharad K., Goel M. K. and Agrawal P. K., Reservoir operation studies of Sabarmati system, India, *Journal of Water Resources Planning and Management*, **124** (1998) pp. 31-38.
15. Lall, Upmanu, Yield model for screening surface-and ground-water development, *Journal of Water Resources Planning and Management*, **121** (1995) pp. 9-22.
16. Malek-Mohammadi, Ehsanolah, Irrigation planning: Integrated approach, *Journal of Water Resources Planning and Management*, **124** (1998) pp. 272-279.
17. Mayya, S. G. and Rama Prasad, Systems analysis of tank irrigation: I. Crop staggering, *Journal of Water Resources Planning and Management*, **115** (1989) pp. 384-405.
18. Michalland, B. Parent E. and Duckstein L., Bi-objective dynamic programming for trading off hydropower and irrigation, Applied Mathematics and Computation, **88** (1997) pp. 53-76.
19. Mizyed, Numan R. Loftis J. C. and Fontane Darrel G., Operation of large multireservoir systems using optimal-control theory, *Journal of Water Resources Planning and Management*, **118** (1992) pp. 371-387.
20. Mohan, S. and Arumugam N., Expert system applications in irrigation management: an overview, *Computers and Electronics in Agriculture*, **17** (1997) pp. 263-280.
21. Mujumdar, P. P. and Narulkar S., Optimization models for multireservoir planning and operation, *Jal Vigyan Sameeksha*, Indian National Committee on Hydrology Publication on Reservoir Operations, **8** (1993) pp. 29-52.
22. Mujumdar P. P. and Teegavarapu Ramesh, A short-term reservoir operation model for multicrop irrigation, *Hydrological Sciences Journal*, **43** (1998) pp. 479-494.
23. Mujumdar, P. P. and Ramesh T. S. V., Real-time reservoir operation for irrigation, *Water Resources Research*, **33** (1997) pp. 1157-1164.
24. Mujumdar P. P. and Vedula S., Performance evaluation of an irrigation system under some optimal operating policies, *Hydrological Sciences Journal*, **37** (1992) pp. 13-26.
25. Naresh, R. and Sharma J., Hydro system scheduling using ANN approach, *IEEE Transactions on Power Systems*, **15** (2000) pp. 388-395.
26. Neelaknatan, T. R. and Pundarikanthan N. V., Neural network-based simulation-optimization model for reservoir operation, *Journal of Water Resources Planning and Management*, **126** (2000) pp. 57-64.
27. Panigrahi, D. P. and Mujumdar P. P., Reservoir operation modelling with fuzzy logic, Water Resources Management, **14** (2000) pp. 89-109.

28. Paudyal, Guna N. and Das Gupta Ashim, Irrigation planning by multilevel optimization, *Journal of Irrigation and Drainage Engineering*, **116** (1990) pp. 273-291.
29. Peng, C.-S. and Buras N., Dynamic operation of a surface water resources system, *Water Resources Research*, **36** (2000) pp. 2701-2709.
30. Peralta, R. C., Cantiller R. R. A. and Terry J. E., Optimal large-scale conjunctive water-use planning: Case study, *Journal of Water Resources Planning and Management*, **121** (1995) pp. 471-478.
31. Ponnambalam Kumaraswamy and Adams B. J., Stochastic optimization of multireservoir systems using a heuristic algorithm: case study from India, *Water Resources Research*, **32** (1996) pp. 733-741.
32. Raju, K. S. and Kumar D. N., Multicriterion decision making in irrigation planning, *Agricultural Systems*, **62** (1999) pp. 117-129.
33. Raman, H. and Chandramouli V., Deriving a general operating policy for reservoirs using neural network, *Journal of Water Resources Planning and Management*, **122** (1996) pp. 342-347.
34. Raman, H. and Kumar Sunil, Expert systems for reservoir operation, *Jal Vigyan Sameeksha*, Indian National Committee on Hydrology Publication on Reservoir Operations, **8** (1993) pp. 63-77.
35. Rama Prasad and Mayya S. G., Systems analysis of tank irrigation: II. Delayed start and water deficit, *Journal of Water Resources Planning and Management*, **115** (1989) pp. 406-420.
36. Ramesh, T. S. V. and Mujumdar P. P., Stochastic models for reservoir planning and operation, *Jal Vigyan Sameeksha*, Indian National Committee on Hydrology Publication on Reservoir Operations, **8** (1993) pp. 20-28.
37. Ravikumar, V. and Venugopal K., Optimal operation of South Indian irrigation systems, *Journal of Water Resources Planning and Management*, **124** (1998) pp. 264-271.
38. Shrestha, Bijaya P., Duckstein L., and Stakhiv E. Z., Fuzzy rule-based modeling of reservoir operation, *Journal of Water Resources Planning and Management*, **122** (1996) pp. 262-269.
39. Sinha, A. K., and Vasudeva Rao B. and Lall Upmanu, Yield model for screening multipurpose reservoir systems, *Journal of Water Resources Planning and Management*, **125** (1999) pp. 325-332.
40. Srivastava, R. C., Methodology for optimizing design of integrated tank irrigation system, *Journal of Water Resources Planning and Management*, **122** (1996) pp. 394-402.
41. Sunantara, J. D. and Ramirez J. A., Optimal stochastic multicrop seasonal and intraseasonal irrigation control, *Journal of Water Resources Planning and Management*, **123** (1997) pp. 39-48.
42. Vedula, S. and Mohan S., Real-time multipurpose reservoir operation: a case study, *Hydrological Sciences Journal*, **35** (1990) pp. 447-461.

43. Vedula S. and Mujumdar P. P., Optimal reservoir operation for irrigation of multiple crops, *Water Resources Research*, **28** (1992) pp. 1-9.
44. Vedula S. and Mujumdar P. P., Modelling for reservoir operation for irrigation, in Water Resources Planning and Management, by V. P. Singh and B. Kumar (eds.) pp. (1996) 113-119.
45. Vedula, S. and Nagesh Kumar D., An integrated model of for optimal reservoir operation for irrigation of multiple crops, *Water Resources Research*, **32** (1996) pp. 1101-1108.
46. Vedula, S., Optimal irrigation planning in river basin development: The case of the Upper Cauvery river basin, *Sadhana*, Proceedings of the Indian Academy of Sciences, **8** (1985) pp. 223-252.
47. Vedula, S. and Sreenivasan K., Reliability based reservoir irrigation of multiple crops, Proceedings of the Ninth Congress of the Asian and Pacific Division of the International Association of Hydraulic Research, Aug. (1994) pp. Singapore, **1**, 134-140.
48. Vedula, S. and Rogers P. P., Multiobjective analysis of irrigation planning in river basin development, *Water Resources Research*, **17** (1981) pp. 1304-1310.
49. Vijaya Kumar, V., Rao B. V., and Mujumdar P. P., Optimal operation of a multibasin reservoir system, *Sadhana*, Proceedings of the Indian Academy of Sciences, **21** (1996) pp. 487-502.

OPPOSITION TO LARGE DAMS IN INDIA: AN ANALYSIS

RAMA PRASAD
Department of Civil Engineering, Indian Institute of Science, Bangalore, INDIA
Email: rama@civil.iisc.ernet.in

Large dams have been built in India for more than 60 years. During the past decade, however, there has been increasing opposition to large projects, especially water resources projects. This paper analyses such opposition in the case of hydroelectric projects near the west coast in the state of Karnataka, the multipurpose project at Tehri in the Himalayas in the North and the Sardar Sarovar Project in Central India. Some ways to reduce the intensity of opposition are suggested.

1 Introduction

Small reservoirs with earthen embankments across small streams, called tanks, have been built in India for thousands of years for irrigation. Large dams have been built for irrigation and power production for more than 60 years across major rivers. Notable among such dams are the Krishnarajasagara, Mettur, Nagarjunasagar and Srisailam dams in the south, Koyna and Ukai in the west, Pong and Bhakra in the north, Hirakud in the east and Upper Wainganga and Gandhi Sagar in central India. All these dams are of great importance to the regions they serve. They have also meant displacement of people from their habitations and submergence of valuable lands with their flora and fauna. These issues have been considered carefully by the World Commission on Dams during the past two years, and a public debate by eminent persons has taken place for and against large dams. Till the 1980's, displacement and submergence were seen as a necessary price to pay for assured water and power supply, although the Silent Valley Project in Kerala was aborted in order to preserve the ecology of the region. During the past two decades, however, opposition to large dams is growing on both counts of displacement of people and environmental impact. This paper analyses such opposition in the case of some projects in the Western Ghats of Karnataka State, the Tehri dam on the Bhagirathi in the north and the Sardar Sarovar on the Narmada in Central India.

2 Western Ghats Projects In Karnataka

Close to the west coast of south India, there are high mountain ranges called the Western Ghats (Fig. 1), which deflect the monsoon winds to receive heavy rainfall varying from 2000 mm to 7000 mm annually. The western slopes of these mountains are very steep, and the rivers draining the slopes carry enormous quantities of water, providing favourable conditions for power generation. Because of the heavy rainfall, the mountains are covered with dense forests. Since most of the rainfall occurs from June to September, storage of water by construction of dams is necessary for power production during the dry season. The mountainous terrain offers very good sites for dams. Several hydropower projects were therefore built in this area until the mid-1970's. Thereafter, new projects have been stalled by local opposition. These are the Bedthi, Sharavathi Tail Race, Aghanashini and Barapole hydroelectric projects. The Bedthi Project for power production of 420 MW was approved by the Central Electricity Authority in 1977, and environmental clearance accorded in 1979. Infrastructure for the project such as some of the roads and buildings were established by the end of 1979, spending about 7% of the project cost. At this stage, the local residents formed a committee called the Anti-Hydel Projects Committee and started agitating against the implementation of the Project. Work on the project was stopped and a committee was constituted by the State Government in 1981 to review the Project. After five years, the Review Committee gave its report in 1986 recommending construction of the Project with a reduced FRL and measures to preserve the ecology of the region. Six years later, the State Government gave a go-ahead and tenders were invited in July 1992 for some components of the Project. At this stage, the local people formed another committee called the Bedthi-Aghanashini Valley Protection Committee and started agitating for abandoning the Project. They also filed a suit in the High Court asking for stoppage of the Project on the ground of adverse environmental impact. This suit is yet to be decided.

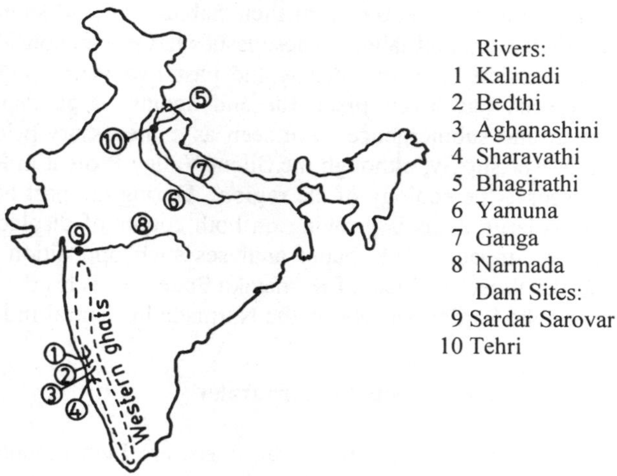

Figure 1. Map showing location of the projects

For the past three decades, a dam on the river Sharavathi at Linganamakki has been generating power of 890 MW. In the 1980's a project was planned to utilise the remaining head between the tail race of this Power house and the sea, to produce additional power of 240 MW. This involves submergence of only 596 ha of land, and no people are displaced. The Government of India gave environmental clearance subject to certain conditions in 1987. Work on the project started in 1989 with the clearing of about 60 ha of forest land, when a suit was filed in the High Court of Karnataka pleading ecological hazards. The Court ordered the Government of India to review the environmental clearance. The Karnataka Power Corporation Ltd, which was implementing the Project, went to the Supreme Court in appeal, which allowed work to be carried out without felling any more trees. In 1992, the Government of India revoked the environmental clearance on the ground that the conditions stipulated by it were not complied with. The Company then made a comprehensive study of the aquatic life, flora and fauna in the catchment and pursued the matter with the Ministry of Environment. The Project was then cleared and is now in the final stages.

The Aghanashini is another river in the region on which a 40m dam was proposed to generate 600 MW of power, involving a total submergence of 2007 ha. Investigation of the site, which was taken up in 1990, was obstructed by the local people, who even burnt the drilling machine deployed on the job. Another Western Ghat river is Barapole which falls by 750m in a very short distance and can produce 600 MW of power. Local people physically obstructed the investigation work taken up here in 1975, and it has been stalled.

In all the above cases, although environmental grounds have been advanced by the agitating people, the real grounds appear to be the reluctance of displaced people to move and lose their plantations or agricultural property. Most of the power produced will not be utilised in the Western Ghats region, where there is not much industrial activity, but will flow mainly to Bangalore, the capital of the State, which is about 300 km away. There is therefore no local support for the projects to counter the agitation of the affected people.

3 The Tehri Dam

The Tehri dam is being built in the Himalayas across the Bhagirathi, a tributary to the Ganga (Fig. 1). It is the first storage project in the Ganga valley, consisting of a 260.5m earth-and-rockfill dam. Planned to irrigate 270 000 ha and produce 2400 MW power, it submerges an area of 42 km^2, including the town of Tehri (population 12000) and 23 other villages besides parts of 72 more villages (population: 34000). A new Tehri town is being built near the reservoir, 25 km from the existing town. It will have the same regional linkages as the present town, *e.g.*, pilgrim traffic from Rishikesh to Yamunotri and Gangotri will pass through the new town, as it does now through the old one. The villagers affected will be resettled near Rishikesh and Haridwar, 100 to 150 km away, with a minimum of two acres of land to each oustee.

Initially, protests against the dam began on the issue of displacement of people, but are now almost wholly focussed on the safety of the dam in the event of an earthquake. The dam is situated in a seismically active region, and its structural failure due to an earthquake could inflict considerable damage downstream. In view of this, seismic aspects of the project have been considered by different bodies in great detail. The University of Roorkee (UOR) has established four permanent observatories operating for more than 20 years, in addition to a number of mobile units [4]. A standing committee of the Central Water Commission recommended preliminary design of the dam for a horizontal acceleration of 0.15*g*. The UOR recommended 0.25*g*, which is what the final design has used. Scientific opinion is divided on what the peak ground acceleration (PGA) would be in an earthquake of the region, as well as on the earthquake magnitude, and the dam is caught in a maze of conflicting predictions about them. The National Geophysical Research Institute has suggested 1.25*g* and 7.5M. The present design based on 0.25*g* and 7.2M was cleared by a committee of experts in 1990. Another high level committee headed by the Director General of the Geological Survey of India concluded that the dam conforms to ICOLD (International Commission on Large Dams) norms on seismic design. Some individual experts have also supported the design, but some other scientists do not agree. Opponents of the dam have picked up the arguments of the latter, which can be summarised as follows: The Central Himalayan region has not experienced a major earthquake in the past 500 years, while all other segments in the Himalayas have had several quakes in the present century. This area is therefore ripe for a major quake in the next 100 years. (This is the 'seismic gap' theory, which assumes that during such a long period of seismic inactivity, energy keeps building up along the thrust fault between the Indian and Eurasian plates, ultimately to be released by a major quake). The reservoir load itself can induce an earthquake. This quake could have a magnitude of 8 + on the Richter scale. For such earthquakes, the recorded PGA has never been as low as 0.25*g* anywhere. Many quakes below 8+M have had PGA much higher than 0.5*g* and even 1*g*. The 1991 Uttarkashi earthquake in the higher Himalayas (50 km from Tehri) was 7M with 0.43*g*. The dam has therefore to be designed for at least 1*g*. Even so, the reliability of earthquake engineering is in doubt because structures designed for 7.9M were destroyed by the 7.2M Kobe earthquake of 1996. Further, even if the dam withstands the quake, the hillsides will fall into the reservoir, which would have the same effect as a dam burst. Hence the dam should not be built.

Proponents of the dam have answers to these points, also based on the opinion of seismologists. According to Valdiya [1], Himalayan earthquakes are due more to movements along tear faults transverse to the hills than along the thrust faults (which run parallel to the hills). This contradicts the seismic gap theory and hence the impending nature of a megaquake. Bolt, Ambraseys and Finn [1] point out that the peak ground acceleration, being associated with a very high frequency, lasts only a short period and is less important in the design than the response spectrum at relatively long periods (of the order of the fundamental period of the dam, typically one or two seconds), and 0.25*g* is therefore more like the relevant figure. For example the Koyna dam, designed for 0.05*g*, withstood the 1967 earthquake (6.5M) with PGA of 0.5*g*. There is also the experience of several dams built in the Himalayas, such as Bhakra, Ramganga, Pong and Pandoh in India and Mangla and Tarbela in Pakistan, some of which are in operation for about three decades. They have not caused any increase in seismic activity in the respective regions. Nor have there been any landslides there of the type feared by opponents of Tehri dam. Chander and Kalpna [2] studied the effect of the reservoir load on the stresses in three seismically active fault planes near the reservoir, and concluded that the reservoir load has a stabilising influence on these faults, i.e., earthquakes would be postponed , as a result of reduction in the stresses. However, they think that the rate of stress build up is not small enough for postponement for a period of the order of the life of the dam to occur. They also conclude that the 1991 Uttarkashi earthquake would have been postponed if the Tehri reservoir had been impounded before 1991. Hence, there will be no reservoir-induced earthquakes in the region. Based on a study of the behaviour of 33 dams near the San Andreas fault during the 1906 San Fransisco earthquake (8.25*g)*, Seed [3] concludes that dams constructed of clay soils on clay or rock foundations have withstood accelerations of 0.35*g* to 0.8*g* with no apparent damage, and that two rockfill dams have withstood moderately strong shocking with no significant damage and if the rockfill is kept dry by means of a concrete facing, such dams should be able to withstand extremely strong shaking with only small deformations. Many defensive design measures are available for ensuring safety of embankments during earthquakes and the dam design incorporates many of these methods [1]. For embankment materials which do not build up large pore pressures, pseudo-static analysis provides in many cases an entirely adequate design procedure, and Finn [1] says that pseudo-static analysis of the Tehri dam suggests that it has seismic resistance at least for base input acceleration of 0.4*g*. He also states that the dam as designed and planned has inherently strong resistance to seismic shaking. Further, hundreds of dams have been built in seismically active areas all over the world and have been quite safe. Recently also, the 325m high earth-and-rockfill Ragun dam and the 300m high rockfill Nurek dam have been built on the Vaksh river in the former USSR adjacent to the Himalayas in an 8+M seismic zone. The former in fact has two faults within the dam foundations [4]. In contrast, there are no faults in the foundation of Tehri dam [4]. Further, the average slope of the hillside around the reservoir, from the ridge to the reservoir level, is not more than the angle of friction of the rock mass. Even though there are local cliffs and steeper gradients on the hill slopes, major foliations and discontinuities have dip directions which favour stability. No area has been found where large scale slides could occur. Also, no large scale movement of soil mass has occurred in the case of Bhakra, Ramganga and other reservoirs which have been functioning for decades in similar regions of the Himalayas. Only sliding or slipping of near-surface overburden and weathered rock is possible, and in the vicinity of the dam and other major structures, stabilisation of slopes with the usual techniques is planned.

It is thus clear that although there is a difference of opinion among seismologists, the balance is strongly supportive of the existing design of the dam. Typically, no opponent of the dam is prepared to believe this group of scientists although it includes some of the leading seismologists in the world.

4 The Sardar Sarovar Project (SSP)

The river Narmada, flowing in a long and narrow valley between the Vindhya and Satpura mountain ranges in Central India from east to west (Fig. 1), has remained more or less untouched till recently. The major part of its catchment lies in the states of Madhya Pradesh and Gujarat, and a small part in the state of Maharashtra. Large areas of the basin are subject to droughts, and in 1969 the Government of India appointed a tribunal (Narmada Water Disputes Tribunal or NWDT) to allocate waters of the river, among the three basin states as well as Rajasthan, a neighbour state. By the time the NWDT was ready to allocate the waters, flow data were available only for 22 years from 1948 (when the Central Water Commission, CWC, started gauging of the river) to 1970. This was considered too short a period to estimate with reasonable accuracy the 75% dependable flow (*i.e.*, flow with 75% probability of exceedence, which is the lower quartile of the flow series), which is the statistic normally used in India for project planning. A longer annual flow series was therefore adopted, which consisted of 'hindcast' flows estimated from annual rainfall for the period from 1891 to 1947, using a regression relation developed for the 15 year period 1948-1962, and measured flows from 1948 onward. The 75% dependable flow was estimated as 27 MAF (Million Acre Feet, one acre foot being equal to 1233 m^3. the MAF unit is used in this paper, since these units have been used by everybody commenting about the Sardar Sarovar Project). With an additional 1 MAF considered to be available from carry-over storage and return flows net of evaporation, a total annual flow of 28 MAF was apportioned by the NWDT: 18.25 MAF to Madhya Pradesh, 9 MAF to Gujarat, 0.25 MAF to Maharashtra and 0.5 MAF to Rajasthan. Madhya Pradesh plans to irrigate 2.76 Mha of land with several reservoirs and produce 1000 MW of power at Narmada Sagar reservoir. Gujarat plans to irrigate 1.8 Mha and produce 1200 MW at Sardar Sarovar. The NWDT award was given in 1979, and envisaged four reservoirs on the main river, at Punasa, Maheshwar and Omkareshwar in Madhya Pradesh and near Garudeshwar in Gujarat, the last being the terminal reservoir, called Sardar Sarovar. All these reservoirs were regarded as mandated by NWDT, and people also appeared to have accepted them as such. There was no opposition to the projects during the years following the NWDT award. In the mid- 1980's, the Forest Conservation Act was passed by the Parliament, which required environmental clearance for large projects, and thus the Narmada reservoirs came under scrutiny. Opposition to the SSP started at the same time, and gathered strength from 1987.

The main focus of the opposition is the oustee problem, but other issues have been advanced as well. These are detailed below.

4.1 The Oustee Problem

Narmada Sagar in Madhya Pradesh (Live storage capacity 7.9 MAF) submerges 249 villages with a total population of 86000. There is little opposition against this reservoir. The Sardar Sarovar in Gujarat (Live Storage capacity 4.72 MAF) submerges 245 villages, 19 in Gujarat, 33 in Maharashtra and 193 in Madhya Pradesh, with a total population of about 100000. Virtually all the oustees in Gujarat and Maharashtra and 40% of the Madhya Pradesh oustees are tribals, whose occupation is cultivating land, grazing animals, collecting forest products (mainly firewood) and fishing. The NWDT, in its order, detailed some aspects of how the states should go about resettlement and rehabilitation (R&R) of the oustees. These steps were further improved as a result of representations by voluntary agencies, discussions with the World Bank which had at that time agreed to fund the Project, decisions of courts etc. Implementation of the R&R plan has been criticised by the Narmada Bachao Andolan (NBA - "Save Narmada Movement") on the following grounds. The tribals, who form the majority of oustees, depend on the forest and the river for their livelihood. At the locations identified for R&R, there is no forest or river. At the rehabilitation sites, the quality of land is not good. The R&R policies followed by the three states are not uniform, Gujarat's being the best, and the option available to oustees under the NWDT Award for choosing the state in which to resettle is therefore only notional. A large number of people other than the oustees, such as tradesmen, suppliers of goods and services to the oustees, shopkeepers, craftsmen, fruit and vegetable growers, small entrepreneurs etc. linked directly or indirectly with the oustee population will be affected by the Project and they have not been taken care of in the R&R policy. Besides, a large number of people will be displaced by the canal network. The reduction of downstream flows will affect a large number of fishermen and boatmen. Resettlement of the oustees at new locations will lead to conflict with the population existing at these locations. Not enough land is available to resettle the oustees. The purchase of land for resettlement itself leads to displacement of landless labourers. Redressal of individual hardship cases is very poor due to the indifference of bureaucracy. All these issues cannot be properly addressed and many other issues cannot be foreseen. Hence the project should be stopped.

4.2 Environment and Ecology

The opponents of the dam have raised various objections to the Sardar Sarovar Project on the ground that it will adversely affect the environment in the neighbourhood of the reservoir as well as downstream: Soil erosion from the catchment will lead to sedimentation of the reservoir cutting down its life. Hence the entire catchment area has to be treated, which makes the project economically unviable. Forest lands will be lost as a result of submergence and resettlement of oustees, with consequent loss of flora and fauna. The dam prevents migration of fish to upstream reaches, resulting in loss of fish population. Also, reduction in flows downstream leads to the same result. Micronutrients will be trapped in the reservoir and will not be available to downstream aquatic life. The reservoir will lead to health problems such as increase in the incidence of Malaria. Water quality in the command area will deteriorate due to the use of fertilisers and pesticides. Waterlogging and salinity will develop in the command area and land will go out of production. The availability of irrigation water will encourage farmers to switch from food crops to cash crops resulting in loss of food production. Because of reduced downstream flow, saline water would intrude into the river reach near the coast.

4.3 Hydrology

A very important point of controversy is the magnitude of the 75% dependable annual flow at Garudeshwar, which was set at 27 MAF by agreement between the basin states and accepted by the NWDT. As already stated, this figure was arrived at using a flow series consisting of observed data from 1948 to 1970 and hindcast data from 1891 to 1947. The hindcasting was done because the 22 year length of observed data was too short to estimate the lower quartile with reasonable accuracy. Now, however, a longer observed series is available. The observed flow for water years from 1948-49 to 1987-88 gives a 75% dependable flow of only 22.9 MAF. In fact even considering the hindcast series also with the observed series up to 1988, the 75% dependable flow comes to only 24.8 MAF.

4.4 Irrigation

A major benefit used to justify the project is that out of the total command of 1.8 Mha, more than half lies in the drought- prone areas of Gujarat, viz. North Gujarat (0.8 Mha), Saurashtra (0.39 Mha) and Kachch (0.04 Mha). Critics say that since the flow is less than assumed, these areas will get less water. Other objections have also been raised: *en route* flows are now taken into account, which was not envisaged earlier. The assumed irrigation efficiency of 60% is too high. Water-intensive crops will be grown in violation of the planned cropping pattern. Waterlogging and salinity will develop in the command area. In the region downstream of the dam, municipal and industrial demands for water will grow, reducing its availability for irrigation, and water will not reach Saurashtra and Kachch. The major part of the irrigation water will go to already prosperous areas like Ahamdabad, Bharuch, Vadodara and Kheda (accounting for 0.89 Mha). Moreover, Saurashtra and Kachch do not need Narmada waters and local water resources can be developed to meet their requirements.

4.5 Drinking Water

Provision of drinking water to 8214 villages and 135 towns is advanced as one of the most important benefits justifying the SSP. Criticisms by the Project opponents on this issue are as follows: Drinking water provision is an afterthought on the part of the Gujarat Government, introduced after opposition to the Project gathered strength. The original allocation for municipal and industrial uses out of Gujarat's share of Narmada waters was 1.06 MAF. Over four years drinking water coverage has been increased from 4720 villages to 8214 villages, and 131 towns to 135 towns, but no increase in water allocation has been made. No plans have been drawn up for the water supply. In any case, it will take around 10 years for the canal network to be built, and in the meantime, alternative measures to meet the drinking water needs are necessary. These measures may be made permanent and hence Narmada waters are not necessary for this purpose!

4.6 Power Generation

Critics say that since the flow is less than assumed, power produced will be correspondingly lower. As irrigation develops over the years, power generation will decrease since less and less water will be released through the river bed power house. Most of the power generated will be used up by the project itself to operate the canal control system and pump water to the Saurashtra and Kachch branch canals. There is therefore no net power production. Anyway, there is no need to generate power at SSP since alternatives are available such as gas that is being flared at present, increasing the power generation from existing reservoirs etc.

4.7 Are the Criticisms Justified?

Any major project, not only reservoir projects, involves displacement of people, and such displacement had been accepted as inevitable in the larger interests of the surrounding region as well as the country. The economy of the tribal inhabitants of the Narmada valley is based on continuous encroachment of forest land for agriculture by a growing population, which is not sustainable in the long run. Ultimately, their economy has to change if forests are to be preserved. Over the last one hundred years, the forests in the submergence area have already got degraded. The quality of land at rehabilitation sites cannot be expected to be uniformly good when thousands of hectares are involved. Some oustees will get good lands and others not so good lands. The oustees will gain the advantage of becoming legal owners of land, instead of their present status as illegal encroachers. One cannot also expect the three different states to follow the same R&R policies, since conditions like availability of land are different. The oustees have the option to choose where to resettle, weighing all the factors together. The providers of goods and services to the oustees, shopkeepers etc will find their business expanding due to improvement in the standard of living of the farmers in the 4.56 Mha of the total Narmada Projects command, instead of depending on the present subsistence economy of the oustees. The increase in wealth and agricultural production will generate thousands of new jobs in the areas of farm labour, transport, marketing etc. Such a result has been seen in the case of irrigation projects already in operation in India. It is not fair to expect the Project to provide for the rehabilitation of such indirectly affected people.

Criticisms of the SSP on environmental grounds would again have been equally applicable to all other reservoir projects built earlier, but in these projects, after several decades of operation, there is very little felt adverse effect on the environment. In fact, there is a felt beneficial effect in many cases. After the Sharavathi and Kali hydro-electric projects near the west coast of Karnataka were commissioned, extremes of flow in the downstream reaches have evened out, and salinity intrusion from the sea has come down considerably. Vast areas in Saurashtra and Kachch now plagued with brackish groundwater will see an improvement in water quality due to recharge of fresh water [5]. If farmers switch to cash crops instead of foodgrains why should they be prevented? More money in their hands will make its own contribution to improvement in the environment. It is often forgotten that a large part of the environmental degradation in India (including the Narmada valley) is due to poverty rather than industrial and irrigation development. Rivers are polluted by raw sewage since there is no money for treating the sewage. Groundwater in villages is polluted due to the same reason. Water supply is unreliable. Forests are denuded (for firewood) because a large number of people cannot afford commercial fuels and the stoves to burn them. Generation of wealth by the SSP will automatically lead to improvement in the environment of towns and villages, although some forests are submerged. If the arguments and methods of calculating benefit cost ratios put forward by opponents of SSP had been applied to existing irrigation projects, none of them would have been built, and today India would have to depend on other countries for its food requirements, besides facing famines of the kind which used to devastate the country earlier. Thanks to these projects, India has been able to build a strong buffer stock of foodgrains which can meet droughts in several parts of the country.

Regarding hydrology, it is true that the observed flow series 1948 to 1988 or later gives less than 27 MAF for the 75% dependable flow. This raises the question whether 40 or 45 years flow record is long enough to estimate the lower quartile with reasonable accuracy. Opponents of the dam are convinced that it is, on the ground that projects have been designed elsewhere on the basis of a record length of that order. However, the accuracy of estimate of a quantile depends on the standard deviation σ and the length of record n. If the fluctuations of flow around the mean are random, the lower quartile has a standard error [6] equal to

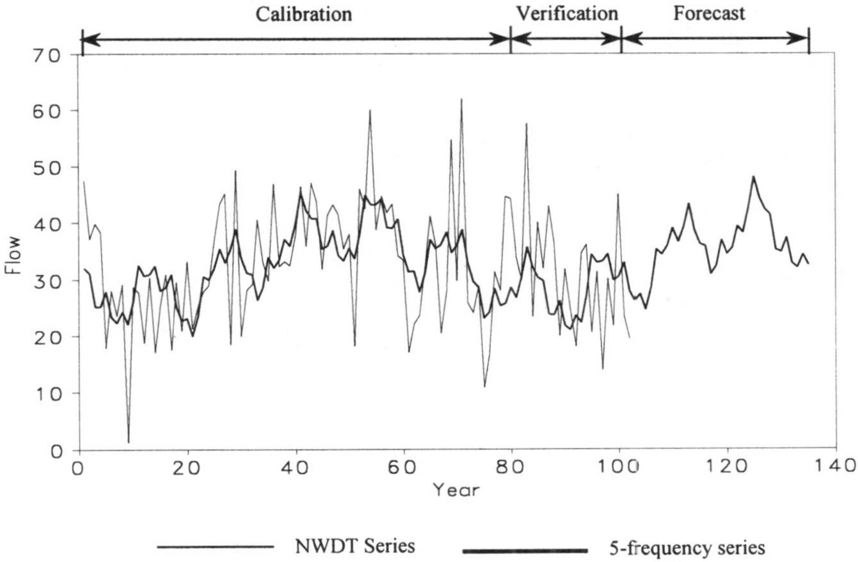

Figure 2. Narmada Flow at Garudeshwar

$1.363\sigma/\sqrt{n}$. For the 45 year observed flow series from 1948-'49 to 1992-'93 the lower quartile is 23.1 MAF, σ is 11.3 MAF and the standard error is 2.29 MAF, which is about 10% of the quartile. If the admissible standard error is 5% (1.15 MAF), then the length of the series should be 180 years. Student's t_{05} for a 45-year record is 2.014, so that the 95% confidence interval for the 75% dependable flow is from 18.5 MAF to 27.7 MAF. It is obvious that the uncertainty in the 75% dependable flow estimated from the 45-year observed record is high, and the NWDT figure of 27 MAF lies within the 95% confidence interval. The case for changing this design parameter is therefore weak. Further, the assumption that fluctuations of the flow about the mean are random is open to doubt, since there is some indication of cyclic behaviour. The Narmada Valley Development Authority, Madhya Pradesh [7] showed that quantiles calculated for a 79-year period positioned anywhere in the interval 1891-92 to 1988-89 were remarkably constant and in particular the lower quartile varied in the small range between 27.6 and 27.9 MAF. A look at the graphical representation of the flow series (Fig. 2) also suggests cyclic behaviour. A Fourier analysis of the series shows the presence of several frequencies, important among them corresponding to periods of 3, 6, 10, 14 and 75 years. Short time Fourier transform analysis shows that the first four periods are present throughout the flow series. A sinusoidal series consisting of five terms with these periods, the amplitude and phase being calibrated for the first 80 years (1891-'92 to 1970-'71), simulates the remaining 22 years (1971-72 to 1992-93) also rather well (Fig. 2). The last 45 years appear to have captured the trough region of the cycle, resulting in lower estimate of the 75% dependable flow. Quantile estimates over a part of the cycle are obviously in error. The expected life of the dam being a few hundred years, the record length considered for a proper estimate should cover an integral number of cycles. Thus, the Narmada Valley Development Authority's proposal of between 27.6 and 27.9 MAF appears to be a closer estimate of the 75% dependable flow than others. This being more than the design basis of 27 MAF, there is no case for revising the design. Finally, even if one accepts the figure of 22.9 MAF based only on the observed record, all it means is that 27 MAF has a lower dependability (65%). This only means that the project will meet the full demands in 65% of the years instead of 75%. Even then, no revision of the design is called for, since the shortfall in the remaining years will be in the Rabi season and confined to a small area.

There is thus no ground for concluding that the irrigation benefits will be less than what is planned. Although significant quantities of water will go to already prosperous areas, the drought-prone Saurashtra, Kachch and North Gujarat account for more than half the command area. The distribution of water between different areas is dictated by topography. Many other criticisms are based on imagination rather than reasonable grounds. Increase in the number of villages proposed to be supplied with drinking water is no serious threat to irrigation water availability, the additional water requirement being a very small fraction of the irrigation water.
Power generation at SSP is planned to meet the peaking requirement. In the initial years before the command area is fully developed, surplus water is available so that continuous power can be generated. The opponents' criticism that power generation declines as the command area develops is thus a topsy-turvy way of looking at the Project. At present load shedding is regularly resorted to because of shortage of power, and any additional generation should be welcomed. Even with SSP producing power, there is need for additional power plants.

The NBA nevertheless filed a petition in the Supreme Court praying for abandonment of the Project. After six years of proceedings, the Court delivered its judgement in October 2000, rejecting the prayer. Significantly, the Court observed that large dams should not be equated with polluting industries in respect of environmental impact, and that past experience in India showed that large dams in fact improved the environment. The Court was also satisfied with the steps taken for rehabilitation of the oustees, and gave the go-ahead for construction of the Dam as per original design. Typically, however, the opponents of the Dam maintained that their own stand was right, the Court was wrong, and that they would continue to obstruct the construction.

5 Analysis of Opposition to Large Dams

An overall view of the opposition to various large dams suggests that there are various factors at work. The natural reluctance of people to move their homes and loss of private agricultural lands and plantations provide the nucleus. There is also in some cases a feeling of alienation that the Project would benefit not the local people, but others living far away. In the case of irrigation projects, land prices in the command area would appreciate and returns from agriculture increase, but the oustees feel that their wealth or standard of living goes down, at least immediately. Delays in implementing different components of a project and lack of coordination between different departments of the government complicate matters. An example of this is provided [8] by the Kalinadi hydro-electric project in the Western Ghats of Karnataka, operating for several years now. The dam is situated near Supa village, which is now submerged. Before the dam was built, the resettlement area for the oustees was identified on the hillslope catchment area. These slopes, forming part of reserve forests, were deforested and handed over to the project authorities by the forest department. In the mean time, a major fault was detected in the dam foundation and construction was delayed by a few years. The oustees did not move to the new location, which got severely eroded without forest cover and became unfit for cultivation. This is an example where a wholly avoidable environmental problem developed. But even otherwise, environmental activists tend to oppose any large project, and throw in various issues, real or imagined, to strengthen the opposition, as illustrated in the case of SSP. Often, projects are made to look uneconomical by the opponents, by loading the costs and ignoring corresponding benefits. For example, catchment area treatment to minimise erosion (and hence sedimentation of the reservoir) is included on the cost side, but the improvement in groundwater resources and agricultural production as well as retention of fertile topsoil are not counted among the benefits. When forest land is submerged in a hydroelectric project, the loss of annual yield of the forest is used to boost the cost [9], but the increase in industrial production due to the power generated is not included in the calculation of benefits. To take another example from the SSP itself, in 1970 the Narmada experienced a peak flood of 69,400 m^3/s, affecting 285,000 people and killing 349. In 1994, a little higher peak flood of 70,800 m^3/s occurred, but because of flood attenuation by the partially built Sardar Sarovar dam, only 35,000 people were affected and 42 lost their lives. Damage to property was also a fraction of that in 1970. Such direct benefits of the dam are ignored in the benefit-cost ratio computation, while all kinds of indirect or imagined costs are included in computing the benefit-cost ratio. None of the existing large dams in India would have passed such logic, but the benefits due to them are there for all to see. All these dams have improved the standard of living of people in the respective regions, including those of the oustees. When parts of Gujarat experienced severe drought in the summer of the year 2000, there was a widespread feeling that had Sardar Sarovar been allowed to be built, the drought would never have occurred.

This is not to say that all is well with the way the oustee problem is handled. The system in place for implementing the R&R policies needs improvement and response to cases of individual hardships must become much quicker. Since availability of land for resettlement is becoming less and less, non-land based rehabilitation schemes have to be developed. Benefits of the project should be available to the oustee population also, *e.g.*, by way of resettlement in command areas in the case of irrigation projects or jobs in the case of hydroelectric projects. Where the power produced does not lead to corresponding increase in industrial activity in the neighbourhood of the project, the sense of alienation of the local people must be reduced, *e.g.*, by a royalty scheme, part of the revenue from the project being used to develop the infrastructure of the surrounding region.

6 Conclusions

The issue of rehabilitation of oustees lies at the heart of large reservoir projects in India. This issue is exploited by vocal environmentalist groups which tend to oppose any large project. Most of the other objections by these groups are based on imaginary grounds, and are raised in order to give an environmental veneer to the opposition. If genuine environmental impact is addressed and R&R is carried out efficiently, the benefits from reservoir projects which the silent majority of the society looks forward to can become a reality.

References

1. INTACH, Earthquake hazard and large dams in the Himalaya, *Proc. Workshop*, Indian National Trust for Art and Cultural Heritage (1993).
2. Chander R. and Kalpna, Probable influence of Tehri Reservoir load on earthquakes of the Garhwal Himalayas, *Current Science* **70** (1996).
3. Seed H. B., Earthquake-resistant design of earth dams, *Proc. Symp. Seismic Design of Embankments and Caverns,* Philadelphia (1983).
4. Kumar D., Environmental aspect of Tehri Dam Project, *Bhagirath* **37** (1990).
5. Irrigation Commission, Report of the Irrigation Commission, Ministry of Irrigation and Power, Government of India, New Delhi (1972).
6. Yule G. U. and Kendall M. G., An introduction to the Theory of Statistics, *Griffin*, London (1950).
7. Narmada Valley Development Authority, Narmada Projects - Controversy on available flows - Concept of hydrometeorological cycle, Bhopal (1979).
8. Gadgil M., Hills, dams and forests, *Proc. Indian Acad. Sci.,* **C2**(3) (1979).
9. Paranjpye V., Dams: Are we damned?, *Proc. Seminar on Protection of Environment - A Key to sustained Growth*, Ahmadnagar (1980).

SUBJECT INDEX